CERTIFICATION MANUAL FOR WELDING INSPECTORS

THIRD EDITION

Published By
American Welding Society
Education Department

American Welding Society

DISCLAIMER

The American Welding Society, Inc. assumes no responsibility for the information contained in this publication. An independent, substantiating investigation should be made prior to reliance on or use of such information.

International Standard Book Number: 0-87171-421-3
Copyright © 1993 by American Welding Society, Miami, Florida.
Third Edition 1993
Revised 10/95
All rights reserved. No part of this book may be reproduced in any form or by any means, electronic or mechanical, including photocopying, recording, or by any information storage or retrieval system, without permission in writing from the publisher.

Printed in the United States of America

PREFACE
CERTIFICATION MANUAL FOR WELDING INSPECTORS

In 1976, the American Welding Society introduced a much needed certification program specifically for those individuals who perform visual welding inspection. Shortly thereafter, the AWS Qualification and Certification Committee initiated the development of a single publication which could serve as a valid reference for those individuals interested in becoming qualified as a Certified Welding Inspector. Prior to the initial publication of *Certification Manual for Welding Inspectors* in 1977, relevant information on the subject could be found scattered among various documents such as *Welding Inspection* and *Guide to Nondestructive Testing of Welds*.

Dr. Hallock C. Campbell was selected to organize existing information on the topic in a manner that would be most helpful to those individuals who wished to become better prepared for both the certification examination as well as their jobs as welding inspectors. Dr. Campbell's ability to present this technical information in a form that inspectors of many different backgrounds could easily comprehend has made the *Certification Manual for Welding Inspectors* one of the most utilized documents on the subject. The initial printing was followed by a revised second printing in 1979, and finally a Second Edition, released in 1980. This is, I feel, adequate proof of the publication's value to the industry.

Since that time, numerous changes have occurred in both the AWS Certified Welding Inspector program and examination as well as the technology related to welding inspection. Consequently, the AWS Continuing Education Committee sought to update the information contained in the Certification Manual. Much of the information contained herein was drawn from other AWS publications, including: *Welding Inspection*, Second Edition, *Welding Inspection Technology*, AWS B1.11-88, *Guide for the Visual Inspection of Welds*, and AWS B1.10-86, *Guide for the Nondestructive Inspection of Welds*. The reader is encouraged to review these and other documents for even more detailed descriptions of much of the information contained in this new edition.

As Dr. Campbell pointed out in previous editions, welding inspectors are employed in a variety of industries. As a result, their duties will differ somewhat from one situation to the next. This Manual has been developed under the assumption that a welding inspector will be performing quality control duties of a general nature. Some inspectors, for example, may be working at a field construction site where he or she is in charge of overall welding quality. At the other extreme, in a large organization, a quality assurance department may make many of the decisions that the Manual assigns solely to the inspector. As noted by Dr. Campbell, in either case, the welding inspector will always perform a key role. The individual inspector's specific role in the quality control activity must therefore mesh with many other activities and personnel, as outlined in the pages which follow.

In this third edition, there has been an attempt to update the technical information, where appropriate. One of the areas where readers of previous editions will note changes is in the terminology used for describing various weld characteristics. There is an ongoing effort to use standard terminology when talking about welding operations and related weld characteristics. Other important changes include the addition of more illustrations and photographs to better describe much of the information. A final major change involves the questions appearing at the end of each chapter. The number of questions has been significantly increased to provide those individuals who are preparing for the CWI examination with numerous examples of the types of questions that appear on the test. To further enhance the value of the questions, they all appear in the same format (multiple choice with five options) as the questions on the CWI examination. While this is intended to specifically aid those studying for the test, it should also be beneficial to others from the standpoint of improving their comprehension of the information presented in the text.

The fourteen chapters included in previous editions have been rearranged and combined into twelve chapters, much like the AWS Course Welding Inspection Technology. It is hoped this change will result in better understanding by the reader. As with previous editions, the target audience remains the same: practical people in welding inspection, technicians, and the novice welding inspector who are not specifically interested in a highly technical and academic treatment of the topics presented. That does not mean, however, that technical topics have been omitted; only that they are presented in a manner and language that should allow for their comprehension by individuals having a moderate amount of education and experience.

I hope this presentation will prove helpful to those interested in becoming welding inspectors and eventually becoming qualified as an AWS Certified Welding Inspector. The job of welding inspector is a tremendously challenging and important one, and those seeking the CWI qualification should be commended and encouraged. My personal desire is that this Manual will assist you in reaching that goal. I wish you the best in that endeavor.

Richard L. Holdren, P.E.
July 1993

TABLE OF CONTENTS

Chapter	Title	Page
	Preface	iii
1	The Welding Inspector	1-1
2	Welding Inspector Responsibilities	2-1
3	Codes, Standards and Specifications	3-1
4	Weld Joint Geometry and Welding Terminology	4-1
5	Welding Symbols	5-1
6	Weldability, Welding Chemistry and Welding Metallurgy	6-1
7	Destructive Testing	7-1
8	Welding Procedure and Welder Qualification	8-1
9	Welding, Brazing and Cutting Processes	9-1
10	Weld and Base Metal Discontinuities	10-1
11	Nondestructive Examination Processes	11-1
12	Inspection Reports	12-1
	Appendix A - Additional Resources & References	A-1
	Appendix B - Sample Forms	B-1
	Appendix C - Answer Key	C-1
	Index	xi

CHAPTER 1: THE WELDING INSPECTOR

Introduction

Welding Inspectors function as quality representatives of organizations that may be the manufacturer, the purchaser, an insurance company, or a government agency. The inspector is responsible for judging the acceptability of a product according to a written specification. The inspector must understand the specification both as to its limitations and intent. Keep in mind that the goal is to strive for the required quality without delaying completion and delivery without proper cause.

Welding inspectors find themselves working in dozens of different industries, with each situation having slightly different job responsibilities. Among those industries employing welding inspectors are: energy production, chemical processing, petroleum product refining and distribution, transportation, and bridge and building construction.

The welding inspector is a composite person, a specialist, highly qualified in the field of welding. Welding inspectors can be classified as:

- Code or governing agency inspector

- Purchaser's, customer's, or owner's inspector

- Fabricator's, manufacturer's, or contractor's inspector

- Architect's or engineer's inspector

Even though the in-house inspector may have different duties from the outside inspector, within this manual, only a single inclusive category of inspector is considered. Hence, the contents of this manual sometimes may apply to all the categories above or limited to one or more of them. In all cases, the inspector has the necessary qualifications and is competent to make the examinations appropriate for the type of weldment being inspected.

Nippon Steel

Lukens Steel Company

Cessna Aircraft Corporation

Western Oceanic Inc.

Important Qualifications for the Welding Inspector

For a person to become a welding inspector, there are a number of important qualifications. Any person who wishes to do his or her inspection job conscientiously and professionally must take these qualifications seriously.

Physical Condition

A welding inspector's physical condition must allow one to be an active inspector. Inspection requires examination before, during and after fabrication. Climbing around large fixtures and assemblies can be a job requirement. Inspection conditions are frequently difficult. Work is sometimes positioned for the convenience of welders and welding operators, but not necessarily for the inspector.

Short of unnecessary endangerment, the inspector must be able to see the weld to perform the visual inspection. Inspecting completed welds in the shortest possible time requires the same means of access to the weld that the welder had.

Vision

Good vision is vital. The ability to examine weld surface conditions and judge their acceptability according to written quality requirements are the primary functions of a welding inspector. An AWS Certified Welding Inspector (CWI) is required to have 20-40 vision, as determined by corrective eye charts, and Jaeger J-1 near vision acuity, with or without corrective lenses. The required eye examination may also include a color perception test for red/green and blue/yellow differentiation. Color perception for most visual welding inspection jobs is not a requirement. According to AWS QC 1, *"It is the employer's responsibility to determine and enforce any color perception requirement."*

Professional Attitude

It is extremely important that the welding inspector maintain a professional attitude. Attitude determines the degree of success or failure. Success will be dependent upon the cooperation of associates in all departments, and the welding inspector must have their respect to obtain their help. It is important that the welding inspector strive to be impartial and consistent in all decisions. The inspector should develop a definite method for inspection procedures. Remember, the welding inspector must be neither stubborn nor readily swayed by persuasive arguments. Under no circumstances may the welding inspector seek favor or incur obligation through personal decisions or pending decisions.

Knowledge of Welding and Inspection Terminology

The improper use of welding terminology by a welding inspector could create an embarrassing situation if it becomes apparent to others in job related conversations. Consequently, the welding inspector must know and communicate correctly the language of welding. The job of welding inspector requires communication of findings to the shop people, who created the welds and make repairs, and the engineers who planned the work and accept the final structure. The vocabulary used in speaking and writing must be in terms understandable to all involved.

Inspectors should make ready use of AWS A3.0, *Welding Terms and Definitions*. The latest edition provides AWS approved terminology used to describe the various aspects of welding. It is advisable that welding inspectors study and consult the standard until the terms become part of their natural vocabulary.

Included in this manual will be explanations of: types of joints and welds, parts of welds and weld application terminology (Chapter 4); terminology related to destructive testing (Chapter 7) and nondestructive examination (Chapter 11); names and descriptions of weld and base metal discontinuities (Chapter 10); terms related to welding metallurgy (Chapter 6); and terminology related to various welding processes (Chapter 9).

Knowledge of Drawings and Specifications

An inspector must be familiar with engineering drawings and able to understand specifications. Welding inspectors must be able to read and understand blueprints and drawings and must know welding and nondestructive examination symbols. It is not necessary to memorize the various standards and specifications that may be in effect. Rather, they should be available for reference whenever the welding inspector needs information contained in these documents. The welding inspector should be familiar with the contents so that it doesn't take long to find information.

Knowledge of Testing Methods

Numerous destructive and nondestructive testing methods are available for use in determining whether a base metal, weld metal, and/or a weldment meets certain specification requirements. While perhaps not performing the testing, the welding inspector must be aware of the test basics, including: application technique, obtainable information, and the advantages and limitations.

If others are performing the tests, the welding inspector must be certain the NDT technicians have the proper credentials and be familiar enough with the method to determine if the test results obtained meet prescribed requirements.

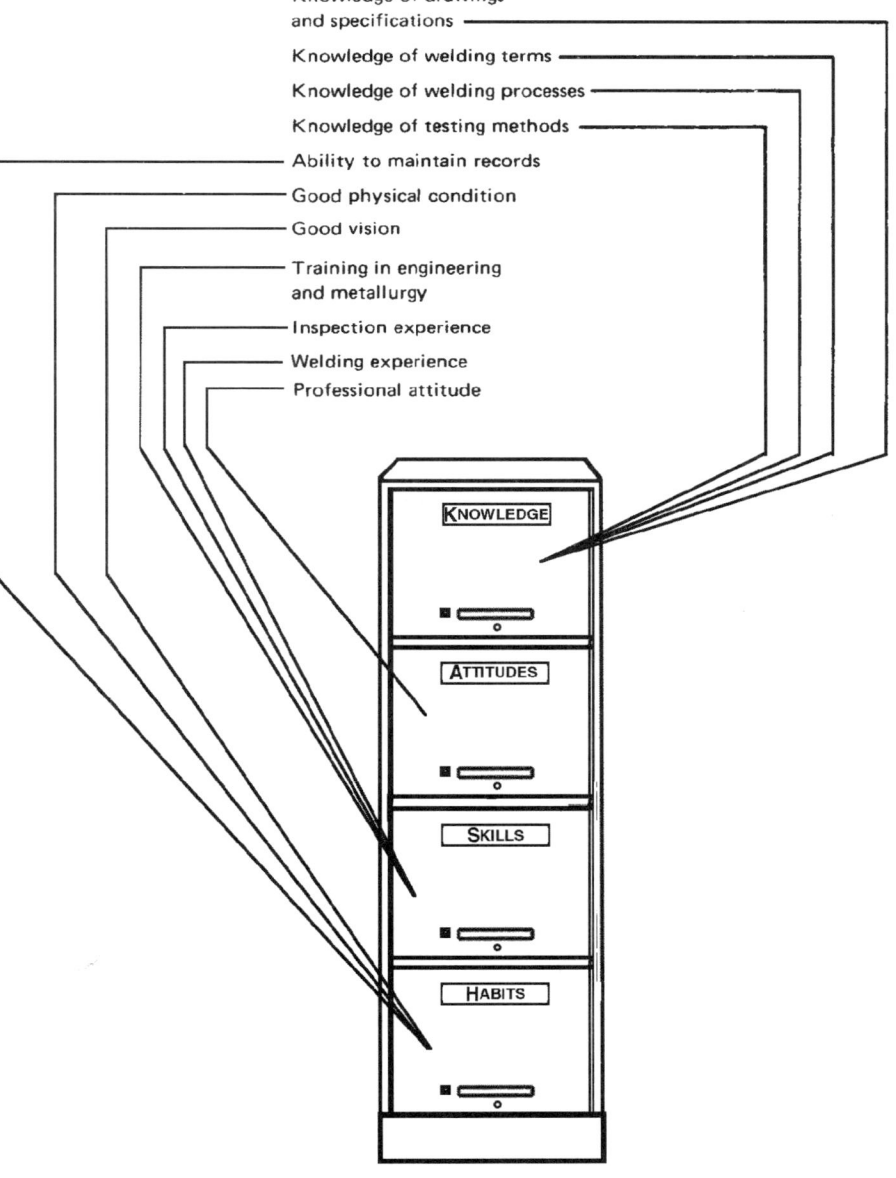

Ability to Produce and Maintain Records

A welding inspector should be able to develop and maintain inspection records. He must be able to write concise, accurate reports. The report should be simple and understandable to anyone familiar with the project. Reports should, at the same time, be complete enough so that the reason for decisions will be clear months or years later.

It is important to remember that well-known facts at the time of the writing often are not remembered as clearly, completely, or accurately later. Records should include not only all results of inspections and tests but also supporting records relating to: welding procedures, welder qualifications, drawing or specification revisions, etc. Good records also protect the welding inspector's reputation.

Knowledge of Welding Processes

Since the welding inspector spends the major portion of his time evaluating welds, knowledge of the various welding processes is essential. Further, actual experience as a welder or welding operator is valuable to a welding inspector, but is not mandatory. Welding experience broadens the inspector's welding knowledge, commands respect, and gives his opinions more credibility when weld quality is being evaluated. There are some employers who require actual welding experience as a prerequisite to becoming a welding inspector; however, welding experience is not a prerequisite to certification as an AWS CWI or CAWI. The inspector who is familiar with the advantages and limitations of the various welding processes is able to identify problems when, or even before, they occur.

Ability to be Trained

Welding inspectors are expected to possess knowledge in a number of different areas to be considered effective, and to be trained in areas of unfamiliarity. Many employers select persons as potential welding inspectors based on their ability to study and gain the necessary knowledge. Training in fundamental engineering can apply toward partial satisfaction of the experience requirements for becoming a CWI. AWS QC1, *Standard for Qualification and Certification of Welding Inspectors* outlines these limits for the CWI program.

Inspection Experience

Only through inspection experience is the attitude and point of view of a good inspector acquired. Even experience in inspecting unwelded materials is extremely helpful in the inspection of weldments, since a good inspector has developed a distinct way of thinking and working. Those learning the ropes should observe the behavior and techniques of experienced inspectors.

To comply with the experience requirement for AWS certification, the welding inspector must show evidence of having performed the functions of a welding inspector. Other job functions that have a close relationship to welding inspection also provide this evidence. Periods of qualifying experience are counted by the actual number of calendar months employed at jobs (not the number of employers). The jobs must have a close relationship to fabrication of weldments according to a code, standard, or specification, and directly involve one or more of the following:

- Design: Preparation of plans and drawings for weldment construction/fabrication.

- Production: Planning and control of welding materials, welding procedures and welding operations for weldment fabrication.

- Construction: Fabrication and/or erection of weldments.

- Inspection: Detection and measurement of weld discontinuities; verification of fabrication requirements, as described in Section 4.2 of AWS QC1.

- Repair: Repair of welds that are defective.

Ethical Requirements for the Welding Inspector

Introduction

This manual will discuss the technical methods, procedures, processes, and functions of the welding inspector. However, effective weld inspection requires not only the performance of duties consistent with the specification requirements, but also the practice of professional conduct and ethical principles. Below is the "Code of Ethics" for welding inspectors, which is included in AWS QC1, *Standard for Qualification and Certification of Welding Inspectors*. The latest edition of AWS QC1 should be consulted for the most recent requirements.

Preamble

In order to safeguard the public's health and well-being and to maintain integrity and high standards of skills, practice, and conduct in the occupation of welding inspection, the AWS Certified Welding Inspector (CWI) and Certified Associate Welding Inspector (CAWI) shall be cognizant of the following principles and the scope to which they apply, with the understanding that any unauthorized practice is subject to the AWS Qualification and Certification Committee's review and may result in suspension, reprimand or revocation of certification.

Integrity

The CWI and CAWI shall act with complete integrity (honesty) in professional matters and to be forthright and candid to the Committee or its representatives on matters pertaining to AWS QC1.

Responsibility to the Public

The CWI and CAWI shall act to preserve the health and well-being of the public by performing the duties required of weld inspection in a conscientious and impartial manner to the full extent of the inspector's moral and civic responsibility and qualification. Accordingly, the CWI and CAWI shall:

- Undertake and perform assignments only when qualified by training, experience and capability.

- Present credentials upon request.

- Neither falsely represent current status nor seek to misrepresent certification level (CWI/CAWI) by modification of certification documents false verbal or written testimony of current level or status.

- Be completely objective, thorough and factual in any written report, statement or testimony of the work and include all relevant or pertinent testimony in such communiques or testimonials.

- Sign only for work the inspector has inspected, or for work over which the inspector has personal knowledge through direct supervision.

- Neither associate with nor knowingly participate in a fraudulent or dishonest venture.

Public Statements

The CWI or CAWI shall issue no statements, criticisms or arguments on weld inspection matters connected with public policy which are inspired or paid for by an interested party, or parties, without first identifying the party, the speaker, and disclosing any possible financial interest.

The CWI or CAWI shall publicly express no opinion on welding inspection subjects unless it is founded upon adequate knowledge of the facts in issue, upon a background of technical competence pertinent to the subject, and upon honest conviction of the accuracy and propriety of the statement.

Conflict of Interest

The CWI or CAWI shall avoid conflict of interest with the employer or client and will disclose any business association, or circumstance that might be so considered.

The CWI or CAWI shall not accept compensation, financial or otherwise, from more than one party for services on the same project, or for services pertaining to the same project, unless the circumstances are fully disclosed and agreed to by all interested parties or their authorized agents.

The CWI or CAWI shall not solicit or accept gratuities, directly or indirectly, from any party, or parties dealing with the client or employer in connection with the CWI's and CAWI's work.

The CWI or CAWI shall, while serving in the capacity of an elected, retained or employed public official, neither inspect, review nor approve work in the capacity of CWI or CAWI on pro-projects also subject to his administrative jurisdiction as a public official, unless this practice is expressly dictated by a job description and/or specification and all affected parties to the action are in agreement.

Solicitation of Employment

The CWI and/or CAWI shall neither pay, solicit, or offer, directly or indirectly, any bribe or commission for professional employment with the exception of the usual commission required from licensed employment agencies.

The CWI and/or CAWI shall neither falsify, exaggerate, nor indulge in the misinterpretation of personal academic and professional qualifications, past assignments, accomplishments, and responsibilities, or those of the inspector's associates. Misrepresentation of current CWI/CAWI certification status at the time of, or subsequent to, submission of requested employment information, or in the solicitation of business contracts wherein current certification is either required or inherently beneficial (advertisements for training courses, consulting services, etc.) shall be a violation of this section.

The CWI and/or CAWI shall not function as an independent inspector in public fields out of his or her capability, without first investigating for possible industry or public requirements and additional education/experience requirements (e.g., industrial labs, in the concrete and soil testing field, etc.).

CAUTION:
While the CWI has established excellent credentials, certification alone may not legally qualify the Inspector to provide inspection services to the public. Contract documents,

Building or Jurisdiction Laws may require inspection to be performed under the direction and responsibility of others, such as a Registered Professional Engineer.

Unauthorized Practice

Any violation of any part of the standard of conduct prescribed by AWS QC1 if related to a CWI's or CAWI's occupation, including any violation of the Code of Ethics contained in AWS QC1, shall constitute an unauthorized practice subject to the imposition of sanctions.

Establishing Lines of Communication

The welding inspector must possess the physical, technical, and ethical qualifications mentioned earlier, as well as the skill of a communicator. The welding inspector's success will be affected by his ability to convey information to others as well as understand what others are trying to explain. Communication can occur in numerous forms, including: spoken words, written words, pictures, numbers, or gestures. Each provides an effective means of conveying information. The inspector should always be cautious regarding verbal communication. It is often important to have verbal communication supported by written information, especially when it relates to changes in the inspection requirements or results. Any form of communication should be a continuous loop. That is, the receiver should have the opportunity to respond to the sender. Effective communication must be considered as a two-way proposition.

The welding inspector often finds himself as the central figure in many fabrication situations; consequently, he must be capable of communicating effectively with many individuals involved in the project. The welding inspector must be able to establish lines of communication with a varied number of associates to accomplish

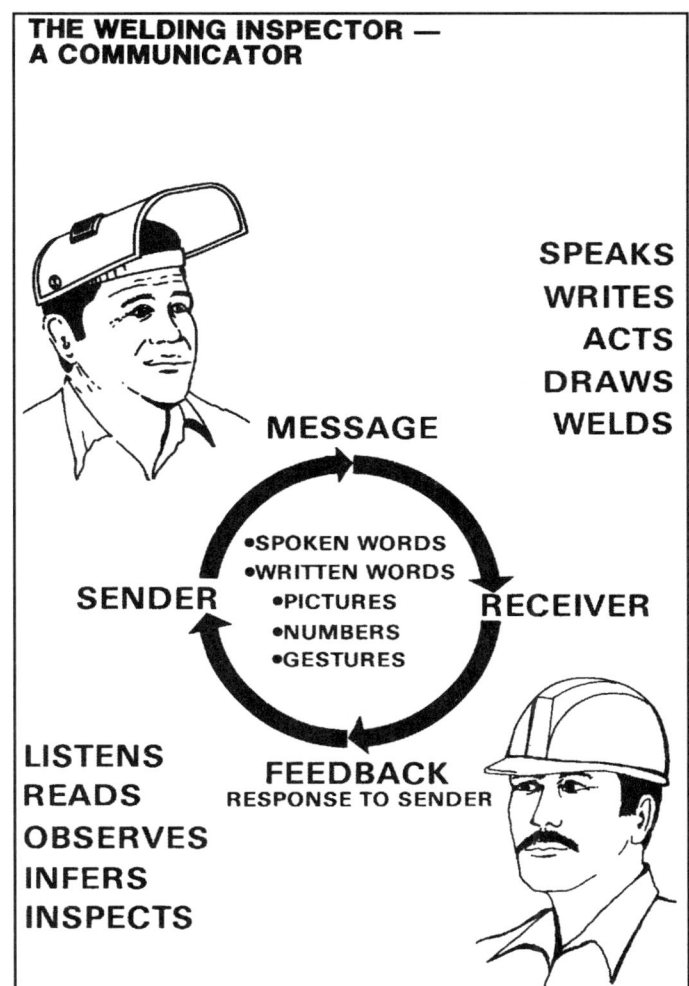

the necessary tasks and responsibilities efficiently and professionally. For example, the associates of a fabricator's inspector will often include the following:

Reporting Supervisor

Virtually all welding inspectors will report to someone; in some cases, it will be the chief inspector. In other situations, it could be the project engineer, a plant manager, an architect, or government official. Regardless of who this individual is, the welding inspector should be able to refer questions or job difficulties to him for answers or guidance. Through effective communication, that supervisor should be able to help the inspector become more knowledgeable of the job, and therefore more effective.

Welders

The relationship between the inspector and the welder is of utmost importance. The welder often knows which welds are of borderline quality. The welder may know where the joint fitup was improper or not as specified. A welder who looks on the inspector as an enemy is not concentrating on making each weld a good one. At any rate, he will not be trying to make the inspector's job any easier. In most cases, good rapport between the welder and inspector will result in improved quality and higher efficiency because problems are identified and corrected when they occur instead of later when repairing it is costly.

Welding Foremen or Supervisors

The welding foreman or supervisor is highly important to the inspector. Given a group of welders equally qualified to meet specifications, the foreman still decides which welder is best suited for more difficult welding jobs. Both the welding inspector and the foreman should be in agreement as to which welders are considered qualified to produce satisfactory welds. At times, the fabricator's representative may suggest that the welding inspector tell the welder what he wants. This would make the welding inspector a party to the operation, and perhaps eliminate the fabricator's responsibility to meet the specification requirements. Remember, the inspector's authority extends only to the determination as to whether the weld is acceptable or rejectable. Volunteering unsolicited information may overstep an inspector's responsibility and should therefore be limited.

These supervisory people are also concerned with production, so some communication must occur to indicate when parts or assemblies will be ready for inspection. After that inspection, they will need to know if the welding has been found to be acceptable. If rejects are noted, then more communication will be necessary to describe and locate the defect so that repairs can be accomplished.

Shop or Field Superintendent

There may also be the need to discuss weld quality matters with the shop or field superintendent. This is most likely because some aspect of the welding quality is having an effect on the overall production schedule of a project. In other cases, the welding inspector may simply have encountered some problem which is beyond the responsibility of the welding foreman or supervisor.

Plant Manager

Communication with the plant manager is much the same as that with the shop or field superintendent. They are responsible for production, as well as the product quality. Consequently, they are very concerned about learning what the status of some part may be when it is subjected to inspection. A single rejected weld could result in delays for the entire project. The welding inspector needs to keep him informed so he is aware of the acceptability or rejectability of some item so appropriate scheduling can be accomplished.

Design/Project Engineers

The design engineer is responsible for the details of any welded fabrication. His intentions are communicated through drawings and specifications. The project engineer is then responsible for the interpretation of these requirements when the work is being performed. The welding inspector may need to communicate with both of these individuals regarding these welding requirements. An overview of the job in consultation with the project engineer will bring out any fabrication or inspection tasks that may require additional planning. For example, in a massive, complex weldment, weld sequencing may be necessary to assure weld soundness and minimize distortion, or some design detail may need modification to facilitate successful fabrication and inspection.

Welding Engineer

The welding inspector should have access to the welding engineer, welding technician or welding specialist so that possible welding-related construction problems can be brought to his attention before they become inspection problems. The welding inspector can be thought of as the welding engineer's "eyes" as the welding is being performed. When such an arrangement exists, the welding inspector can communicate with the welding engineer to describe fabrication problems so that corrective action can be instituted. Proper communication is important to the inspection process. The welding inspector's primary function is to inspect the fabricator's work to see that it meets the requirements of the contract. The quality of work being accomplished is the substance of the welding inspector's reports. Whether the fabricator takes advantage of this information may depend upon how clearly it has been presented in the report. The fabricator still has full responsibility for the quality of the final product.

Proper communications will allow the welding inspector to keep in touch with the activities of the production organization. Early correction of a fault results in producing a satisfactory product instead of one that would otherwise have to be rejected and subsequently repaired.

Summary

The job of welding inspector requires a wide variety of talents and physical capabilities. The individual must be both mentally and physically prepared for the many tasks at hand. The inspector's day-to-day existence dictates that he has the proper knowledge and training, can accurately report his findings and maintain that information for future reference, and can conduct himself in an ethical manner.

Good welding inspectors are invaluable to a company. When permitted to act as specified by an effective quality control system, the welding inspector can often save a company money by identifying problems when or before they occur to minimize correction costs.

REVIEW - CHAPTER 1
THE WELDING INSPECTOR

Q1-1 Of the following, which is considered an important duty of the welding inspector?
   a. It is a welding inspector's responsibility to judge the quality of the product in relation to some form of written specification.
   b. A welding inspector functions as a judicial representative of an organization.
   c. A welding inspector must be able to interpret the specification limitations and intent.
   d. all of the above
   e. none of the above

Q1-2 Of the following, which is not considered an important attribute of a welding inspector?
   a. welding experience
   b. inspection experience
   c. professional attitude
   d. engineering experience
   e. ability to be trained

Q1-3 What document describes the important requirements of the AWS Certified Welding Inspector program?
   a. AWS D1.1
   b. AWS A5.1
   c. AWS QC 1
   d. AWS D14.1
   e. none of the above

Q1-4 As a welding inspector, must you know how to weld?
   a. yes, according to AWS D1.1
   b. yes, according to AWS QC 1
   c. yes, if inspecting highway bridges
   d. no, according to AWS D1.1
   e. not mandatory, according to AWS QC 1

Q1-5 When may you, as a welding inspector, speak out on matters of public policy?
   a. when paid by an interested party or parties
   b. anytime, if no money interest is involved
   c. after disclosing all possible financial connections of the statement, criticism, or argument presented
   d. when the statement is based upon adequate facts, upon a background of technical competence pertinent to the subject, and upon honest conviction of the accuracy and propriety of the statement
   e. Welding inspectors should never make public statements

Q1-6 Which of the following are important ethical requirements for the welding inspector?
   a. integrity
   b. professional ability
   c. good physical condition
   d. volunteering public statements regarding an inspection for personal exposure.
   e. all of the above

Q1-7 Of those attributes considered to be important to the welding inspector, which is probably most influential in his gaining the cooperation and respect of others with which he works?
   a. ability to be trained
   b. professional attitude
   c. ability to complete and maintain inspection records
   d. good physical condition
   e. ability to interpret drawings and specifications

Q1-8 The welding inspector is likely to work in which of the following industries?
   a. shipbuilding
   b. automotive
   c. bridge construction
   d. pressure vessel construction
   e. all of the above

Q1-9 According to the requirements of the AWS CWI program, what is the necessary visual acuity of a welding inspector?
a. 20/20 natural vision
b. 20/20 corrected vision
c. 20/40 natural vision
d. 20/40 corrected vision
e. 20/40 natural or corrected vision

Q1-10 Which of the following could be considered essential knowledge for a welding inspector?
a. nondestructive testing
b. welding symbols
c. welding processes
d. destructive testing
e. all of the above

Q1-11 When a weld requires repair due to some deficiency, to whom should your inspection report be directed?
a. to the welder whose mark is on the weld
b. to another welder, better trained
c. to the project engineer
d. to the welding engineer
e. to the welding foreman or supervisor

Q1-12 What professional attributes are most helpful in performing inspection duties?
a. being informed, impartial and consistent in your decisions
b. being close friends with welders and superiors
c. being a former welder
d. being a non-union employee
e. being a nondestructive examination technician as well as CWI

Q1-13 With whom may the welding inspector communicate during the performance of his or her inspection responsibilities?
   a. welding engineer
   b. welding foreman
   c. welders
   d. inspection supervisor
   e. all of the above

Q1-14 What document defines the proper terminology for use by the CWI?
   a. AWS QC 1
   b. AWS A3.0
   c. AWS D1.1
   d. AWS A5.1
   e. none of the above

Q1-15 With regard to drawings and specifications, the CWI must:
   a. be familiar with engineering drawings and able to understand specifications.
   b. memorize the content.
   c. memorize those portions of these documents applicable to a particular job.
   d. all of the above
   e. none of the above

CHAPTER 2: WELDING INSPECTOR RESPONSIBILITIES

Introduction

The welding inspector holds a position of responsibility. This responsibility demands a professional person with good character, ability, and common sense. A welding inspector may work at various fabrication plants and at various job sites. In all cases the welding inspector should observe the working hours of the fabricating organization. Strict observance of all rules and regulations especially those pertaining to personal conduct, safety, and security are mandatory. Never should the welding inspector consider himself entitled to special privileges. In dealing with the fabricating organization, the welding inspector should be impartial, render decisions promptly, and be tolerant of the opinion of others during communications. Remember, however, to stick to the facts when making decisions. Do not be easily swayed by differing opinions.

To perform visual inspection effectively, it is necessary to observe as many of the individual stages of fabrication as possible. Consequently, the various responsibilities of the welding inspector are categorized when they occur, specifically: before, during and after welding. Typical inspection requirements might be:

Inspection Responsibilities Before Welding

- Review all applicable drawings and standards.
- Check purchase orders to ensure that base and filler materials are properly specified.
- Check and identify materials as they are received against the purchase specifications.
- Check the chemical compositions and mechanical properties shown on mill test reports against specified requirements.
- Check the condition and storage of filler metals
- Check the condition and adequacy of equipment to be used.
- Check weld joint edge geometries.
- Check joint fit.
- Check joint cleanliness.
- Check the welding procedures and welder qualifications.
- Check preheat temperature.

Inspection Responsibilities During Welding

- Check welding parameters and technique for compliance with welding procedure.
- Check quality of individual weld passes.
- Check interpass cleaning.
- Check interpass temperature.
- Verify that in-process NDE is performed, if required.

Inspection Responsibilities After Welding

- Check finished weld appearance.
- Check finished weld sizes and lengths.
- Check dimensional accuracy of completed weldment.
- Select production test samples.
- Evaluate test results.
- Verify that additional NDE has been performed, if required.
- Verify that postweld heat treatment has been done satisfactorily, if required.
- Prepare and maintain inspection reports.

Inspection Responsibilities Before Welding

Knowledge of Drawings and Standards

Drawings, designs, standards, contracts, etc., should be studied in advance so that the welding inspector is aware of the construction details, the proposed use of subassemblies, and the specifics of the welding operation. Note which materials are to be used in the welded structure and whether any of them require special treatment for satisfactory welding. This information should be clearly stated in the standards or welding procedures. If it is not there, the project engineer should be contacted for clarification.

During fabrication of a welded structure or component, the welding inspector may be called upon to interpret drawings or standards on the spot. Prior study of the drawings and design requirements will enhance one's ability to make clear and concise decisions. Clear and concise decisions will speed completion of the work under contract, increase the inspector's professional image, and greatly aid in exercising authority.

Situations that require a deviation from the drawing or detailed standard may arise during the fabrication of any structure. It is the welding inspector's responsibility to alert the project engineer or quality assurance (QA) personnel. They will decide whether the deviation in question should be permitted or rejected.

Sometimes acceptance or rejection of a large welded structure will be involved. It may be the inspector's duty, after careful study, to recommend whether the error can be corrected and whether the method of correction to be used will still insure a satisfactorily completed product in accordance with the drawings and standards. In all cases, he should exercise extreme caution in accepting deviations. Deviations from drawings should be referred to the design agency for approval. Their approval should be received "in writing."

It is not always possible to write an all-inclusive standard containing all the detailed information needed to provide an answer for any question that might arise. If parts of the standard have requirements that are not fully defined, the inspector is often responsible for determining the meaning and intent of that document. Communication with engineering and design personnel may be necessary before responding to a fabricator.

Purchase Specifications Check

The specifications for the job should identify all the materials that will be used. This should include all consumable material such as welding electrodes, welding or brazing fluxes, shielding gases, consumable inserts, and backing bars or rings. The inspector should review the purchase order or contract to see that the materials ordered meet the specification requirements. For example, commercial specifications for steel, such as ASTM A572, frequently include more than one grade of the product, which must be individually identified on the purchase order to obtain the correct grade.

As another example, specifications for welding electrodes, such as AWS A5.1, will cover a dozen or more electrode classifications in the one document. In A5.1, the six E70XX electrodes are all equivalent in strength, but they are not interchangeable when low hydrogen grades are required. If low hydrogen types are required, either E7015, E7016, E7018, or E7028 must be called out on the purchase order.

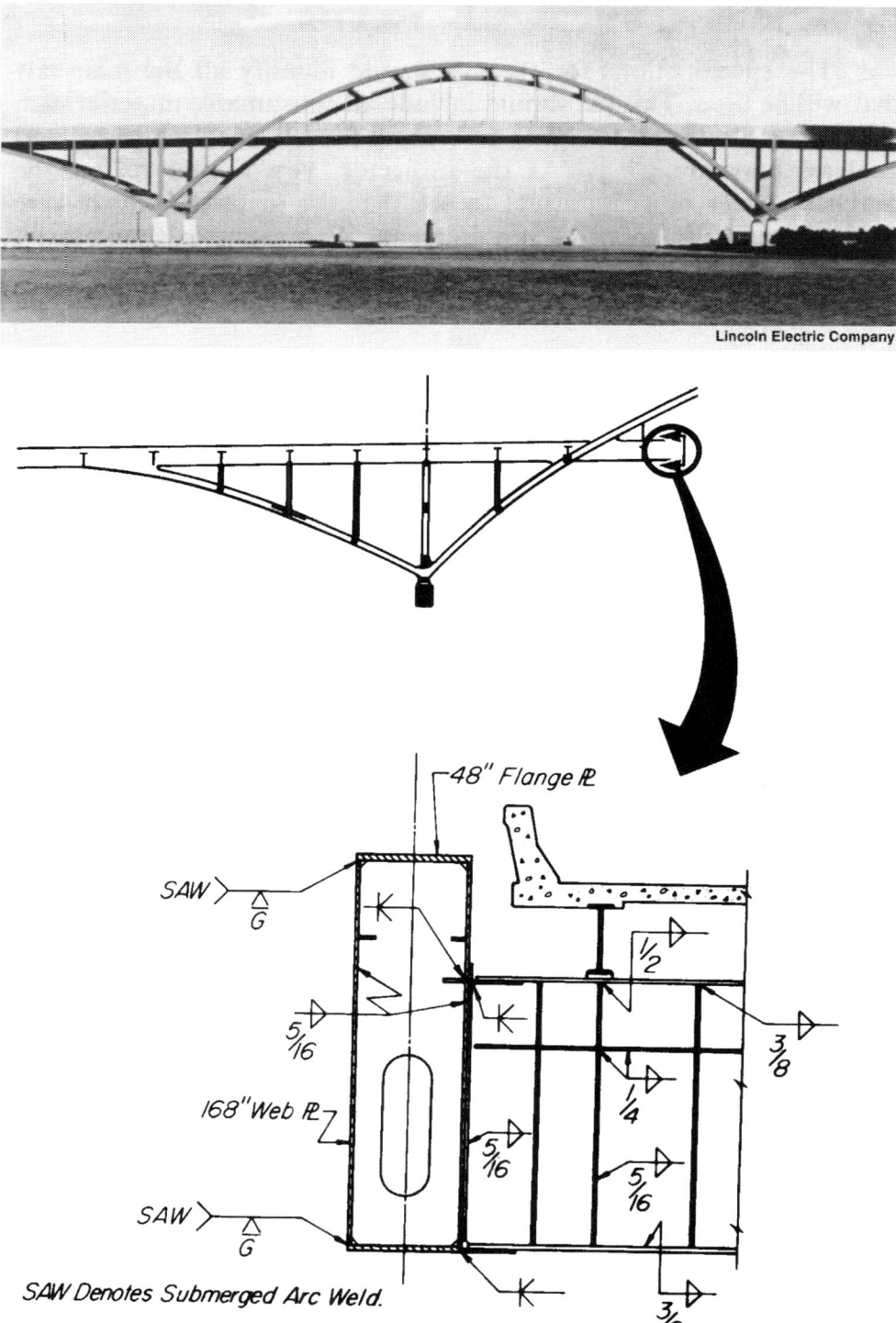

Job Material Verification

Mistakes do happen. Many metals look alike, leading to the possibility of inadvertent mix-ups. To prevent this, the inspector must verify that the materials supplied match the applicable purchase order, when received. In some cases, there are additional requirements for marking the proper identification visibly on each piece, preferably at multiple locations. Identification is lost for the rest of the plate if the first slice cut off by the shop removes the labeled end with the grade stamp on it. Good practice calls for remarking remnants produced in cutting operations prior to the actual cutting while the identity is still verifiable.

Identification of low hydrogen welding electrodes is sometimes lost when stockroom clerks prematurely remove electrodes from their containers to store them in holding ovens. The individual electrodes will still bear classification numbers on their coverings (at the stub end), but the manufacturer's control number only appears on the now discarded container. In nuclear and other critical work, that manufacturer's control number may be needed for each weld. The preferred container for a welding electrode is its original unopened package, or stored in its opened package in an oven or suitable location.

Chemical Analysis and Mechanical Properties Tests

Confirming tests of incoming materials are desired for many reasons. The number of tests required will depend on the inspector's judgment and past experience, unless everything is called out under quality assurance procedure requirements. If the material is in question, the inspector should request tests on representative samples of critical incoming materials to verify chemical composition and mechanical properties. Where material test reports from reliable suppliers (that is, suppliers accredited through an approved quality assurance program) give assurance of conformance, the welding inspector should ask for check tests when errors are strongly suspected.

Base Metal Defects Investigation

The quality of mill products supplied for a job must equal or exceed the quality specified for the final weldment or structure. Base metals almost always contain many small discontinuities. Their effect depends on the thickness of the metal, the type of loading and the criticality of the design. In some instances, they may be cause for rejection; while in other instances, they may not.

The welding inspector has the responsibility to see that discontinuities in the base metal are detected, identified, evaluated, and repaired properly, where necessary, so that they will not be incorporated into the welded product. That requires that the inspector be aware of the acceptable limits for these discontinuities. At times there may be a need to communicate with the responsible engineer in deciding what to do about a major defect.

Some specifications may require that a defective piece be rejected and replaced unless it can be repaired. In fact, some defects may require rejecting the piece altogether. On the other hand, minor repairs such as flame straightening of members accidentally bent in shipping or handling may be permitted by the specification. As another example, the presence of laminations will rule out placing that plate where it must withstand tensile stresses in the through thickness direction, but other locations can be found where the same plate may be used with complete safety. Engineering should make that decision. The welding inspector must then inspect any replacement materials.

Condition and Storage of Filler Materials Check

The welding inspector should check the condition of filler metals to be used. This is

CERTIFICATION OF TESTS

CUSTOMER TECNIWELD, INC.   AML Q.M.   RECEIVED MAY 19 1990

PURCHASE ORDER 4563   **DATE SHIPPED** 5-14-90

DESCRIPTION
1 3/4" THK., 10 PCS. 39-1/2 O.D. X 9-5/8 I.D. TEMP
2
3
4

SPECIFICATION: ASTM-A-516 GR. 70 PVQ - FINE GRAIN

PROPERTIES	1	2	3	4
HEAT NO.	91E060			
C	.22			
MN	1.05			
P	.009			
S	.022			
SI	.23			
NI	.02			
CR	.02			
MO	.005			
CU	.02			
V	.004			
YIELD PSI	48.2 KSI			
TENSILE PSI	76.3 KSI			
%ELONGATION	8" = 25.0			
% RED OF AREA				
BEND TEST				

CB 004

WE HEREBY CERTIFY THAT THE ABOVE MATERIALS HAVE BEEN INSPECTED AND TESTED IN ACCORDANCE WITH THE APPLICABLE SPECIFICATION AND THAT THE ABOVE FIGURES ARE CORRECT AS CONTAINED IN THE RECORDS OF THIS COMPANY.

SWORN TO AND SUBSCRIBED BEFORE ME THIS _____ DAY OF _____ 19 ___

JANE DOE
Notary Public, State of Ohio
Recorded in Cuyahoga County
My Commission Expires 2/22/2001

NOTARY PUBLIC

GENERAL STEEL SHEET & PLATE, INC.
ANYTOWN, U.S.A.

By _J B Manager_

Typical Example of a Mill Test Report

PLATE LAMINATION

especially true in the case of shielded metal arc welding electrodes that have a flux coating that can be easily damaged. Other types of filler metals that may be stored out in the open could also deteriorate with time. For example, solid wire and flux cored electrodes can develop rust, which could result in the production of porosity in the weld.

When low hydrogen type shielded metal arc welding electrodes are being used, they must be stored in electrically heated, vented and thermostatically-controlled storage ovens to maintain their low moisture content once they have been removed from their shipping containers.

Besides these filler metals, some fluxes and flux cored electrodes also require special protection from moisture, whether it be rain or high humidity. Some submerged arc fluxes even require heated storage containers.

ELECTRODE OVEN

Welding Equipment Check

All welding equipment, including that to be used for testing, should be checked periodically for operational capability, calibration, and safety. For example, always check the ammeters and voltmeters if present. These meters on welding machines may not always be accurate because of mistreatment, shop contamination, and overloads. Periodic calibration is recommended.

The equipment should also be checked to make certain it has the necessary output capacity to satisfy the welding procedure requirements. Welding leads, gas hoses and wire feed apparatus must also be examined to ensure their good condition and operability.

Weld Joint Edge Geometry Check

Specific tolerances are listed for weld joint edge geometries in various codes and specifications. The prequalified joints found in Section 2 of AWS D1.1 are typical examples. Inspection responsibilities include examination of the unwelded joint for edge geometry, including root face dimensions and groove angles. The suitability of the joint for the welding process to be used will be discussed in greater detail in Chapter 5.

Weld Joint Fit Check

Again, looking at the unwelded joint, the welding inspector should observe the fit of the parts. For butt joints, alignment and root opening are important. Keep in mind that prestressing or cambering may be needed for welds that will be subject to distortion as a result of weld shrinkage stresses. The fabricator should not attempt such welds without the necessary knowledge.

Inspector Checking Equipment

Procedure modifications may give an acceptable weld, but the welding inspector may offer suggestions to the project engineer, not to the foreman. If the foreman later says, "You told me to do it this way," the inspector has more difficulty in turning down an unacceptable result.

The fit of backing preparations needs particular attention. Backing bars and rings should fit tightly against the pieces to be joined. Transverse joints between segments of a backing material are undesirable, because they induce cracking in the root pass. For this reason, when welding in accordance with AWS D1.1, steel backing on groove welds is required to be made continuous for the entire length of that backing member.

Weld Joint Cleanliness Check

In welding, the cleanliness of the base metal surfaces in and adjacent to the joint is a critical factor. Welding over contamination such as oil, grease, paint, moisture, rust, etc. will likely result in porosity in the completed weld. In many cases, this contamination could also lead to the occurrence of incomplete fusion, or even cracking. Consequently, it is imperative that the welding inspector check the cleanliness of the weld zone prior to welding.

Welding Procedure Qualification Check

The fabricator must pre-scribe the details of the welding procedure that will be followed in producing weldments. They should produce welded joints with acceptable mechanical properties as required by the particular specification or code. Chapter 10 of this manual describes the basic elements of a welding procedure specification, the reasons for its use, its qualification, and your responsibilities as welding inspector in verifying proper application of the welding procedure on the job.

Welder Qualification Check

Codes and specifications that apply to the fabrication of weldments usually require qualification of all welders and welding operators. It is the welding inspector's duty to verify that every welder and welding operator who works under the code or specification has been properly qualified in accordance with those requirements. Verification can be made either by personally witnessing each test or a review of verified test results. It is important for the welding inspector to monitor the welders and welding operators to ensure that they are working within the scope of their qualifications with respect to such variables as: base metal type and thickness, position of welding, welding process, electrode type and size, etc.

Inspector observing welder qualification testing.

Welding codes and specifications do not normally require requalification of procedures, welders or welding operators for each new contract or design. The welding inspector should review the requirements of the contract specification or code to make this determination. Remember that the main objective of qualification tests is to insure that procedures and welders or welding operators are adequate for the intended purpose. To be fair to the fabricator and purchaser, the inspector should make every effort to avoid unnecessary qualification tests.

Preheat Temperature Check

Most codes and specifications require that certain materials are preheated prior to welding. In Chapter 6, there is a discussion of some of the reasons why preheat is necessary. For carbon steels, preheat will be required as the base metal alloy content or thickness increases. Most often, preheat temperature is verified using temperature indicating crayons that are formulated to melt at the temperature noted on their coatings. However, digital contact pyrometers are an effective alternate. Since preheat is necessary to prevent degradation of the base metal properties during the welding operation, the preheat temperature should be measured approximately 1 inch from the edge of the weld preparation, unless the part is greater than 3 inches in thickness. In that case, the temperature measurement should be made at a distance from the weld approximately equal to the part thickness. Normally, the temperature should be maintained during all welding of the joint.

For most carbon steels, the preheat is specified as a minimum. However, for some types, such as the quenched and tempered steels, the preheat temperature is expressed as a range of temperatures having a minimum and maximum value.

Taper Gauge

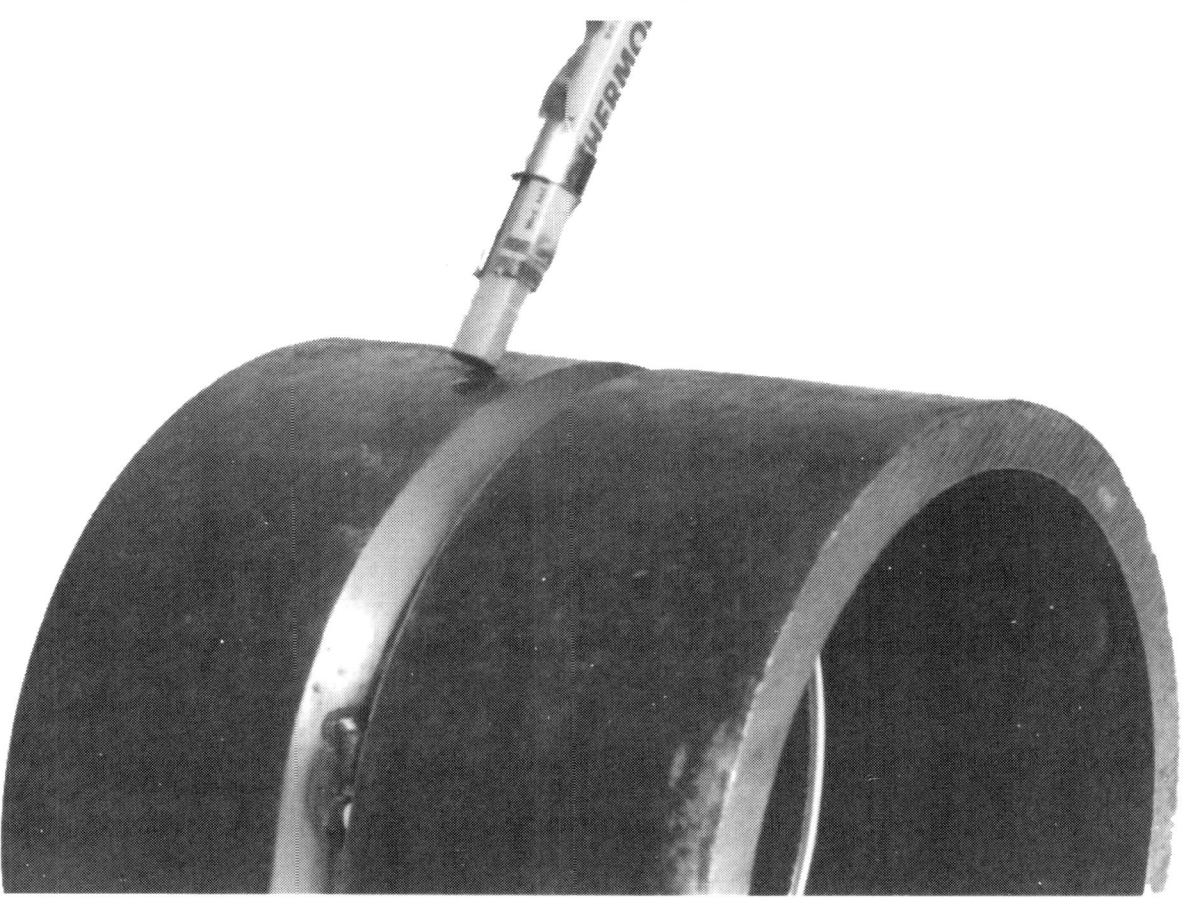

Temperature Sensitive Crayon for Measuring Metal Temperature

Welding Inspector Responsibilities During Welding

In order to continue the ongoing welding quality control, the welding inspector has numerous items to check as the welding is actually being performed. As was the case for inspections performed prior to welding, these checks can hopefully detect problems when they occur so they can be more easily corrected. During this phase of the fabrication process, the inspector's knowledge of welding will be extremely beneficial, since part of the inspection will involve the evaluation of the actual welding technique as well as the resulting weld quality. It is realized that it is unrealistic to think that the welding inspector can observe the deposition of each and every weld pass. Therefore, the experienced welding inspector should be able to select those aspects of the welding sequence which are considered to be critical enough to warrant his presence.

Following are some of the aspects of this phase of visual welding inspection which you may need to perform.

Check Production Welding for Compliance with Welding Procedure

When conducting welding inspection during the production welding, the inspector must rely on the welding procedure to guide that inspection. This document will specify all of those important aspects of the welding operation, including: welding process, materials, specific technique, preheat and interpass temperature, plus any additional information which describes how the production welding should be performed.

Tong Test Ammeter

The welding inspector's job will essentially consist of monitoring the production welding to assure that it is being performed in accordance with the appropriate procedure.

Check the Quality of Individual Passes

One of the aspects of the welding inspection during production welding is the visual examination of the individual weld passes as they are deposited. At that time, any surface discontinuities can be detected and corrected, if necessary. It is also important to note any weld profile irregularities which may hinder subsequent welding. An example of such a situation may occur during the welding of a multipass groove weld. If one of the intermediate passes is deposited such that it exhibits a very convex profile which creates a deep notch at its toe, that configuration may prevent a subsequent pass from properly fusing at that location. If noted, the welding inspector could ask that some grinding be done to assure that thorough fusion can be attained on the next pass.

Checking the in-process quality is especially critical in the case of the root bead. In most situations, this portion of the weld cross section represents the most difficult welding condition, especially in the case of an open root joint. In conditions of high restraint, the shrinkage stresses from welding may be sufficient to fracture the root pass if it is not thick enough to resist those stresses. The welding inspector should be aware of these problems and thoroughly check the root pass prior to any additional welding so that any irregularities can be found and corrected when they occur.

Check Interpass Cleaning

Another feature which should be evaluated during the welding operation relates to cleanliness of intermediate weld passes. If the welder fails to thoroughly clean the weld deposit between individual passes, there is a possibility that slag inclusions and/or incomplete fusion could result. This is especially critical when using a welding process which uses a flux for protective shielding. However, careful interpass cleaning is still recommended for those processes using gas shielding. Proper cleaning may be hindered when the deposited weld exhibits a convex profile which prevents sufficient access to the slag coating. As indicated above, it may then be necessary to

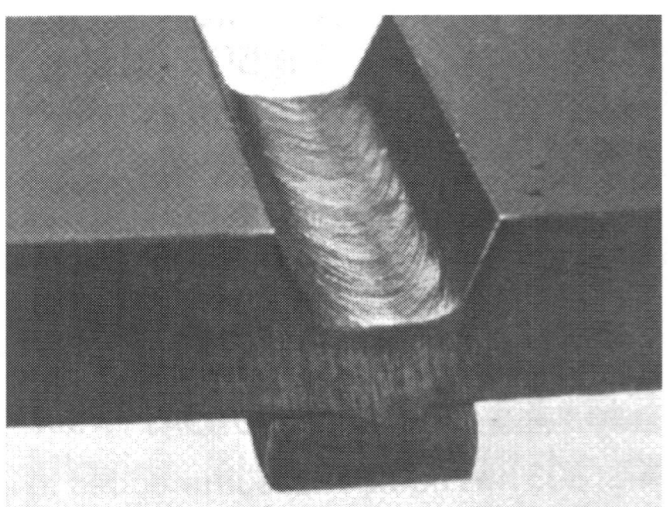

Partially Filled Groove Welded Joint With Backing Strip

perform additional grinding to remove the objectionable profile and facilitate proper cleaning.

Check Interpass Temperature

For welding procedures requiring interpass temperature control, the welding inspector may need to monitor this aspect of the process. As with preheat, the interpass temperature could be specified as a minimum, maximum or both. The interpass temperature is also measured on the base metal surface near the weld zone. Devices such as temperature indicating crayons and surface contact pyrometers are used for these measurements.

In-Process Nondestructive Examination

For some welds, there may be a requirement for other types of nondestructive examination other than visual examination. This testing may occur at various stages during the production of welds. For example, it is common for the root pass to be evaluated using magnetic particle or penetrant testing to assure that it is free of surface discontinuities or cracking. Discovery of problems at this time will result in a relatively easy and inexpensive repair compared to that required if the problem were not detected until the weld was completed.

Nondestructive examination operations shall be performed by an individual qualified in accordance with the recommendations of ASNT's "SNT-TC-1A," or equivalent. If the welding inspector has this qualification, he can perform this inspection as well as the visual examination. However, often a separate NDE technician will perform the nondestructive test. The welding inspector is required to verify that the proper test has been administered and the results have been properly recorded.

Welding Inspector Responsibilities After Welding

Once a weld has been completed, the welding inspector must examine the finished product to assure that all preceding steps have been performed to produce a quality weld. If all of the preliminary steps have been performed as required, the postweld inspection should simply confirm that the weld is of sufficient quality and size. However, the codes specify the required attributes of the finished weld, so the welding inspector must examine the weld visually to determine if those requirements have been met. Some of the important features of this post-weld inspection are discussed below.

Check Final Weld Appearance

In general, visual inspection after welding consists of looking at the appearance of the finished weld. This visual examination will detect surface discontinuities in the weld and adjacent base metal. Of special importance during this aspect of the welding inspection is the evaluation of the weld's profile. Sharp surface irregularities could result in premature failures of a component during service or create difficulty in film interpretation if the weld is to be radiographed. These visible features are evaluated in accordance with the applicable code which will describe the permissible amount of a certain type of discontinuity.

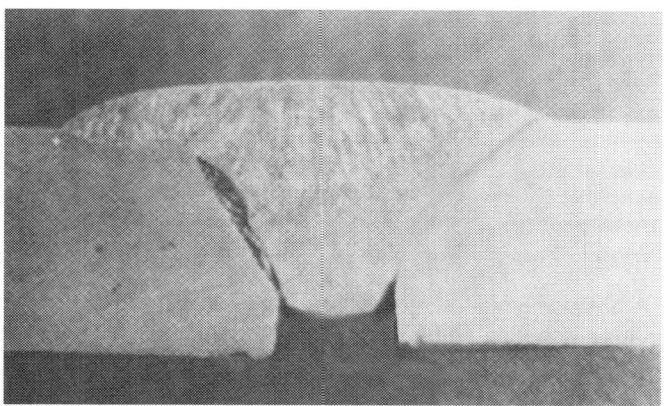
Cross-section of a partial penetration groove weld with heavy slag inclusion

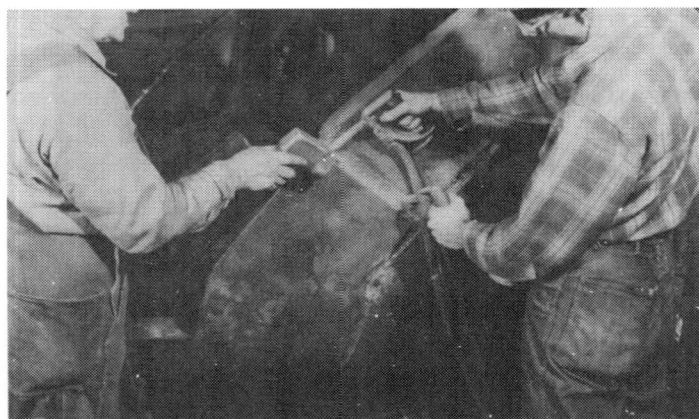
Magnetic Particle Testing of a partially filled groove weld

Check Final Weld Sizes and Lengths

Included in visual examination of a weld is the measurement of the weld to determine if it is as large as required by the drawing. For a groove weld, a primary concern is whether or not the weld groove is filled flush with the base metal surfaces or has the required reinforcement. Any underfill which is present must be corrected by depositing additional weld metal. Excessive weld reinforcement must be removed in such a manner that its height is within specified limits and it blends smoothly with the adjacent base metal.

For fillet welds, the size determination is normally accomplished with the aid of a fillet weld gage. While this measurement could be performed using standard measuring devices, fillet weld gages have been manufactured to facilitate much easier and accurate gaging of the fillet weld size. There are numerous different types of fillet weld gages which can be used, including gages or templates which are specially made for a particular fillet weld configuration.

Since fillet weld sizes are designated as nominal dimensions, there should realistically be some tolerance applied to their measurement. Since commercially-available gages are typically graduated in 1/16 inch increments, it would seem reasonable to gage fillet weld sizes to the closest 1/32 inch. Conditions warranting such an approach include: the difficulty in positioning your eyes properly to view the gage, the fact that weld sizes cannot be thought of in terms of machining precision, gage imprecision, base and weld metal surface irregularities, and the difficulty in determining the exact location of the toe of a convex fillet weld.

Once a weld has been measured to determine if it is of sufficient size, the welding inspector must then evaluate its length to assure that enough weld metal was deposited to satisfy the requirements. This is of special importance where intermittent fillet welds have been specified. Here each segment should be measured as well as their center-to-center, or pitch, distances.

For continuous groove or fillet welds, they are only considered to be of sufficient length if they are filled to their full cross section for the entire length of the shorter of the two members being joined. Normally, minimum lengths are specified, so the presence of extra fillet weld length is not considered rejectable. However, in some cases, excessive weld lengths may be unacceptable.

Check Dimensional Accuracy of Completed Weldment

Other measurements are required to evaluate the overall dimensional accuracy of the completed weldment. This is important since the shrinkage stresses from welding may have caused the size of the part to change. For example, a weld deposited around the outside of a machined bore may cause the diameter of that bore to be reduced, necessitating further machining to provide the appropriate bore size. Some of this dimensional evaluation will be to determine if any distortion resulted from welding. The localized heat of welding could cause members to be distorted or misaligned with respect to other parts of the weldment. These measurements will determine if the amount of distortion which is present is enough to cause the part to be rejectable or unusable.

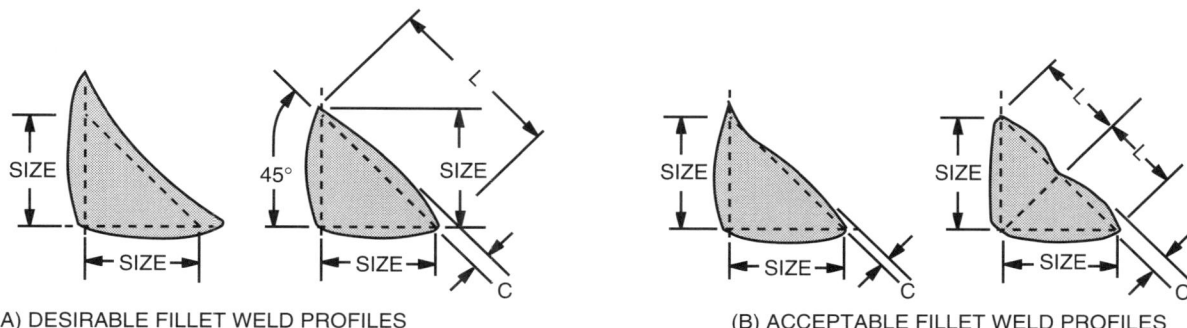

(A) DESIRABLE FILLET WELD PROFILES

(B) ACCEPTABLE FILLET WELD PROFILES

Note: Convexity, C, of a weld or individual surface bead shall not exceed the value of the following table:

Measured Leg Size or Width of Individual Surface Bead, L	Max. Convexity
L ≤ 5/16 in. (8 mm)	1/16 in. (1.6 mm)
5/16 in. < L ≤ 1 in. (25.4 mm)	1/8 in. (3mm)
L ≥ 1 in.	3/16 in. (5mm)

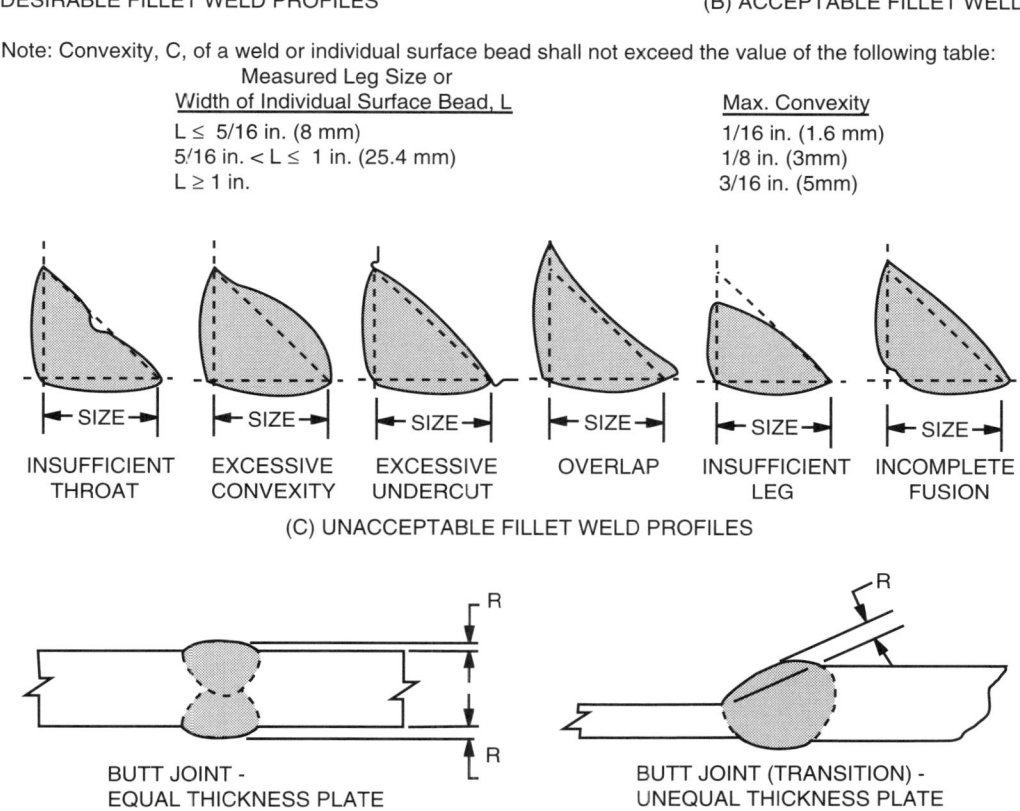

INSUFFICIENT THROAT | EXCESSIVE CONVEXITY | EXCESSIVE UNDERCUT | OVERLAP | INSUFFICIENT LEG | INCOMPLETE FUSION

(C) UNACCEPTABLE FILLET WELD PROFILES

BUTT JOINT - EQUAL THICKNESS PLATE

BUTT JOINT (TRANSITION) - UNEQUAL THICKNESS PLATE

NOTE: Reinforcement = R

(D) ACCEPTABLE GROOVE WELD PROFILE IN BUTT JOINT

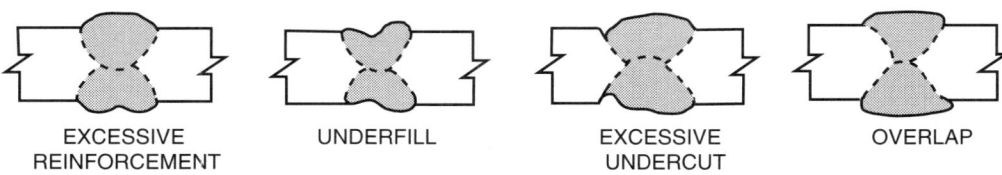

EXCESSIVE REINFORCEMENT | UNDERFILL | EXCESSIVE UNDERCUT | OVERLAP

(E) UNACCEPTABLE GROOVE WELD PROFILES IN BUTT JOINTS

ACCEPTABLE and UNACCEPTABLE WELD PROFILES (Source AWS D1.1-92)

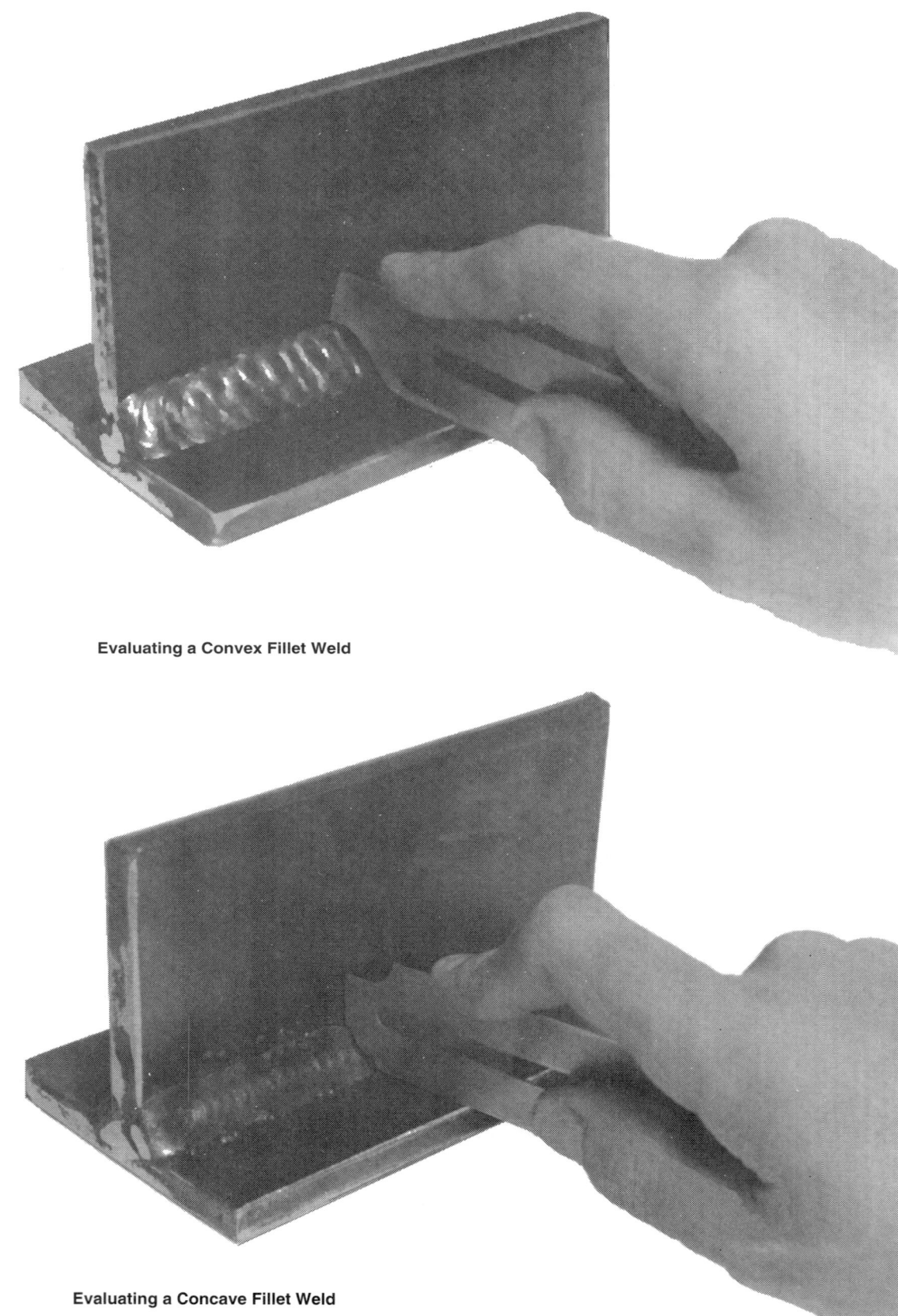

Evaluating a Convex Fillet Weld

Evaluating a Concave Fillet Weld

Selection of Production Test Samples

In welded assemblies, inspection of the product may be performed on samples taken from the production line. These samples may be selected by chance or according to an established order. In either case, witnessing their selection and testing is one of the welding inspector's important duties. Sometimes selection of samples is left to the inspector's judgment and discretion. It is not necessary to take more samples than are necessary to determine conformance. Typical tests include: radiography and other nondestructive tests, hydrostatic tests, chemical analysis, metallurgical examinations, and destructive mechanical testing. It is important to assure that such work is properly carried out. Various sampling plans, test processes and inspection methods are described in Chapters 6 and 9.

Evaluation of Test Results

Sometimes it will be impractical for a welding inspector to witness all tests. However, enough tests should be witnessed to satisfy the inspector that the tests are being performed in the proper manner and that the results are accurately reported. It is important to make certain that the testing equipment calibration is documented. When the tests have been performed, the inspector must evaluate the results and decide whether the product meets the specifications.

Final Nondestructive Examination

Some welds must also be examined upon completion using other nondestructive test methods in addition to the visual inspection. You may also perform this testing, or it could be done by some other nondestructive examination technician. In either case, the individual performing the test shall be properly qualified.

A CWI certification does not qualify a person to perform nondestructive tests. If someone else performs the testing, the welding inspector is required to verify that these nondestructive examinations have been performed by qualified personnel in the specified manner. In that case, the welding inspector's primary responsibility is to review the resulting information to assure that the results are complete. The welding inspector is then responsible for the maintenance of those test records.

Maintaining Records and Reports

Code work always requires record keeping. Whether other jobs call for it or not, complete records should be kept by every inspector. These can be in the form of either detailed notes or a formal inspection report.

The welding inspector should check official records for completeness and accuracy and make certain that they are available. Records that require the fabricator's signature should be prepared by the fabricator, not by the inspector. Records should be as detailed as necessary and should be entered in ink. Errors are to be crossed out with a single line and not erased; otherwise record tampering may be suspected. The final report should comment on the general character of the work, how well it stayed within prescribed tolerances, difficulties that occurred, and any defects that were noted. Repairs should be explained. If there are reports describing the presence of weld defects, there should be accompanying reports which describe the acceptance of those subsequent repairs. Copies should be distributed to all involved parties, and the welding inspector should maintain a copy for his own personal records should some question arise later. Chapter 12 should be reviewed for a more detailed description of this aspect of the welding inspector's responsibilities.

REVIEW - CHAPTER 2
WELDING INSPECTOR RESPONSIBILITY

Q2-1 Of those listed below, which is an acceptable way to correct an error on an inspection report?
- a. Draw a line through the incorrect portion of the report.
- b. Erase the incorrect word or words.
- c. Throw away the report.
- d. Line out the error, make the correction, and initial and date the correction.
- e. none of the above

Q2-2 What records should you keep as a CWI?
- a. copies of reports of all inspections you perform
- b. copies of reports relevant to your areas of responsibility (material test reports, welder qualification paperwork, procedure qualification paperwork, etc.) even though you didn't prepare them
- c. copies of sales literature describing welding equipment
- d. both a and b
- e. all of the above

Q2-3 When a particular type of weld is consistently marginal, with rejects occurring, what action would be appropriate for you as the inspector?
- a. Tell the welder what you want.
- b. Bring the problem to the attention of engineering personnel in order for corrective action to be taken, if possible.
- c. Simply continue to accept or reject the welds according to specified criteria [no more action is appropriate for inspectors].
- d. all of the above
- e. none of the above

Q2-4 A specification for a weld joint which must be immediately accepted or rejected lacks detailed information about that particular joint. Who should rule on the meaning and intent of the specification?
  a. the designer
  b. the welding engineer
  c. the project engineer or quality assurance personnel [if their approval is required by contract]
  d. you, as the CWI
  e. none of the above

Q2-5 How can you identify an individual low hydrogen electrode which a welder is already consuming to make a weld?
  a. Read the classification numbers painted on the covering near the stub end of the electrode.
  b. Ask the welder what it is.
  c. Ask the welding foreman.
  d. Look at the completed weld and identify the type of electrode by the visual appearance of the weld deposit.
  e. Look on the drawing or specification to determine what type of electrode is required for that weld.

Q2-6 How should low hydrogen electrodes be stored before use?
  a. in their original unopened containers
  b. in ovens held at a temperature which assures the maintenance of their low moisture content
  c. in tool room cribs, properly labeled, ready for quick distribution
  d. either a or b
  e. all of the above

Q2-7 What joint fit should you insist on?
  a. within tolerances specified on drawings or specifications
  b. Groove welds should have minimal root openings to reduce distortion.
  c. root openings greater than 1/8" to assure complete penetration
  d. Fillet welds should have root openings so that the resulting weld's effective throat will be greater.
  e. none of the above

Q2-8 Which welders are allowed to work on a "code" job?
  a. only those with certification papers from former jobs
  b. only those tested by the fabricator for this particular job
  c. only those qualified in accordance with job specifications
  d. only those you have requalified for this job
  e. all of the above

Q2-9 How should low hydrogen electrodes out of their original containers be stored?
  a. in their original resealed containers
  b. in heated storage ovens
  c. in open tool crib shelves
  d. in individual welders' electrode pouches
  e. none of the above

Q2-10 How can a CWI verify that the specified material is used on the job?
  a. For code jobs, each piece of material must be correctly marked with its identity.
  b. Perform a quick carbon analysis with a field test kit.
  c. Material must be scrapped if no identification is evident.
  d. Once the material leaves the storage area, the CWI no longer has to verify it.
  e. none of the above

Q2-11 If a mill product has imperfections such as splits, tears or surface irregularities, what action should you as the inspector take?
  a. Reject any imperfect materials.
  b. Judge whether or not the imperfections meet acceptance criteria according to applicable job specifications.
  c. Ignore the irregularities if not in the immediate vicinity of the weld joint.
  d. Wait until the welder finishes the weld to see if any cracking occurs before making any judgment.
  e. none of the above

Q2-12 Which of the conditions below suggests a weldability problem?
   a. One of the welders seems to have inordinate trouble making a sound weld on the grade of steel being used.
   b. Every Monday, five of the welders make poor welds, but the rest of the week all goes well.
   c. One of the welders produces undercut with welding in the vertical position.
   d. Cracking is repetitive when welding a certain steel alloy.
   e. none of the above

Q2-13 Which of the following is a welding inspector's responsibility prior to welding?
   a. Check joint fit.
   b. Check preheat temperature.
   c. Check interpass temperature.
   d. a and b above
   e. b and c above

Q2-14 A 1/4" fillet weld is specified on the drawing. When the CWI inspects the weld, it is measured to be 3/8." What should be done?
   a. Reject the weld for being oversize.
   b. Accept the weld if no weld size tolerances are specified.
   c. Ask for an engineering review of the design.
   d. b and c above
   e. none of the above

Q2-15 Fillet welds should be measured using what tolerance?
   a. + 1/16"
   b. + 1/32"
   c. -1/16"
   d. -1/32"
   e. no tolerance

CHAPTER 3: CODES, STANDARDS AND SPECIFICATIONS

Introduction

The responsibility associated with inspection is quite large. People who inspect for a living must have a critical eye, knowledge of their field, and know how to exercise judgment.

Some welding applications are more critical than others; however, <u>all</u> welding inspections are important, no matter how critical the weld or the application. The importance of welding inspection activities is underscored by the array of codes and specifications that inspectors must use.

In addition to covering the various codes and specifications, this chapter tells how different kinds of weldments are inspected, and also discusses the administrative side of welding inspection........

There are many approaches to inspection. Today's growing technical demands by purchasers and the cost of product liability rule out casual or sloppy inspection. However, there are still areas of interpretation—for instance, where choices of the level of sensitivity are to be selected.

Where life and safety are involved, the quantity and quality of welding inspection usually must conform to a code, standard, or specification. The code may come from a government agency or a private agency such as an engineering society. A code is a body of rules for construction or fabrication of some product. For example, cities have "building codes" which describe the construction requirements for structures in that municipality. Codes are made mandatory by laws or regulations and may be used along with other referenced standards or specifications.

A standard is any document or object used as a basis of comparison. So, once a "standard" has been defined, you as an inspector must then judge the adequacy of product based upon how it compares with that established standard. "Standard" is a general term which could be used to describe a number of documents, including: codes, specifications, procedures, recommended practices, and groups of graphic symbols. Standards can be considered mandatory, such as a code, or nonmandatory, as is the case for a recommended practice.

The third type of document important to the welding inspector is the specification. The specification differs from the code in that it describes the requirements for a particular object, material, service, etc., while the code describes a much larger scope of construction or fabrication. You might more easily remember this difference by realizing that a specification describes a specific part or material that may become an integral part of a product fabricated in accordance with some code. Almost all codes and standards require written welding procedure specifications (WPS's). These specifications provide the welder and inspector with valuable details about the welding processes to be applied. The WPS is a "recipe" for a welded joint.

During the execution of any inspection, the welding inspector must constantly refer to the codes, standards, specifications, or a combination of the three to learn many important requirements for the fabrication or construction taking place. A number of these documents exist that relate to some aspect of weld quality. They are primarily produced by national professional organizations or government agencies. Depending on the type of weldment being constructed, different requirements will be enforced.

The following discussion includes descriptions of these various welding codes, standards and specifications and their areas of coverage.

Structural Welding Code - Steel (AWS D1.1)

AWS D1.1, *Structural Welding Code- Steel*, is published by the American Welding Society, with new editions being scheduled for publication every other year. It describes the welding requirements for steel structures, including statically loaded, dynamically loaded and tubular structures. Sections 1 through 7 contain topics applicable to all structures, while Sections 8, 9 and 10 refer to requirements specific to statically loaded structures, dynamically loaded structures and tubular structures, respectively.

Below is a listing of the various Sections of AWS D1.1 and their areas of coverage:
(1) General Provisions
(2) Design of Welded Connections
(3) Workmanship
(4) Technique
(5) Qualification
(6) Inspection
(7) Stud Welding
(8) Statically Loaded Structures
(9) Dynamically Loaded Structures
(10) Tubular Structures
(11) Strengthening and Repairing Existing Structures
(12) Appendices & Commentaries

In AWS D1.1, section 6 there are two definitions of welding inspectors: fabrication/erection and verification. The fabrication/ erection inspection and testing is the responsibility of the contractor unless otherwise provided in contract documents. However, the verification inspection and testing is the responsibility of the owner or engineer. While AWS does make this distinction, the requirements of the Code are intended to apply equally to the work of both parties.

Sections 6.1 through 6.5 describe the duties and responsibilities of the welding inspector, and Section 6.6 details the responsibilities of the contractor in satisfying the wishes of the inspection personnel. Sections 6.7 and 10.17.2 describe the methods to be used for nondestructive examination in accordance with AWS D1.1. For information pertaining to the various NDT methods, refer to Table 3-1, page 3-3.

Other AWS Structural Welding Codes (D1.2, D1.3, D1.4, and D1.5)

In addition to D1.1, AWS is also responsible for the development of four other codes dealing with structural welding requirements. They include:

D1.2, *Structural Welding Code- Aluminum*;
D1.3, *Structural Welding Code- Sheet Steel*;
D1.4, *Structural Welding Code- Reinforcing Steel*;
D1.5, *Bridge Welding Code*.

All four follow the general format of AWS D1.1; however, as the names imply, they each refer to a specific type of structure. As such, the welding requirements differ slightly.

The newest of these four codes, AWS D1.5 *Bridge Welding Code*, was prepared under the joint guidance of the AWS Structural Welding Committee and AASHTO (American Association of State Highway and Transportation Officials).

ASME Boiler and Pressure Vessel Code

The ASME Boiler and Pressure Vessel Code prescribes those requirements for design and construction of pressure-containing components, as well as the care and operation of heating and

Table 3-1
Nondestructive Testing Methods - D1.1 Code Reference

Test Method	Applicable Code References
Radiographic	Section 6, Part B and Section 10.18
Ultrasonic	Section 6, Part C and Section 10.19
Magnetic Particle	6.7.5 (ASTM E709)
Liquid Penetrant	6.7.6 (ASTM E165)
NDT Personnel Qualification	6.7.7 (SNT-TC-1A)

The acceptance criteria for both the visual inspection and nondestructive examination are found in the following Code locations, depending upon the type of structure being constructed:

Type of Structure	Weld Acceptance Criteria Reference
Statically Loaded	8.15
Dynamically Loaded	9.25
Tubular	10.17

power boilers. Different sections of this Code refer to both fossil and nuclear power equipment. Due to the wide range of coverage offered by the ASME Code, it is contained in several different sections. Table 3-2, page 3-4 describes the various areas of coverage for these Code sections.

Definition of Inspector

Under the ASME Code, the term *Inspector* refers only to an *Authorized Inspector* who is employed by:
1) A state or municipality of the U.S. or a province of Canada;
2) An insurance company authorized to write boiler and pressure vessel insurance
3) A company having vessels made for use by that company exclusively and not for sale (applies to Section VIII, Division I, only)

An Authorized Inspector is not an inspector hired by the fabricator. Qualification of a National Board commissioned Authorized Inspector is by written examination prepared by the National Board of Boiler and Pressure Vessel Inspectors and administered by a city, state or province. The CWI is not an Authorized Inspector. The position of Authorized Inspector is a separate certification.

General Inspection Requirements

General inspection requirements concerning materials, procedures, and personnel are covered in paragraph UG90 of ASME Section VIII, Division I.

Materials

The welding inspector must be sure that the specified materials are used and verify that suitable test examinations proved that filler materials were correct.

Table 3-2 - ASME Boiler and Pressure Vessel - Code Reference

ASME Section	Topic
I	Power Boilers
II	Material Specifications
	Part A: Ferrous Materials
	Part B: Nonferrous Materials
	Part C: Welding Filler Materials
	Part D: Properties
III	Nuclear Components
IV	Heating Boilers
V	Nondestructive Examination
VI	Care and Operation of Heating Boilers
VII	Care of Power Boilers
VIII	Unfired Pressure Vessels
IX	Welding and Brazing Qualification
X	Fiberglass-Reinforced Plastic Pressure Vessels
XI	Inservice Inspection of Nuclear Components

When performing inspection in accordance with the ASME Code, several of the different sections must be referenced, because they contain different types of requirements. As an example, for an unfired pressure vessel constructed from steel in accordance with the ASME Code, the various sections which apply are:

Section	Information Provided
II, Part A	Description, including chemical and mechanical properties, of the steel base metals to be used.
II, Part C	Description, including chemical and mechanical properties, of the welding filler materials.
II, Part D	Material properties
V	Methods for performing various required nondestructive examinations.
VIII	Design, fabrication and inspection requirements for unfired pressure vessels, including weld acceptance criteria.
IX	Requirements for the qualification of welding procedures and welders.

Procedures

It must be verified that the welding procedures have been qualified. Evidence of this qualification can be found by reviewing a written *Welding Procedure Specification (WPS) and a Procedure Qualification Record (PQR)*.

Personnel

All welders, welding operators, brazers, and brazing operators must be qualified.

Nondestructive Examination

You must ensure that nondestructive examinations and tests are carried out as needed.

ASTM Standards

Numerous standards of the American Society for Testing and Materials are referenced in the ASME, AWS, and other codes. Most common are the specifications for materials. However, for ASME Code fabrication, the designations for

ASTM Standards are assigned the prefix "S". Hence, ASTM A285 steel is specified as SA285 for ASME applications. However, when accepted by ASME, the information found in both is identical.

API Standards

The American Petroleum Institute publishes three important standards which deal with welded pipelines and vessels. They include:

API Standard	Title
1104	"Standard for Welding Pipeline and Related Facilities"
620	"Recommended Rules for Design and Construction of Large Welded Low-Pressure Storage Tanks"
650	"Standard for Welded Steel Tank for Oil Storage"

API Standard 1104 contains requirements for qualification of procedures and welders, qualification of radiographers, radiographic techniques, as well as weld quality requirements for production welding.

For procedure and welder qualification, test coupons are subjected to various destructive tests, including: tensile, nick break, root bend, face bend, and side bend. Fillet welds are examined by the nick break test. However, as in AWS D1.1 and ASME Section IX, API 1104 permits the qualification of welders using radiographic testing in lieu of destructive tests. (see specific code for requirements)

Requirements for the qualification of radiographers are given in Paragraph 8.0, but no limitations are placed on the qualification of destructive test personnel. Fractured test specimens speak for themselves, presuming that the welding inspector can verify their source or application.

API Standard 620 is intended to apply to the design and construction of large, welded, low-pressure, carbon steel, above-ground storage tanks that have a single vertical axis of revolution. These tanks may be operated at temperatures up to 200° F and pressures up to 15 psi.

API Standard 650 applies to the material, design, fabrication, erection, and testing requirements for vertical cylindrical aboveground, closed and open-top, welded steel storage tanks in various sizes and capacities for internal pressures approximating atmospheric pressures, except that a small internal pressure is permitted when the additional requirements of Appendix F are met. API 650 applies only to nonrefrigerated service.

Military Standards

The United States Department of Defense produces a staggering quantity of military standards. For a complete list of all military specifications, refer to the *Index to Military Specifications and Standards* that may be available at companies producing products under a military contract. Here are a few examples of the military standards.

Welding electrode specifications have the identifying letter "E" and a multi-digit number, such as MIL-E-22200/1 for "Electrodes, covered low alloy steel"; MIL-E-22200/2 for covered stainless steel electrodes; MIL-E-19933D for bare stainless steel electrode wires; and MIL-E-16053L for bare aluminum electrode wires. Other important filler metal specifications include MIL-R-17131B for "Rods, bare hard surfacing alloys for GTA welding" and MIL-I-2341B for "Inserts, consumable."

ANSI Standards

The American National Standards Institute has either developed or adopted other codes and standards which are referred to as "American National Standards." Many of these documents carry identification referring to both organizations responsible for its development and maintenance.

Two common ANSI standards are ASME B31.1, "Power Piping" and ASME B31.3, "Chemical Plant and Petroleum Refinery Piping." Although developed by the American Society of Mechanical Engineers (ASME), they are best known by their ANSI designation. Both refer to ASME Section IX for welding and brazing qualification requirements.

Many of the AWS standards are also considered to be American National Standards, and can be correctly referred to by the designation prefix ANSI/AWS. A few of the many examples are: ANSI/AWS D1.1, ANSI/AWS B1.10, ANSI/AWS D14.3, etc.

AWS Filler Metal Specifications

Thirty different specifications for filler materials are published by the American Welding Society. Reference is made in ASME, ANSI, and AWS Codes, and indirectly in the API Standards to the AWS filler metal specifications. Certain of these are reviewed and adopted by ASME, adding the letters "SF" preceding the AWS designation. The AWS Filler metal specifications are listed as follows:

A5.1 Carbon Steel Covered Arc Welding Electrodes

A5.2 Iron and Steel Gas Welding Rods

A5.3 Aluminum and Aluminum Alloy Arc Welding Electrodes

A5.4 Corrosion-Resisting Chromium and Chromium-Nickel Steel Covered Electrodes

A5.5 Low Alloy Steel Covered Arc Welding Electrodes

A5.6 Copper and Copper Alloy Covered Electrodes

A5.7 Copper and Copper Alloy Bare Welding Rods and Electrodes

A5.8 Brazing Filler Metal

A5.9 Corrosion Resisting Chromium and Chromium-Nickel Steel Bare and Composite Metal Cored and Stranded Arc Welding Electrodes and Welding Rods

A5.10 Aluminum and Aluminum Alloy Welding Rods and Bare Electrodes

A5.11 Nickel and Nickel Alloy Covered Welding Electrodes

A5.12 Tungsten Arc Welding Electrodes

A5.13 Surfacing Welding Rods and Electrodes

A5.14 Nickel and Nickel Alloy Bare Welding Rods and Electrodes

A5.15 Welding Rods and Covered Electrodes for Welding Cast Iron

A5.16 Titanium and Titanium Alloy Bare Welding Rods and Electrodes

A5.17 Bare Carbon Steel Electrodes and Fluxes for Submerged Arc Welding

A5.18 Carbon Steel Filler Metals for Gas Shielded Arc Welding

A5.19 Magnesium Alloy Welding Rods and Bare Electrodes

A5.20 Carbon Steel Electrodes for Flux Cored Arc Welding

A5.21 Composite Surfacing Welding Rods and Electrodes

A5.22 Flux Cored Corrosion-Resisting Chromium and Chromium-Nickel Steel Electrodes

A5.23 Bare Low Alloy Steel Electrodes and Fluxes for Submerged Arc Welding

A5.24 Zirconium and Zirconium Alloy Bare Welding Rods and Electrodes

A5.25 Consumables Used for Electroslag Welding of Carbon and High Strength Low Alloy Steels

A5.26 Consumables Used for Electrogas Welding of Carbon and High Strength Low Alloy Steels

A5.27 Copper and Copper Alloy Gas Welding Rods

A5.28 Low Alloy Steel Filler Metals for Gas Shielded Arc Welding

A5.29 Low Alloy Steel Electrodes for Flux Cored Arc Welding

A5.30 Consumable Inserts

A5.31 Specification for Fluxes for Brazing and Braze Welding

Alphabetic Index to AWS Filler Metal Specifications

Aluminum	A5.3, A5.10
Brazing	A5.8, A5.31
Carbon Steel	A5.1, A5.17, A5.18, A5.20, A5.25, A5.26
Cast Iron	A5.15
Consumable Inserts	A5.20
Copper	A5.6, A5.7, A5.27
Corrosion Resistance	A5.4, A5.9, A5.22
Electrogas	A5.26
Electroslag	A5.25
Flux Cored	A5.20, A5.22, A5.29
Gas Shielded Arc	A5.18, A5.28
Low Alloy Steel	A5.5, A5.23, A5.25, A5.26, A5.28
Magnesium	A5.19
Nickel	A5.11, A5.14
Stainless Steel	A5.4, A5.9, A5.22
Submerged Arc	A5.17, A5.23
Surfacing	A5.13, A5.21
Titanium	A5.16
Tungsten	A5.12
Zirconium	A5.24

Summary

As a welding inspector, it will be necessary to work under the guidance and requirements of certain codes, standards and specifications. These documents should describe all of the necessary information, including:

- what to inspect,
- where to inspect,
- how to inspect,
- the extent of inspection, and
- the acceptance criteria to be used during these evaluations.

Various organizations are responsible for the development and maintenance of these codes, standards and specifications. As required by contract documents, the welding inspector will refer to those documents which will be in effect for his portion of the inspection.

A number of different codes, standards and specifications exist for different types of fabrication and construction. Their requirements differ depending on the specific needs of that particular industry.

REVIEW - CHAPTER 3
CODES, STANDARDS AND SPECIFICATIONS

Q3-1 Job quality requirements can be found in all but which of the following?
a. codes
b. standards
c. specifications
d. text books
e. a&b only

Q3-2 Of the following documents, which may be considered a "standard"?
a. codes
b. specifications
c. recommended practices
d. a and b above
e. all of the above

Q3-3 The type of document which has legal status is:
a. code
b. standard
c. specification
d. both a and b above
e. all of the above

Q3-4 That type of document which describes the requirements for a particular object or component is referred to as:
a. code
b. standard
c. specification
d. a and b above
e. b and c above

Q3-5 Of the following types of documents, which is the more general type. In fact, the other documents could be considered as more specific types of this classification.
a. codes
b. standards
c. specifications
d. drawings
e. none of the above

Q3-6 The code which covers the welding of steel structures is:
a. ASME Section IX
b. ANSI B31.1
c. API 1104
d. AWS D1.1
e. none of the above

Q3-7 The code which covers the design and fabrication of unfired pressure vessels is:
a. ASME Section IX
b. ASME Section VIII
c. ASME Section III
d. API 1104
e. AWS D1.1

Q3-8 The specification covering the requirements for welding electrodes are designated as:
a. AWS D1.X
b. AWS D14.X
c. AWS A5.X
d. ASTM A53
e. ASTM A36

Q3-9 The standard describing the requirements for welding of crosscountry pipelines is:
a. AWS D1.1
b. ASME Section VIII
c. ASME Section IX
d. API 1104
e. none of the above

CHAPTER 4: WELD JOINT GEOMETRY AND WELDING TERMINOLOGY

Introduction

There was considerable discussion in Chapter 1 that one of the more important aspects of the welding inspector's job is communication with others involved in the fabrication of a weldment. The inspector needs to be capable of effectively stating inspection findings to others as well as understanding their responses or questions. This does not imply that all communication must be oral. It also applies to written or graphic explanations.

As pointed out in Chapter 2, another part of the welding inspector's job is the review and interpretation of various documents relating to the welded fabrication. All aspects of this communication require that the individual have a full understanding of the proper terms and definitions used.

The American Welding Society has long realized the need for standardized terms and definitions for use by those actively involved in the fabrication of welded products. In answer to this need, AWS has published the document AWS A3.0, *Welding Terms and Definitions*. It was developed by the Committee on Definitions and Symbols to aid in the communication of welding information. The standard terms and definitions published in AWS A3.0 are those that should be used in the oral and written language of welding. While these are the standard, or preferred terms, they are by no means the only terms used to describe various situations. Since the purpose here is to educate, it is felt to be important to mention some of these common terms, even though they are not preferred terminology. When these terms are mentioned, they will appear in parentheses after the preferred words.

While most of the terms used apply to the actual welding operation, it is important for the welding inspector to understand other definitions which apply to other related operations. For example, the welding inspector should understand how to describe the various weld joint configurations and elements of the fitting process. After welding is completed, the welding inspector may need to describe the location of some welding discontinuity which has been discovered. If such a discontinuity requires further attention, it is important that the inspector accurately describe the location of the problem so that the welder will know where the repair is to be made.

Types of Joints

Before welding begins, the welding inspector may be required to examine the weld joint configuration and fit. This is one of the most important aspects of welding inspection, because it is possible to detect potential problems. When discovered at this stage, these problems can be corrected more economically.

When a welding inspector is performing a joint inspection, it is necessary that he know the differences between the various types of weld joints. A *joint* is "the junction of members or edges of members which are to be joined or have been joined." There are five basic types of joints: butt, corner, T-, lap, and edge.

These five joint types get their names from their basic configurations. The butt joint describes the configuration when two members to be joined lie in the same plane and are connected at their edges.

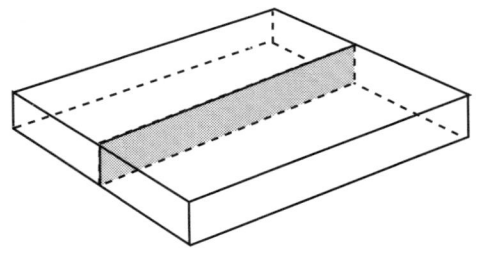

APPLICABLE WELDS
BEVEL GROOVE	U-GROOVE
FLARE BEVEL GROOVE	V-GROOVE
FLARE V-GROOVE	EDGE FLANGE
J-GROOVE	BRAZE
SQUARE GROOVE	

(A) BUTT JOINT

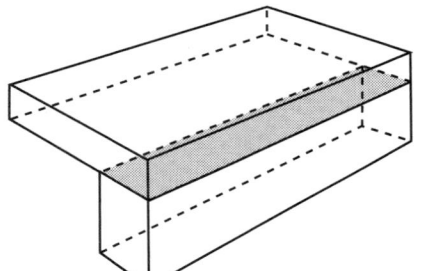

APPLICABLE WELDS
FILLET	CORNER FLANGE
BEVEL GROOVE	EDGE FLANGE
FLARE BEVEL GROOVE	PLUG
FLARE V-GROOVE	SLOT
J-GROOVE	SPOT
SQUARE GROOVE	SEAM
U-GROOVE	PROJECTION
V-GROOVE	BRAZE

(B) CORNER JOINT

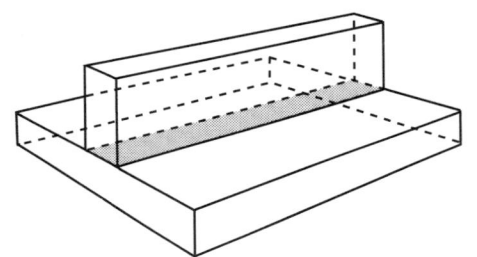

APPLICABLE WELDS
FILLET	SLOT
BEVEL GROOVE	SPOT
FLARE BEVEL GROOVE	SEAM
J-GROOVE	PROJECTION
SQUARE GROOVE	BRAZE
PLUG	

(C) T-JOINT

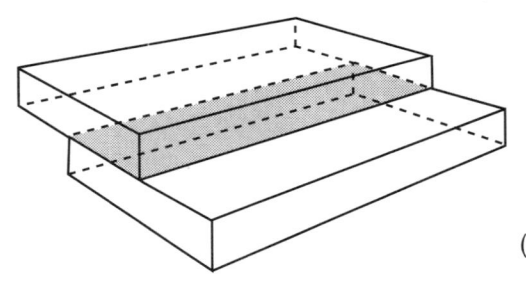

APPLICABLE WELDS
FILLET	SLOT
BEVEL GROOVE	SPOT
FLARE BEVEL GROOVE	SEAM
FLARE V-GROOVE	PROJECTION
J-GROOVE	BRAZE
PLUG	

(D) LAP JOINT

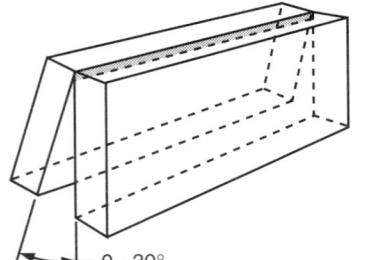

APPLICABLE WELDS
BEVEL GROOVE	V-GROOVE
FLARE BEVEL GROOVE	EDGE
FLARE V-GROOVE	CORNER FLANGE
J-GROOVE	EDGE FLANGE
SQUARE GROOVE	SEAM
U-GROOVE	

(E) EDGE JOINT

0 - 30°

JOINT TYPES (Source: ANSI/AWS A3.0)

With a corner joint, the two members to be joined lie in perpendicular planes and their edges are connected. The T-joint is similar in that the two members lie in perpendicular planes, except the edge of one member is joined to the planar surface of the other. In a lap joint, the two members lie in parallel planes, but not the same plane. The joint occurs where the two members overlap each other to form a double thickness region. The final joint configuration, the edge, also has the two members lying in parallel planes, but the two members lie with their planar surfaces in contact so that the actual welding occurs around the perimeter, or outside, of the joint.

Parts of the Weld Joint

Once the type of joint has been identified, it may be necessary to further describe the exact configuration required. To do this, the welding inspector must be capable of naming the various features of that particular joint. Some of these elements include: joint root, groove face, root face, root edge, root opening, bevel, bevel angle, groove angle, and groove radius. Depending upon the particular type of joint configuration, these features may take on slightly different shapes.

A perfect example of this is the *joint root*, or "that portion of a joint to be welded where the members approach closest to each other. In cross section, the joint root may be either a point, line, or an area."

Some other related terms are: groove face, root face and root edge. By definition, *groove face* is "that surface of a member included in the groove." The *root face* (also commonly called the land, nose or flat) is "that portion of the groove face adjacent to the joint root." The *root edge* is defined as "a root face of zero width."

Terms describing actual shapes and dimensions are some of the other features which may require description by the welding inspector. These elements are often essential variables for welding procedure specifications, so the welding inspector may be required to actually measure them to judge their compliance with applicable drawings or other documents. The *root opening* (gap) is described as "the separation between the workpieces at the joint root." The *bevel* (also referred to as chamfer) is "an angular edge preparation." The *bevel angle* is defined as "the angle formed between the prepared edge of a member and a plane perpendicular to the surface of the member." The *groove angle* is "the total included angle of the groove between workpieces." For a single-bevel-groove-weld, the bevel angle and the groove angle are equal. The final term, groove radius, applies only to J- and U-groove-welds. It is described as "the radius used to form the shape of a J- or U-groove weld." Normally, a J- or U-groove weld configuration is specified by both a bevel (or groove) angle and a groove radius.

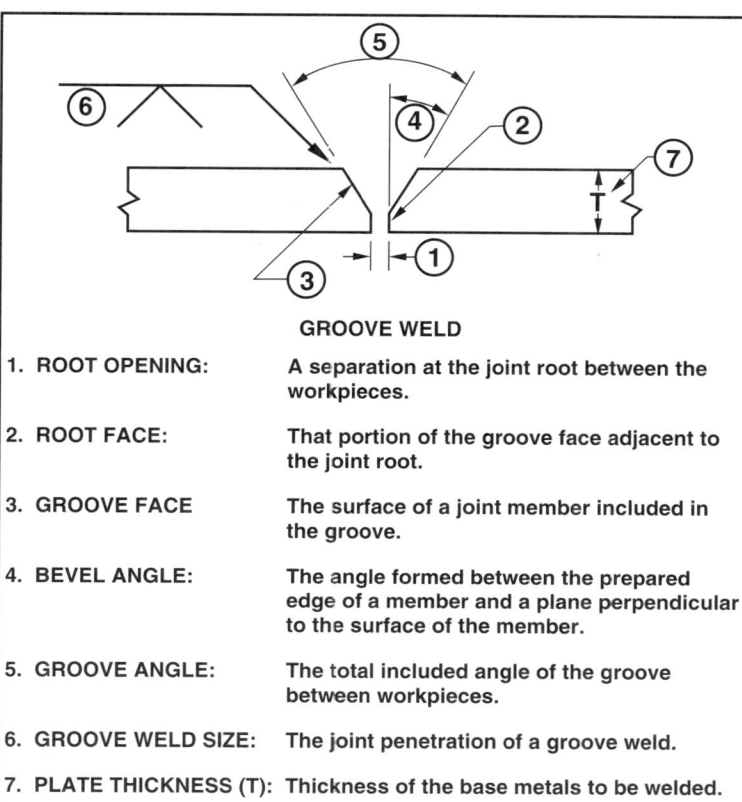

GROOVE WELD

1. **ROOT OPENING:** A separation at the joint root between the workpieces.
2. **ROOT FACE:** That portion of the groove face adjacent to the joint root.
3. **GROOVE FACE:** The surface of a joint member included in the groove.
4. **BEVEL ANGLE:** The angle formed between the prepared edge of a member and a plane perpendicular to the surface of the member.
5. **GROOVE ANGLE:** The total included angle of the groove between workpieces.
6. **GROOVE WELD SIZE:** The joint penetration of a groove weld.
7. **PLATE THICKNESS (T):** Thickness of the base metals to be welded.

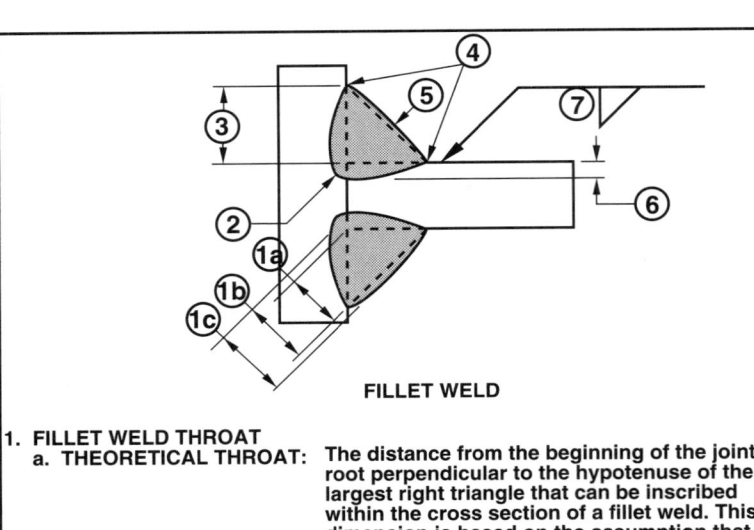

FILLET WELD

1. **FILLET WELD THROAT**
   a. **THEORETICAL THROAT:** The distance from the beginning of the joint root perpendicular to the hypotenuse of the largest right triangle that can be inscribed within the cross section of a fillet weld. This dimension is based on the assumption that the root opening is equal to zero.
   b. **EFFECTIVE THROAT:** The minimum distance minus any convexity between the weld root and the face of a fillet weld.
   c. **ACTUAL THROAT:** The shortest distance between the weld root and the face of the fillet weld.
2. **WELD ROOT:** The points, shown in a cross section, at which the root surface intersects the base metal surfaces.
3. **FILLET WELD LEG:** The distance from the joint root to the toe of the fillet weld.
4. **WELD TOE:** The junction of the weld face and the base metal.
5. **WELD FACE:** The exposed surface of a weld on the side from which welding was done.
6. **DEPTH OF FUSION:** The distance that fusion extends into the base metal or previous bead from the surface melted during welding.
7. **FILLET WELD SIZE:** For equal leg fillet welds, the lengths of the largest isosceles right triangle that can be inscribed within the fillet weld cross section. For unequal leg fillet welds, the leg lengths of the largest right triangle that can be inscribed with the fillet weld cross section.

The figure above illustrates the various parts of a fillet weld.

Types of Welds

There are numerous welds which can be applied to the various types of joints. According to AWS A3.0, there are 18 basic types of welds used in arc welding, including:

1) Square-groove weld
2) Bevel-groove weld
3) V-groove weld
4) J-groove weld
5) U-groove weld
6) Flare-bevel-groove weld
7) Flare-V-groove weld
8) Fillet weld
9) Edge weld
10) Edge-flange weld
11) Corner-flange weld
12) Spot weld
13) Seam weld
14) Plug weld
15) Slot weld
16) Surfacing weld
17) Back weld
18) Backing weld

With this variety of weld geometries available, the welding fabricator can choose the one which best suits his needs. This choice could be based on considerations such as: accessibility, type of welding process being used, method of joint preparation, and adaptation to particular designs of the structure being fabricated. The previous figure showing the various types of weld

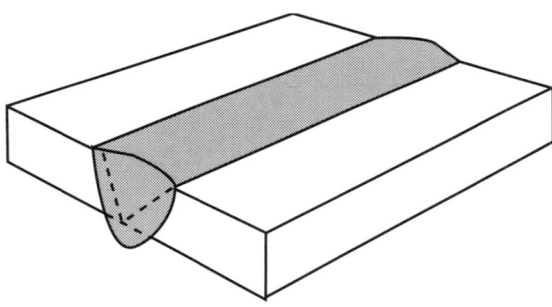

SINGLE V-GROOVE WELD

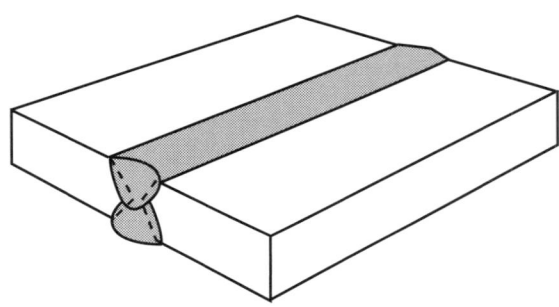

DOUBLE V-GROOVE WELD

joints also indicates which of the above welds can be applied to each of the five types of weld joints.

The first seven types of welds refer to different groove configurations. Their names imply what the actual configurations look like when viewed in cross section. All of these groove weld types can be applied to joints which are welded from a single side or both sides. As would be expected, a *single-welded joint* is "a fusion welded joint that is welded from one side only," while a *double-welded joint* is "a fusion welded joint that is welded from both sides."

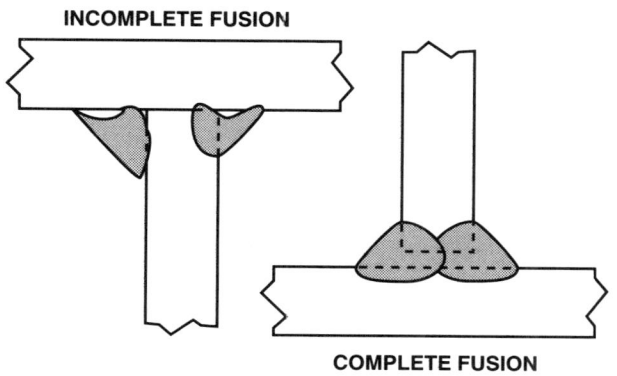

The next category of weld is the fillet weld. This weld type is possibly the most commonly used. An important thing to remember is that a fillet weld is not a type of joint. It is a particular type of weld which can be applied to a lap, T- or corner joint. AWS A3.0 defines a *fillet weld* as "a weld of approximately triangular cross section

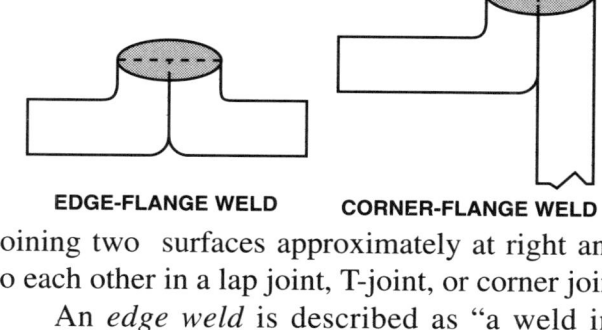

joining two surfaces approximately at right angles to each other in a lap joint, T-joint, or corner joint."

An *edge weld* is described as "a weld in an edge joint." Two modifications of the edge weld are used for flange welds. A flange weld is "a weld made on the edges of two or more members to be joined, at least one of which is flanged." The two common types are the edge-flange weld and the corner-flange weld. The edge-flange weld has both of the members flanged, while the corner-flange weld is used to join two members where only one of the members is flanged.

The next type of weld to be discussed is the spot weld. We most commonly associate spot welds with resistance welding, which is used extensively in the automotive and aerospace industries. However, a very effective way to join a lap joint configuration is through the use of an arc spot weld. In general, a *spot weld* is "a weld made between or upon overlapping members in which coalescence (the act of combining or uniting) may start and occur on the faying surfaces or may proceed from the outer surface of one member." A *faying surface* is "the mating surface of a member

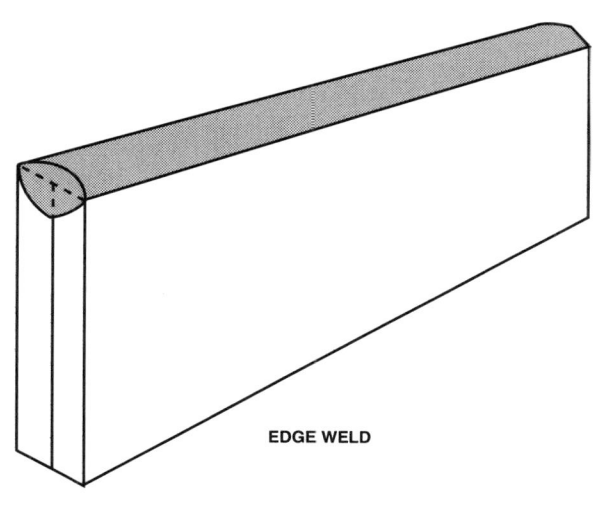

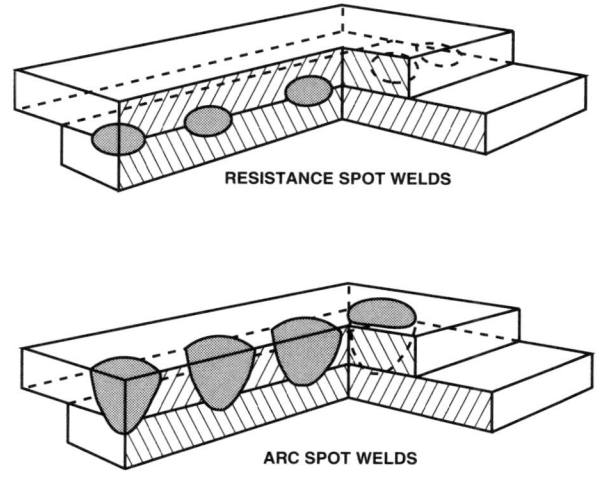

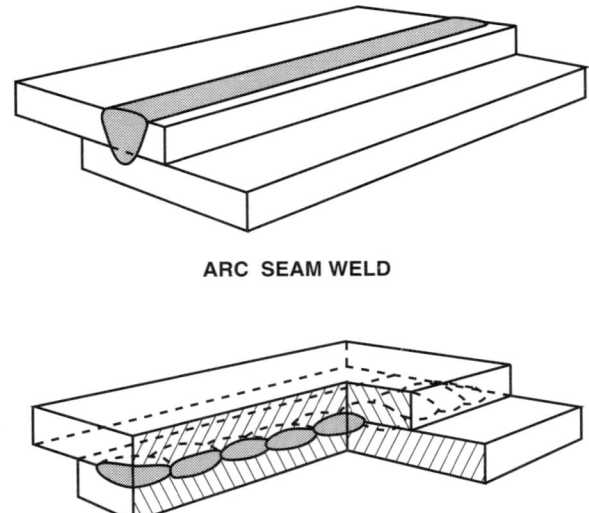

ARC SEAM WELD

RESISTANCE SEAM WELDS

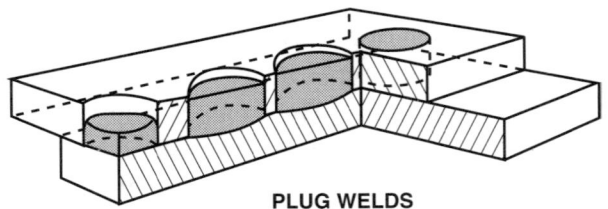

PLUG WELDS

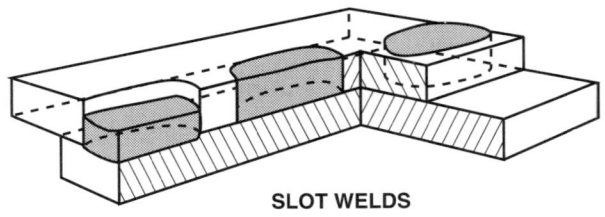

SLOT WELDS

that is in contact with or in close proximity to another member to which it is joined." In the case of arc spot welding, the weld is accomplished by melting through the top member using one of the arc welding processes such that fusion occurs between it and the member which it overlaps.

A seam weld is a similar type of weld except that instead of the weld being applied in a single spot, it forms a continuous weld having some length. As was the case for spot welds, seam welds can be accomplished using resistance or arc welding methods. In either case, the *seam weld* is defined as "a continuous weld made between or upon overlapping members, in which coalescence may start and occur on the faying surfaces, or may have proceeded from the outer surface of one member." Like the arc spot weld, the arc seam weld extends through one member to provide fusion to the member which is overlapped.

Two other types of welds which are used for the joining of overlapping members are plug and slot welds. They are different than spot and seam welds in that the near side member has a hole or slot cut to provide access to the member which it overlaps. The *plug weld* is defined as "a weld made in a circular hole in one member of a joint fusing that member to another member.

Similarly, a *slot weld* is "a weld made in an elongated hole in one member of a joint fusing that member to another member. The hole may be open at one end." In either case, the hole or slot is normally filled completely flush with the top surface. A fillet weld applied in a circular hole or slot is not considered to be either a plug or slot weld.

The next weld type of interest is the surfacing weld. As one might expect, the particular type of weld is applied to the surface of a metal. Normally, the primary reason for this application is to provide some barrier against abrasion or corrosion. Often, this approach is more economical than the use of a full thickness of some more expensive material. AWS A3.0 defines a *surfacing weld* as "a weld applied to a surface, as opposed to making a joint, to obtain desired properties or dimensions."

The final weld types to be discussed are called back and backing welds. From the names, it is apparent that these welds are meant to be applied to the back side of a weld joint. Although

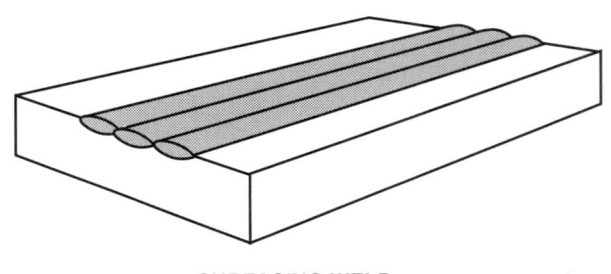

SURFACING WELD

they are applied to the same location, they differ depending upon when they are deposited. AWS A3.0 describes a *back weld* as "a weld made at the back of a single groove weld," and a *backing weld* as "backing in the form of a weld." Therefore, a back weld is applied after the front side has already been welded, while the backing weld is deposited prior to the welding of the front side.

Parts of Completed Welds

So far, the discussion has been limited to the description of weld joints and types of weld configurations. However, the welding inspector must also be aware of terms used to describe conditions or features of completed welds. When a completed weld is being inspected, the inspector should describe the conditions which exist when reporting inspection findings. Groove welds have several primary components. The first part, the *weld face*, is "the exposed surface of a weld on the side from which welding was done." "The junction of the weld face and the base metal" is referred to as the *weld toe*. Opposite the weld face is the weld root. The *weld root* is defined as "the points, as shown in cross section, at which the root surface intersects the base metal surfaces. The root surface is "the exposed surface of the weld opposite the side from which welding was done." Therefore, the root surface is bounded by the weld root on either side.

Other terms relate to *weld reinforcement*, which is "weld metal in excess of the quantity required to fill a joint." The *face reinforcement* (also commonly called weld crown) is "the weld reinforcement at the side of the joint from which welding was done." Conversely, the root reinforcement is "the weld reinforcement opposite the side from which welding was done." In both cases, this represents that portion of the weld metal which extends beyond the surface of the base metal.

The previous explanations assume a single-welded joint, or all welding was performed from one side. In the case of a double-welded joint, both sides of the joint will have a weld face, and the amount of reinforcement present on each side will be referred to as the face reinforcement. This is illustrated in the figure below showing the use of a back weld.

Just as groove welds have names for various parts, standard terminology exists for parts of fillet welds. As with the groove weld, the surface of the weld which the inspector will evaluate is

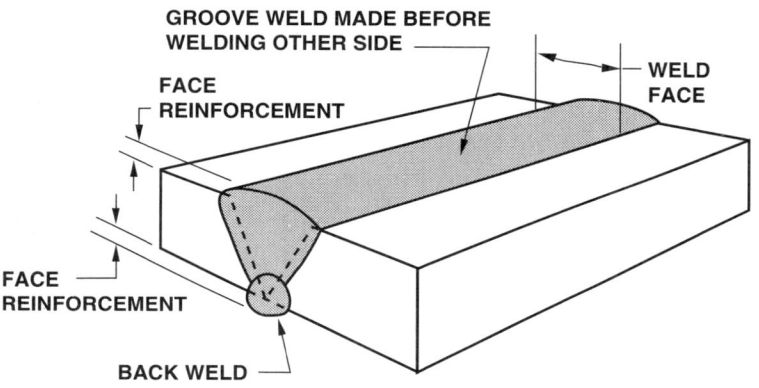

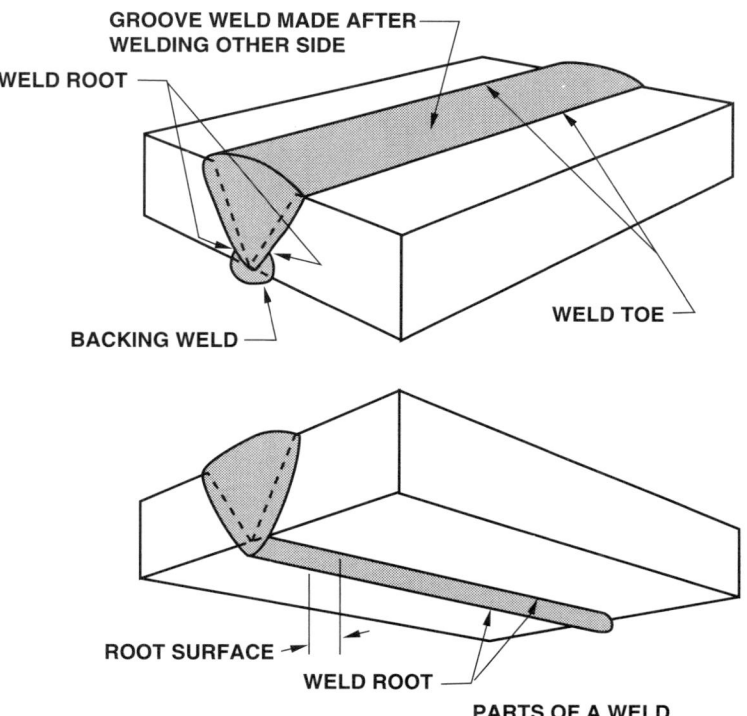

PARTS OF A WELD

referred to as the weld face. The junctions of the weld face with the base metal are the weld toes. The furthest penetration of the weld metal into the joint is the weld root.

The distance from the weld toe to the joint root is called the *fillet weld leg*. One other feature of a fillet weld is the weld throat. In general, this is the shortest distance through the cross section of the weld. Various types of fillet weld throats will be discussed in more detail under the topic of sizing convex and concave fillet welds.

Fusion and Penetration Terminology

There are also terms relating to the fusion and penetration of the weld metal into the base metal. Although these are features which are difficult for the visual inspector to check without further destructive or nondestructive examination, it is still important to understand what the various terms mean.

In general, fusion refers to the actual melting together of the filler metal and base metal, or of

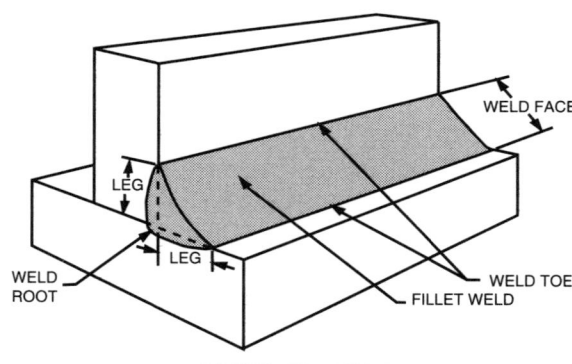

PARTS OF A WELD

the base metal only. Penetration is a term which relates to the distance that the weld metal has progressed into the joint. The degree of penetration achieved has a direct effect on the strength of the joint and is therefore related to the weld size.

Numerous terms exist which describe the degree or location of either fusion or penetration. During the welding operation, the original groove face is melted such that the final boundary of the weld metal is deeper than the original surface. The groove face is referred to as the fusion face since it will be melted during welding. The

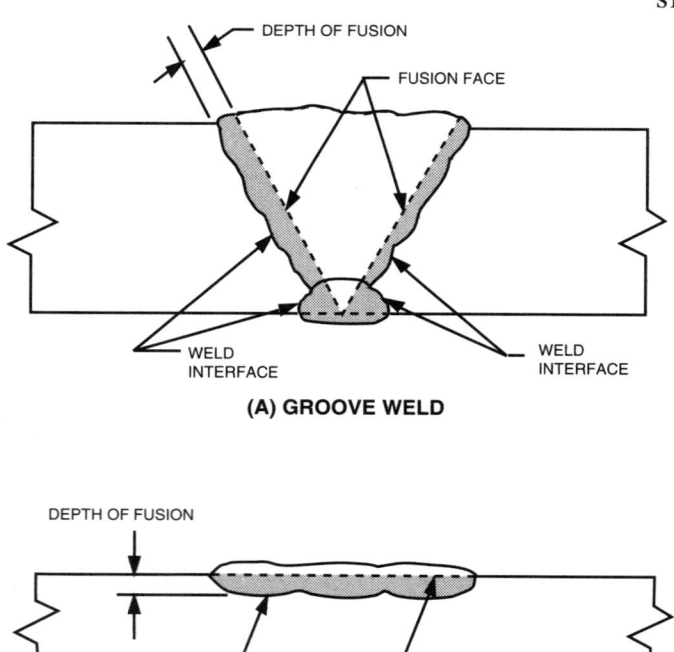

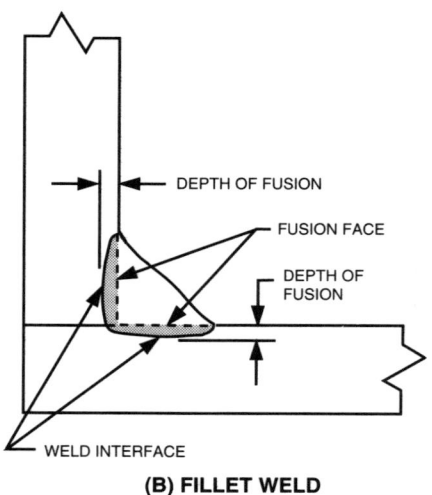

FUSION WELDS

boundary between the weld metal and base metal is referred to as the weld interface. The depth of fusion is the distance from the fusion face to the weld interface. The depth of fusion is always measured perpendicular to the fusion face. These terms are applied similarly for other types of welds such as fillet and surfacing welds.

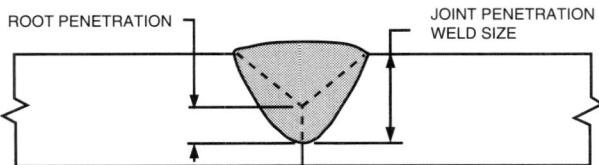

INCOMPLETE JOINT PENETRATION OR PARTIAL JOINT PENETRATION

There are also several terms which refer to penetration of the weld. *Root penetration* is the distance that the weld metal has melted into the joint beyond the joint root. The *joint penetration* is the distance from the furthest extension of the weld into the joint to the weld face, excluding any weld reinforcement which may be present. For groove welds, this same length is also referred to as the weld size (sometimes improperly referred to as effective throat).

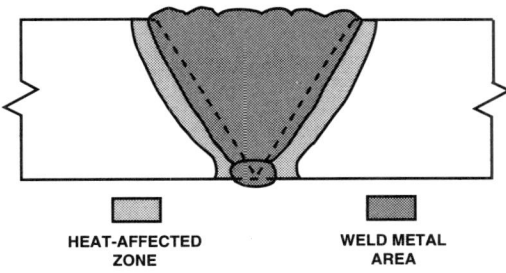

Another related term is *heat-affected zone*. This region is defined as "that portion of the base metal that has not been melted, but whose mechanical properties or microstructure have been altered by the heat of welding, brazing, soldering, or cutting."

Weld Size Terminology

The previous discussion describes joint penetration, and therefore the weld size, for single-groove weld configurations.

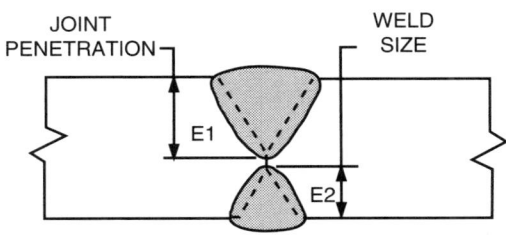

WELD SIZE, E, EQUALS E1 PLUS E2

INCOMPLETE JOINT PENETRATION OR PARTIAL JOINT PENETRATION

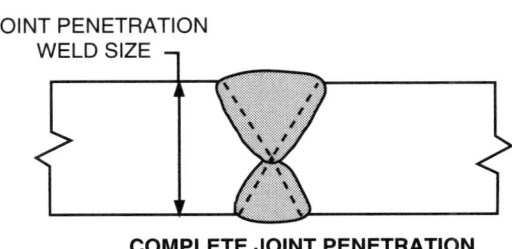

COMPLETE JOINT PENETRATION

For the case of a double-groove weld configuration where the joint penetration is less than complete, the weld size is equal to the sum of the joint penetrations from both sides.

For a complete penetration groove weld, the weld size will be equal to the thickness of the thinner of the two members joined, since there is no credit given for any weld reinforcement present.

To determine the size of a fillet weld, we must first know whether the final weld configuration is convex or concave. By convex, we mean that the weld face exhibits some buildup causing it to appear slightly curved outwardly. This is referred to as the amount of convexity. Convexity in a fillet weld is analogous with weld reinforcement in a groove weld. If a weld has a concave profile, we mean that its face is "dished in."

For either configuration, the fillet weld size for equal leg fillet welds is described as the leg lengths of the largest isosceles (two legs of equal length) right triangle which can be inscribed within the fillet weld cross section."

These inscribed isosceles right triangles are shown with dotted lines in the two illustrations

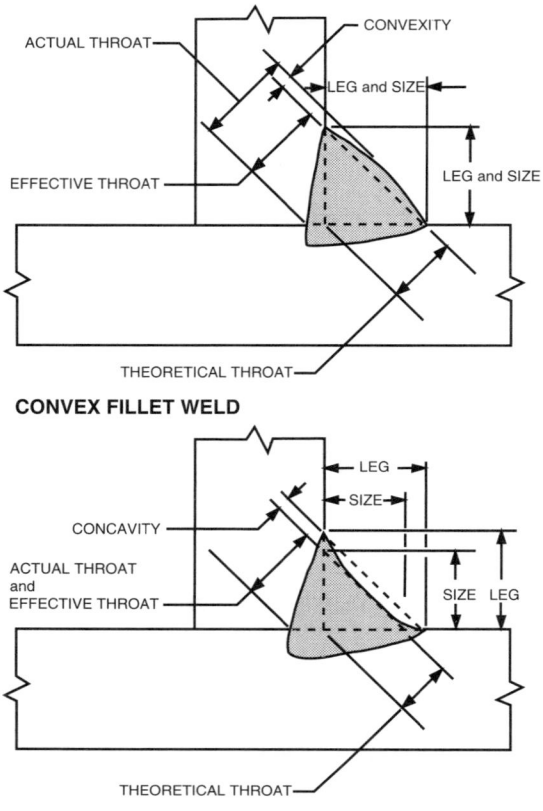

CONVEX FILLET WELD

CONCAVE FILLET WELD

above. So, for the convex fillet weld, the leg and size are equal. However, the size of a concave fillet weld is slightly less that its leg length.

For unequal leg fillet welds, the fillet weld size is defined as "the leg lengths of the largest right triangle that can be inscribed within the fillet weld cross section.

It can be noted that there are additional notations on these illustrations which refer to fillet weld throats. There are really three different types of weld throats with which we may be concerned. The first is the theoretical throat. This is the minimum amount of weld which the designer counts on when he originally specifies a weld size.

The *theoretical throat* is described as "the distance from the beginning of the joint root perpendicular to the hypotenuse (side of the triangle opposite the right angle) of the largest right triangle that can be inscribed within the cross section of a fillet weld. This dimension is based on the assumption that the root opening is equal to zero."

The *effective throat* takes into account any additional joint penetration which may be present. So, the effective throat can be defined as "the minimum distance minus any convexity between the weld root and the face of a fillet weld." The final throat dimension, the actual throat, takes into account both the joint penetration as well as any additional convexity present at the weld face.

Technically, the *actual throat* is "the shortest distance between the weld root and the face of a fillet weld." For a concave fillet weld, the effective throat and actual throat are equal, since there is no convexity present.

The welding inspector may also be asked to somehow determine the sizes of other types of welds. One example might be a spot or seam weld, where the weld size is equal to the actual nugget size or diameter.

For an edge or flange weld, the weld size is equal to the total thickness of the weld from the weld root to the weld face.

Weld Application Terminology

To complete this discussion of welding terms and definitions, it seems appropriate to mention some of the terminology associated with the actual application of welds. Some welding procedures will refer to these details, so the welding inspector should be familiar with their meanings. The first aspect is the difference

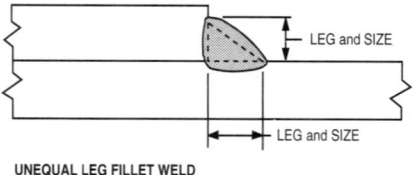

UNEQUAL LEG FILLET WELD

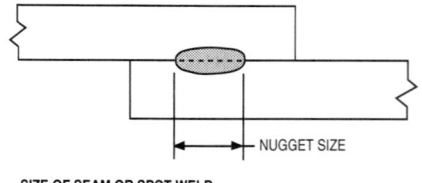

SIZE OF SEAM OR SPOT WELD

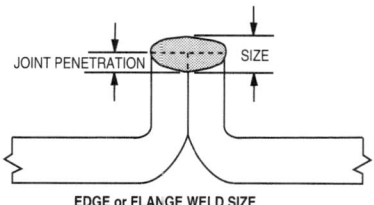

EDGE or FLANGE WELD SIZE

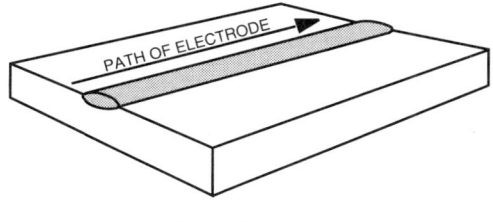

STRINGER BEAD

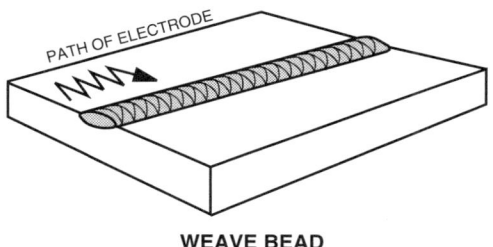

WEAVE BEAD

among the terms weld pass, weld bead and weld layer. A *weld pass* is a single progression of welding along a joint. The *weld bead* is that weld which results from a weld pass. A *weld layer* is a single level of weld within a multiple-pass weld. A weld layer may consist of a single bead or multiple beads.

When a weld bead is deposited, it could have a different name, depending upon the technique which the welder uses. If the welder progresses along the joint with little or no side-to-side motion, the resulting weld bead is referred to as a *stringer bead*. A *weave bead* results when the welder manipulates the electrode laterally, or side to side, as the weld is deposited along the joint.

The weave bead is typically wider than the stringer bead. Due to the amount of lateral motion used, the travel speed, as measured along the longitudinal axis of the weld, is less than would be the case for a stringer bead.

There are several terms which describe the actual sequence in which the welding is to be done. This is commonly done to reduce the amount of distortion caused by welding. Three common techniques are: *backstep sequence, block sequence and cascade sequence*. The backstep sequence is a technique where each individual weld pass is deposited in the direction opposite that of the overall progression of welding.

A *block sequence* is defined as "a combined longitudinal and cross sectional sequence for a continuous multiple pass weld in which separated increments are completely or partially welded before intervening increments are welded." With the block sequence, it is important that each subsequent layer is slightly shorter than the previous one so that the end of the block has a gentle slope. This will provide the best chance of obtaining adequate fusion when the adjacent block is filled in later.

A *cascade sequence* is described as "a combined longitudinal and cross sectional sequence in which weld passes are made in overlapping layers." This method differs from the block sequence in that each subsequent pass is longer than the previous one.

When fillet welds are required, there will be some cases where the design does not warrant the use of continuous welds. The designer may therefore specify intermittent fillet welds. If there are intermittent fillet welds specified on both sides

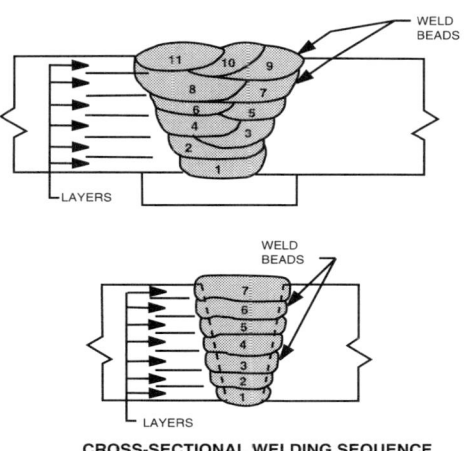

CROSS-SECTIONAL WELDING SEQUENCE

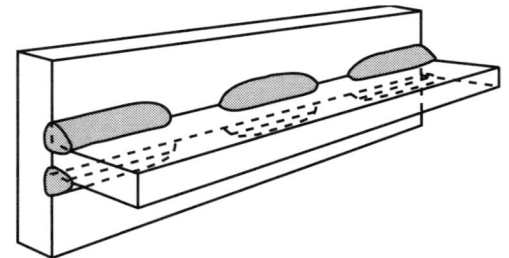

CHAIN INTERMITTENT FILLET WELD

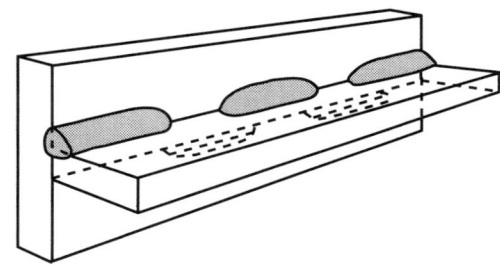

STAGGERED INTERMITTENT FILLET WELD

of a particular joint, they can be detailed as either chain intermittent or staggered intermittent fillet welds. The *chain intermittent fillet weld* has the increments on either side of the joint directly opposite each other.

Similarly, a *staggered intermittent fillet* weld is an intermittent fillet weld on both sides of a joint in which the weld increments on one side are alternated with respect to those on the other side.

One final term related to the actual welding operation is boxing (commonly referred to as end returning). *Boxing* is defined as "the continuation of a fillet weld around a corner of a member as an extension of the principal weld."

Summary

While numerous terms have been discussed here, that does not imply that these are the only ones which are applied to welding. This does provide some basis upon which the inspector can begin to understand how to describe a weld or some feature of that weld. These written explanations and illustrations are simply a beginning. As the you gain experience, you will learn to correlate these "textbook" terms with actual physical characteristics. It is only after working with and using these terms that you will gain full understanding of how to describe various welding attributes.

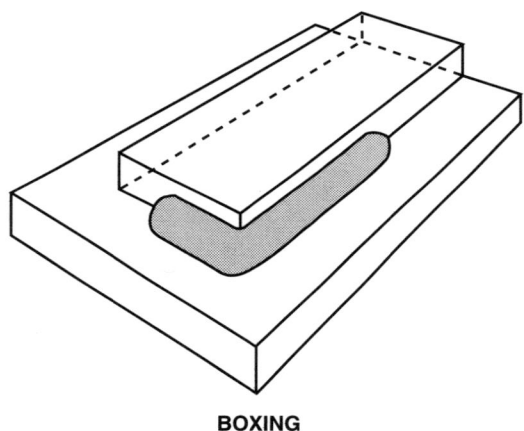

BOXING

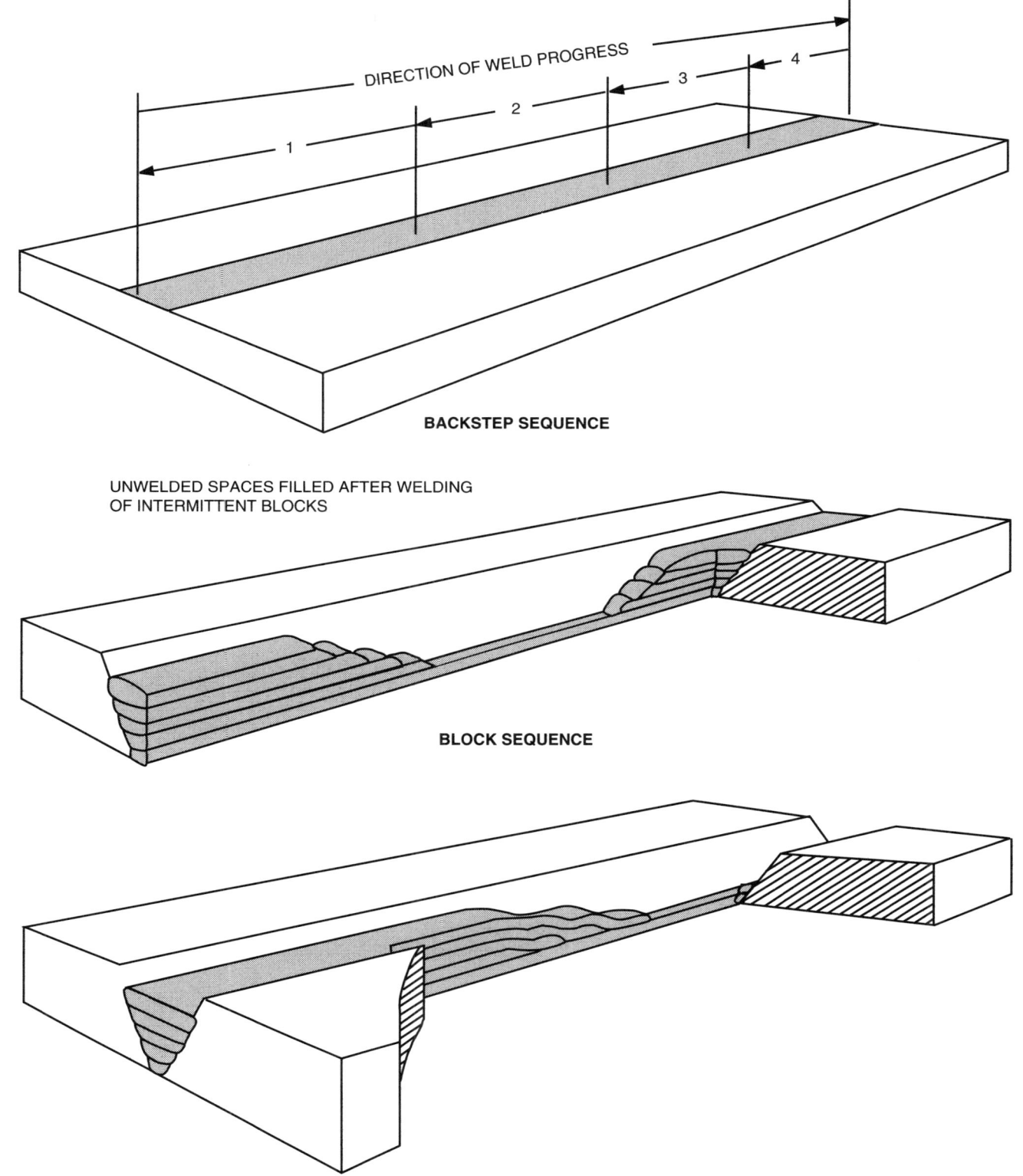

REVIEW – CHAPTER 4
WELD JOINT GEOMETRY AND WELDING TERMINOLOGY

Q4-1 Which of the following is not considered a type of joint?
 a. butt
 b. T
 c. fillet
 d. corner
 e. edge

Q4-2 The type of joint formed when the two pieces to be joined lie in parallel planes and their edges overlap is called:
 a. corner
 b. T
 c. edge
 d. lap
 e. butt

Q4-3 That portion of the joint where the two pieces to be joined come closest together is referred to as the:
 a. bevel
 b. joint root
 c. groove angle
 d. root face
 e. both b & d

Q4-4 In a single-V-groove weld, the sloped surfaces against which the weld metal is applied are called:
 a. root face
 b. joint root
 c. groove face
 d. groove angle
 e. bevel angle

Q4-5 The type of weld produced by filling an elongated hole in an overlapping member fusing it to the member beneath is called a:
a. plug weld
b. spot weld
c. seam weld
d. slot weld
e. none of the above

Q4-6 The type of weld having a generally triangular cross section and which is applied to either a T-, corner or lap joint is called a:
a. flange weld
b. flare weld
c. fillet weld
d. slot weld
e. spot weld

Q4-7 The type of weld used to build up thinned surfaces, provide a layer of corrosion protection, provide a layer of abrasion-resistant material, etc. is referred to as a:
a. edge weld
b. flare weld
c. flange weld
d. slot weld
e. surfacing weld

Q4-8 The type of weld applied to the opposite side of a joint before a single-V-groove weld is completed on the near side of a joint is called a:
a. melt-through weld
b. backing weld
c. back weld
d. root weld
e. none of the above

Q4-9 In a completed groove weld, the surface of the weld on the side from which the welding was done is called the:
a. crown
b. weld reinforcement
c. weld face
d. root
e. none of the above

Q4-10 In a completed weld, the junction between the weld face and the base metal is called the:
   a. root
   b. weld edge
   c. weld reinforcement
   d. leg
   e. weld toe

Q4-11 The height of the weld above the base metal in a groove weld is called the:
   a. crown
   b. buildup
   c. face
   d. weld reinforcement
   e. none of the above

Q4-12 In a fillet weld, the leg and size are the same for what type of configuration?
   a. equal leg
   b. concave
   c. convex
   d. unequal leg
   e. oversize

Q4-13 When looking at the cross section of a completed groove weld, the difference between the fusion face and the weld interface is called the:
   a. depth of fusion
   b. depth of penetration
   c. root penetration
   d. joint penetration
   e. effective throat

Q4-14 For a concave fillet weld, which throat dimensions are the same?
   a. theoretical and effective
   b. actual and effective
   c. theoretical and actual
   d. all of the above
   e. none of the above

Q4-15 In a partial penetration single-V-groove weld, the dimension measured from the joint root to the weld root is called the:
a. joint penetration
b. effective throat
c. root penetration
d. depth of fusion
e. weld interface

Q4-16 The size of a spot weld is determined by its:
a. depth of fusion
b. spot diameter
c. depth of penetration
d. thickness
e. none of the above

Q4-17 In the performance of a vertical position weld, the type of weld progression having a side-to-side motion is called:
a. stringer bead technique
b. stagger bead technique
c. weave bead technique
d. unacceptable
e. none of the above

Q4-18 Which of the following is only found in a single-welded groove?
a. weld face
b. weld root
c. weld toe
d. root reinforcement
e. face reinforcement

Q4-19 The technique used to control distortion of a long joint where individual passes are applied in a direction opposite the general progression of welding in the joint is called:
a. backstepping
b. boxing
c. staggering
d. cascading
e. blocking

Q4-20 A technique used in a multiple layer weld deposit where each successive layer is longer than the previous one is called:
a. block sequence
b. box sequence
c. cascade sequence
d. backstep sequence
e. stagger sequence

Questions **Q4-21** through **Q4-28** refer to Figure 1 below:

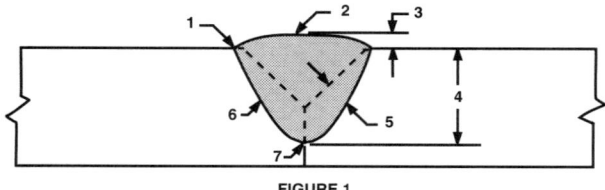

FIGURE 1

Q4-21 The weld face shown in Figure 1 is labeled:
a. 1
b. 2
c. 3
d. 6
e. 7

Q4-22 The weld root shown in Figure 1 is labeled:
a. 1
b. 2
c. 3
d. 6
e. 7

Q4-23 The type of weld shown in Figure 1 is a:
a. double-bevel-groove
b. single-bevel-groove
c. double-V-groove
d. single-V-groove
e. none of the above

Q4-24 The weld reinforcement height shown in Figure 1 is labeled:
   a. 1
   b. 2
   c. 3
   d. 6
   e. 7

Q4-25 The weld toe shown in Figure 1 is labeled:
   a. 1
   b. 2
   c. 3
   d. 6
   e. 7

Q4-26 Number 6 shown in Figure 1 is the:
   a. weld root
   b. fusion face
   c. groove face
   d. weld interface
   e. depth of fusion

Q4-27 Number 5 (between arrows) shown in Figure 1 is the:
   a. weld root
   b. fusion face
   c. groove face
   d. weld interface
   e. depth of fusion

Q4-28 Number 4 shown in Figure 1 is the:
   a. weld size
   b. joint penetration
   c. actual throat
   d. theoretical throat
   e. a and b above

Questions **Q4-29** through **Q4-40** refer to Figure 2 below:

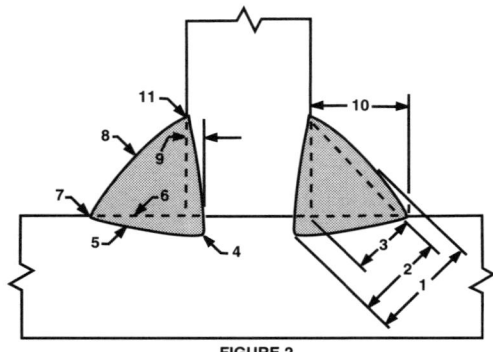

FIGURE 2

Q4-29 The weld face shown in Figure 2 is labeled:
a. 7
b. 8
c. 6
d. 11
e. 10

Q4-30 The weld root shown in Figure 2 is labeled:
a. 6
b. 4
c. 5
d. 9
e. 3

Q4-31 The welds shown in Figure 2 are:
a. concave fillets
b. conical fillets
c. convex fillets
d. T-fillets
e. fillet of fish

Q4-32 The actual throat shown in Figure 2 is labeled:
a. 1
b. 2
c. 3
d. 10
e. 9

Q4-33 The weld toe shown in Figure 2 is labeled:
   a. 11
   b. 8
   c. 10
   d. 7
   e. both a and d

Q4-34 Number 6 shown in Figure 2 is the:
   a. weld root
   b. fusion face
   c. groove face
   d. weld interface
   e. depth of fusion

Q4-35 Number 9 shown in Figure 2 is the:
   a. weld root
   b. fusion face
   c. groove face
   d. weld interface
   e. depth of fusion

Q4-36 Number 5 shown in Figure 2 is the:
   a. weld root
   b. fusion face
   c. groove face
   d. weld interface
   e. depth of fusion

Q4-37 Number 4 shown in Figure 2 is the:
   a. weld root
   b. fusion face
   c. groove face
   d. weld interface
   e. depth of fusion

Q4-38 Number 2 shown in Figure 2 is the:
a. weld size
b. effective throat
c. actual throat
d. theoretical throat
e. a and b above

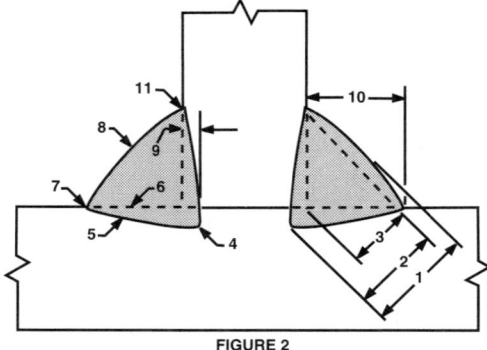

FIGURE 2

Q4-39 Number 3 shown in Figure 2 is the:
a. weld size
b. effective throat
c. actual throat
d. theoretical throat
e. a and b above

Q4-40 Number 10 shown in Figure 2 is the:
a. weld size and leg size
b. weld size
c. leg
d. theoretical throat
e. actual throat

Questions **Q4-41** through **Q4-45** refer to Figure 3 below:

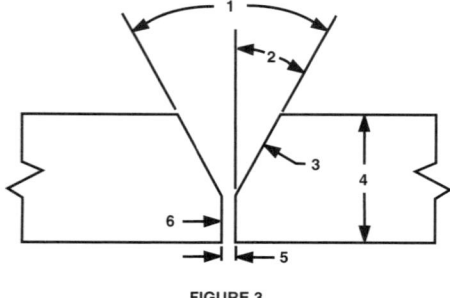

FIGURE 3

Q4-41 The groove angle shown in Figure 3 is labeled:
a. 1
b. 2
c. 3
d. 4
e. 5

Q4-42 The bevel angle shown in Figure 3 is labeled:
   a. 1
   b. 2
   c. 3
   d. 4
   e. 5

Q4-43 Number 3 shown in Figure 3 is the:
   a. groove angle
   b. bevel angle
   c. groove face
   d. fusion face
   e. c and d above

Q4-44 Number 6 shown in Figure 3 is the:
   a. groove face
   b. fusion face
   c. bevel face
   d. root face
   e. bevel

Q4-45 Number 5 shown in Figure 3 is the:
   a. fusion face
   b. groove face
   c. root opening
   d. root face
   e. weld root

Questions **Q4-46** through **Q4-54** refer to Figure 4 below:

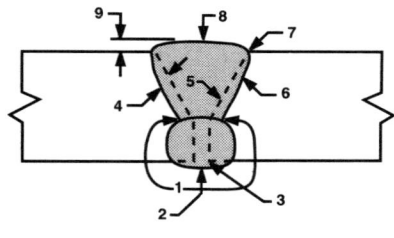

FIGURE 4

Q4-46 The weld faces shown in Figure 4 are labeled:
   a. 8 & 2
   b. 9 & 7
   c. 7 & 4
   d. 1 & 6
   e. 3 & 9

Q4-47 The weld root shown in Figure 4 is labeled:
   a. 1
   b. 2
   c. 3
   d. 7
   e. a and c above

Q4-48 The weld shown in Figure 4 includes a:
   a. backing weld
   b. back weld
   c. double-V-groove
   d. double-bevel-groove
   e. none of the above

Q4-49 The weld size shown in Figure 4 is labeled:
   a. 9
   b. 8
   c. 7
   d. 2
   e. none of the above

Q4-50 The weld toe shown in Figure 4 is labeled:
a. 1
b. 2
c. 3
d. 6
e. 7

Q4-51 Number 6 shown in Figure 4 is the:
a. weld root
b. fusion face
c. groove face
d. weld interface
e. depth of fusion

Q4-52 Number 5 shown in Figure 4 is the:
a. weld root
b. fusion face
c. groove face
d. weld interface
e. depth of fusion

Q4-53 Number 4 (between arrows) shown in Figure 4 is the:
a. weld root
b. fusion face
c. groove face
d. weld interface
e. depth of fusion

Q4-54 Number 2 shown in Figure 4 is the:
a. root surface
b. fusion face
c. weld face
d. weld interface
e. depth of fusion

Questions **Q4-55** through **Q4-58** refer to Figure 5 below:

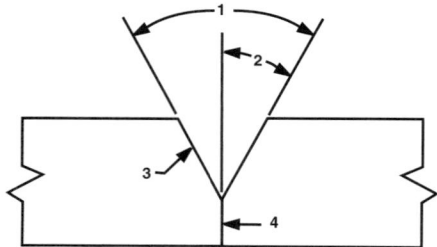

FIGURE 5

Q4-55 The bevel angle shown in Figure 5 is labeled:
- a. 1
- b. 2
- c. 3
- d. 4
- e. c and d above

Q4-56 The joint root shown in Figure 5 is labeled:
- a. 1
- b. 2
- c. 3
- d. 4
- e. none of the above

Q4-57 The groove face shown in Figure 5 is labeled:
- a. 1
- b. 2
- c. 3
- d. 4
- e. c and d above

Q4-58 The root face shown in Figure 5 is labeled:
- a. 1
- b. 2
- c. 3
- d. 4
- e. none of the above

Questions **Q4-59** through **Q4-69** refer to Figure 6 below:

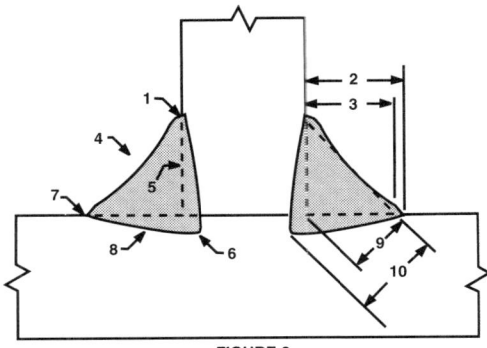

FIGURE 6

Q4-59   The weld face shown in Figure 6 is labeled:
   a.   1
   b.   4
   c.   7
   d.   3
   e.   2

Q4-60   The weld root shown in Figure 6 is labeled:
   a.   6
   b.   4
   c.   5
   d.   9
   e.   10

Q4-61   The welds shown in Figure 6 are:
   a.   concave fillets
   b.   conical fillets
   c.   convex fillets
   d.   T-fillets
   e.   fillets of fish

Q4-62   The actual throat shown in Figure 6 is labeled:
   a.   9
   b.   10
   c.   3
   d.   2
   e.   5

Q4-63 The weld toe shown in Figure 6 is labeled:
- a. 1
- b. 8
- c. 10
- d. 7
- e. both a and d above

Q4-64 Number 6 shown in Figure 6 is the:
- a. weld root
- b. fusion face
- c. groove face
- d. weld interface
- e. depth of fusion

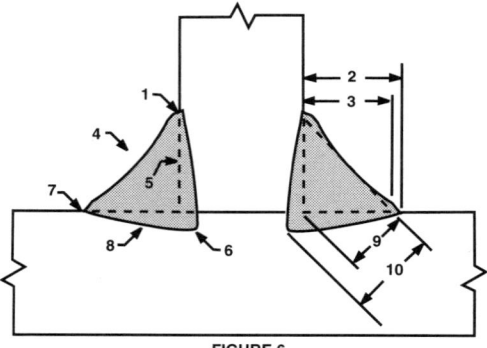

FIGURE 6

Q4-65 Number 8 shown in Figure 6 is the:
- a. weld root
- b. fusion face
- c. groove face
- d. weld interface
- e. depth of fusion

Q4-66 Number 5 shown in Figure 6 is the:
- a. weld root
- b. fusion face
- c. toe
- d. weld interface
- e. depth of fusion

Q4-67 Number 2 shown in Figure 6 is the:
- a. weld size
- b. leg size
- c. leg and weld size
- d. theoretical throat
- e. actual throat

Q4-68 Number 3 shown in Figure 6 is the:
  a. weld size
  b. leg
  c. leg and weld size
  d. theoretical throat
  e. actual throat

Q4-69 Number 9 shown in Figure 6 is the:
  a. effective throat
  b. weld size
  c. leg
  d. theoretical throat
  e. actual throat

CHAPTER 5: WELDING AND NONDESTRUCTIVE EXAMINATION SYMBOLS

Introduction

In the previous sections, there has been discussion relative to the importance of the welding inspector being capable of effectively communicating with others involved in the fabrication of some welded product. Much of this communication is achieved through the use of various types of documents which describe what attributes that product must exhibit. While these documents provide the basis upon which the inspection will be performed, confusion could occur if there is a tremendous amount of material involved. If the welding inspector must spend a great deal of time studying this information, it may detract from the actual inspection.

One method for reducing the mass of information contained in documents (especially drawings) is through the use of symbols. This practice replaces written words and detailed graphic illustrations with specific symbols to convey the same information in an abbreviated manner. To provide continuity, the American Welding Society has developed a standard which describes the construction and interpretation of all types of welding and nondestructive examination symbols. This document, AWS A2.4, *Symbols for Welding and Nondestructive Examination*, details all requirements relating to the use of these symbols.

Welding and nondestructive examination symbols can be thought of as a "shorthand" method for conveying information pertinent to these operations. While quite simple, this system provides a powerful method of describing detailed information. The designer can communicate a vast amount of information to fabrication and inspection personnel regarding numerous aspects of some welding project.

The welding or examination symbol can be used to provide a great deal of information; however, they must be used properly to be effective. If misapplied or misinterpreted, the symbols may tend to cause confusion rather than aid in the understanding of some welding or testing detail. For that reason, it is important to understand how the welding and nondestructive examination symbols are used.

To aid in the description of interpretation of a particular welding or nondestructive examination symbol, the following is a detailed description of the steps used in the construction of the symbol. This is convenient, because there are numerous elements of the symbol which have some specific meaning due to their location with respect to other parts of that symbol. Once it is understood how a symbol was constructed, that information can be applied in reverse to gain insight as to what is actually required for a weld to be in compliance with a symbol. Therefore, the following discussion takes one through the various steps associated with the construction of a welding or nondestructive examination symbol.

Elements of the Welding Symbol

Before describing the various parts of a welding symbol, it is important to understand some of the terminology relating to symbols. A basic distinction which must be pointed out is the difference between the terms "weld symbol" and "welding symbol." As stated in AWS A2.4, "the weld symbol indicates the type of weld."

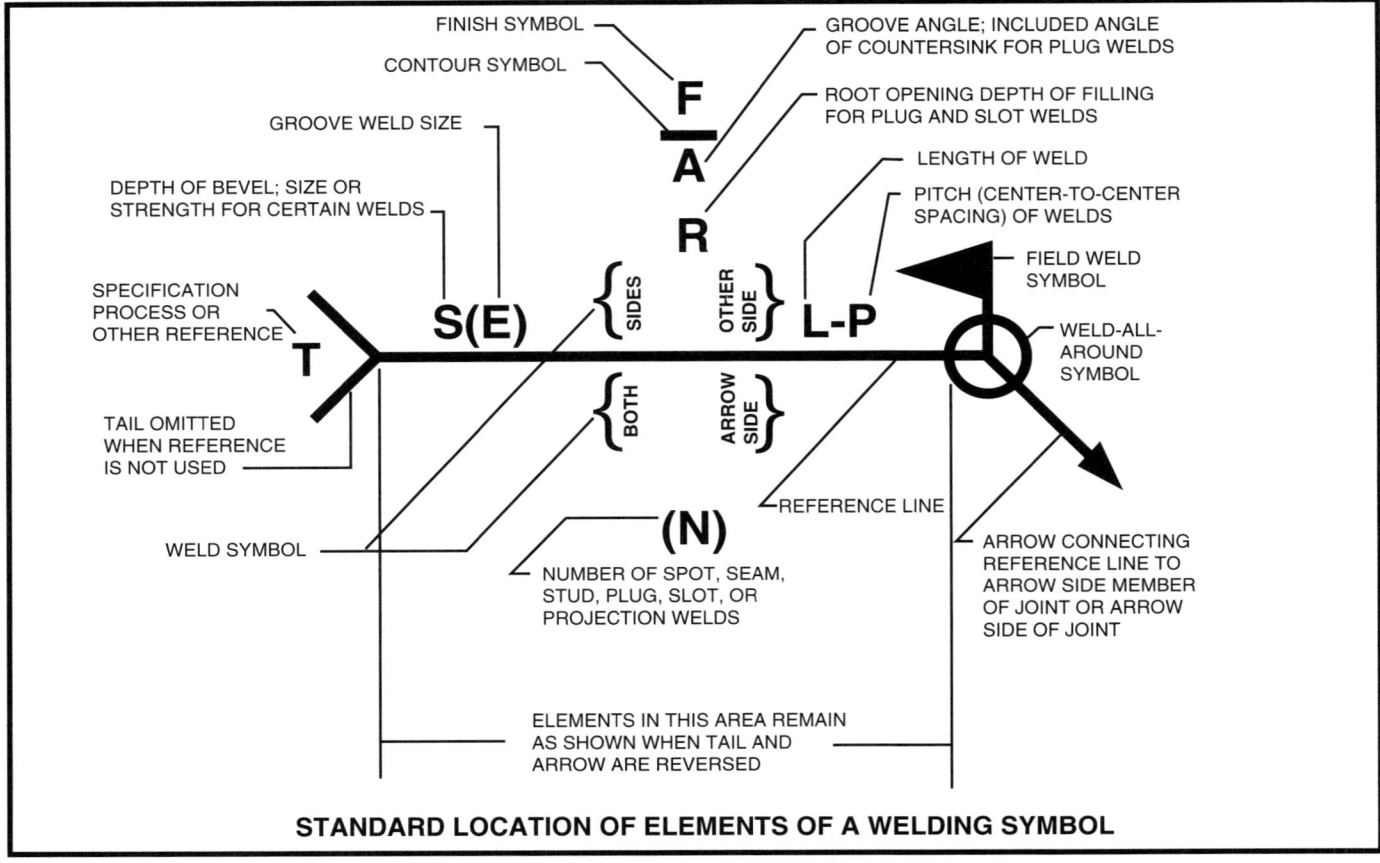

STANDARD LOCATION OF ELEMENTS OF A WELDING SYMBOL

The welding symbol is a method of representing the weld symbol on drawings, and includes supplementary information and consists of the following eight elements. Not all elements need be used unless required for clarity.

1) Reference line (shown horizontally)
2) Arrow
3) Basic weld symbols
4) Dimensions and other data
5) Supplementary symbols
6) Finish symbols
7) Tail
8) Specification, process, or other reference

In the construction of a welding symbol, the primary element which is always included is the reference line. This is simply a horizontal line segment which provides the basis for all other parts of the symbol. It must appear on the drawing as a horizontal line, because there is a significance whether information lies above or below the line.

The next element of the welding symbol is the arrow. This line segment is connected to one end of the reference line and points to one side of the weld joint. This gives significance to the terms arrow side and other side. That is, the side to which the arrow points is referred to as the arrow side, while the opposite side is called the other side. Once the arrow side and other side have been assigned by the placement of the arrow, it is now possible to specify information relating to either or both sides.

The AWS rule is that any information placed below the reference line relates to the arrow side of the joint, and that information above the reference line describes what will occur on the other side of the joint.

This rule will never change, no matter which end of the reference line is attached to the arrow or which direction the arrow may point. Even with

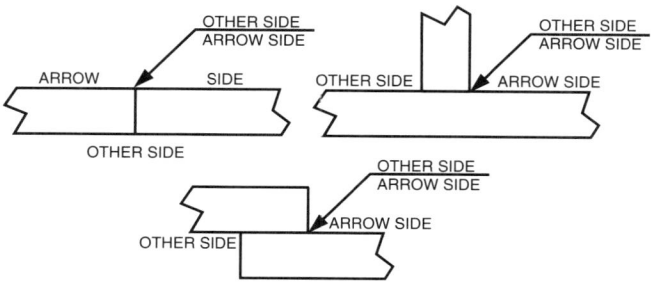

ARROW - LOCATION SIGNIFICANCE

the arrow oriented in different directions and at either end of the reference line, the operations will be performed on the side of the joint to which the arrow points.

Once the reference line and arrow are in place, the next element of the welding symbol can be added. This particular part, the weld symbol, describes what the actual weld configuration will be. Weld symbols depicting arrow side welds will appear below the reference line and those referring to other side welds will appear above the reference line. It is also interesting to note that some of the weld symbols are placed such that the reference line splits them in half (e.g., spot, projection and seam welds). This simply implies the weld has no side significance, meaning that it makes no difference which side is called the arrow side. With the exception of the surfacing weld, which always appears as an arrow side weld, all other types can be shown as arrow side, other side or both sides.

Most of the weld symbols appear much like the actual weld configuration, which makes it easier to remember exactly what type of weld is specified by a particular weld symbol.

Another feature which should be noted for all of those weld symbols which represent welds having only one of the two members prepared is that the perpendicular side of the symbol will always appear on the left side (e.g., bevel, J- and flare-bevel grooves; fillet; and corner-flange welds). For these groove welds, the designer can designate which of the two members actually receives the preparation by using an arrow with a break in the line. The rule is that the last segment of the arrow points to that member receiving the specified preparation.

GROOVE							
SQUARE	SCARF	V	BEVEL	U	J	FLARE-V	FLARE-BEVEL
∥	∥	∨	⋁	∪	⌐	)(	)(

FILLET	PLUG OR SLOT	STUD	SPOT OR PROJECTION	SEAM	BACK OR BACKING	SURFACING	FLANGE	
							EDGE	CORNER
△	▭	⊗	○	⊖	⌣	⌣⌣	‖	‖

WELD SYMBOLS (NOTE: THE REFERENCE LINE IS SHOWN DASHED FOR ILLUSTRATIVE PURPOSES.)

Groove Weld Detailing

After designating the type of groove weld required and at which side or sides of the joint it will be deposited, other information is necessary. Most of this data relates to dimensional requirements. Groove weld features needing dimensions include: the joint configuration, weld size and the extent of welding.

Groove Weld Dimensions: Some of the groove weld dimensions are placed within or slightly outside the weld symbol. A dimension appearing within the weld symbol indicates the required root opening, while a dimension appearing just outside (above or below) the weld symbol refers to the necessary groove angle.

Another important piece of information for the preparation of the groove is the depth of preparation. This dimension is always shown to the <u>left</u> of the groove weld symbol. This depth is measured from the base material surface. The specified depth of preparation in each case is that dimension outside of the parentheses. In general, dimensions appearing to the left of the weld symbol refer to the weld size required.

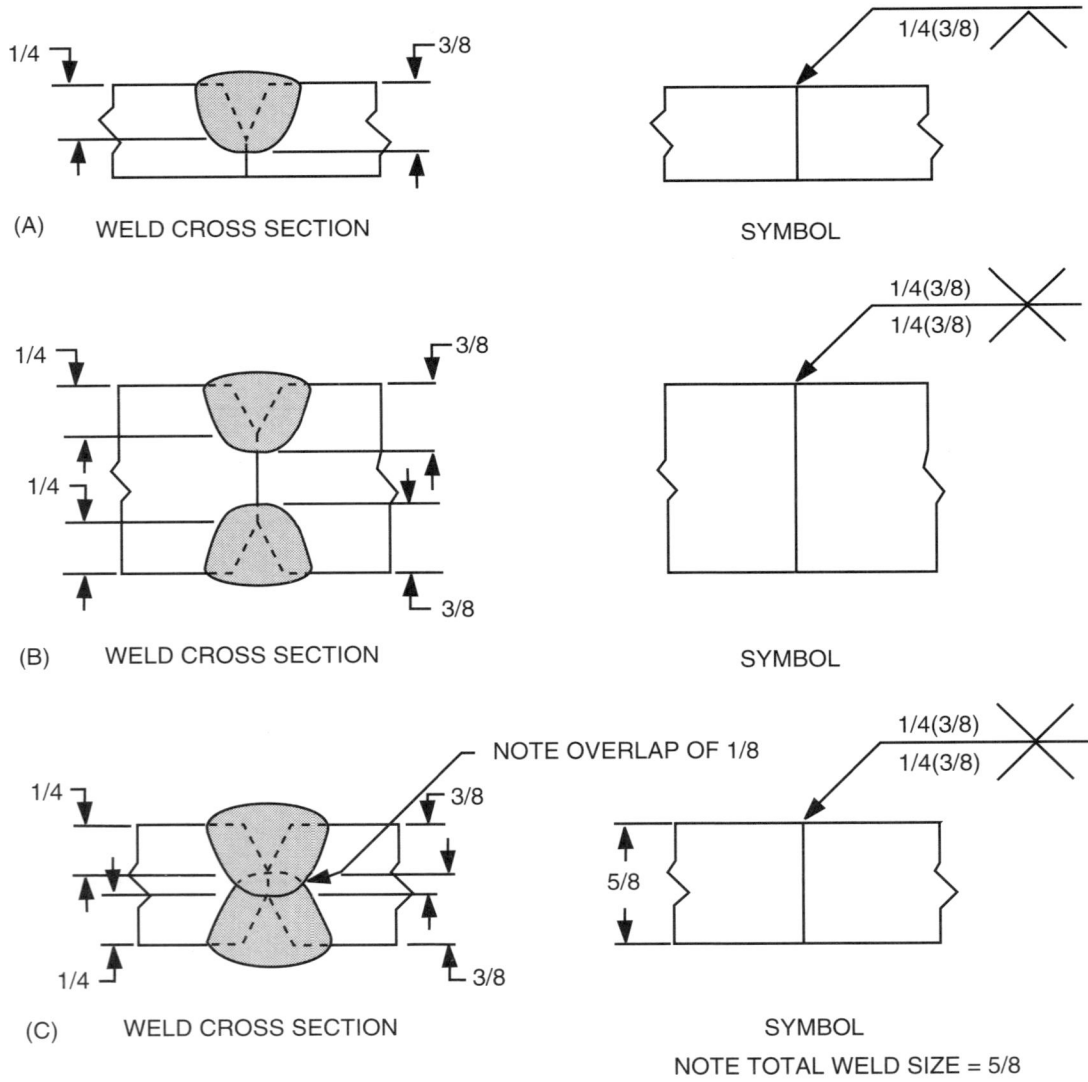

SPECIFICATION OF GROOVE WELD SIZE AND DEPTH OF BEVEL

The dimensions which are <u>enclosed in parentheses</u> refer to the groove weld size (or joint penetration) required. For groove welds, absence of dimensions for depth of preparation or weld size implies that the required weld is to have complete joint penetration.

The final piece of dimensional information necessary for a groove weld is the required length. This detail is always shown on the welding symbol to the <u>right</u> of the weld symbol. If no dimension is shown, it is assumed that the specified weld is to be the entire length of the joint. If a dimension is present to the right of the weld symbol, it refers to the length of groove weld segment required.

Fillet Weld Detailing

There is also dimensional information pertinent to fillet welds. As was the case for groove welds, the size of a fillet weld is dimensioned to the *left* of the weld symbol.

Another feature identical to the groove weld application is that the length of a fillet weld is dimensioned to the *right* of the weld symbol.

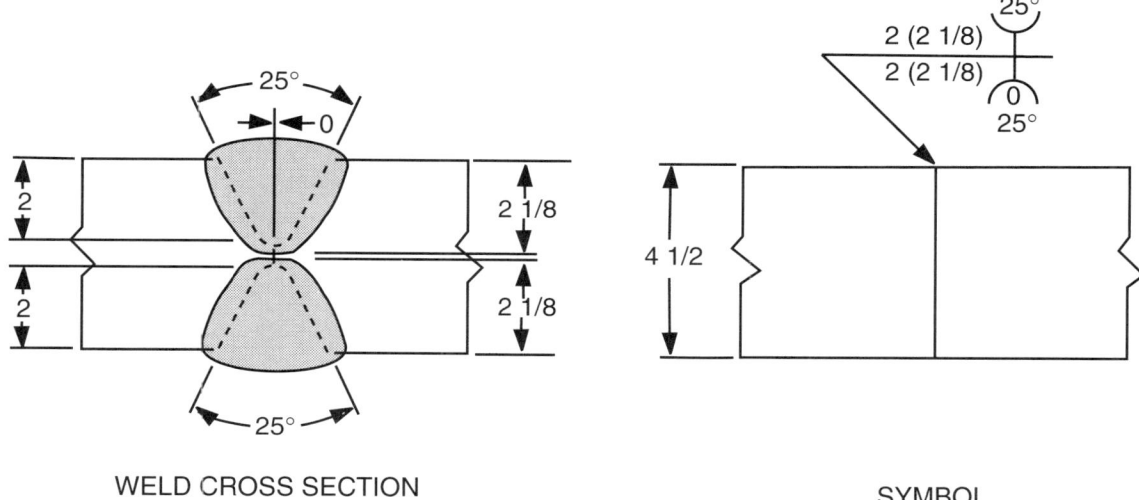

GROOVE WELD SYMBOL WITH COMBINED DIMENSIONS

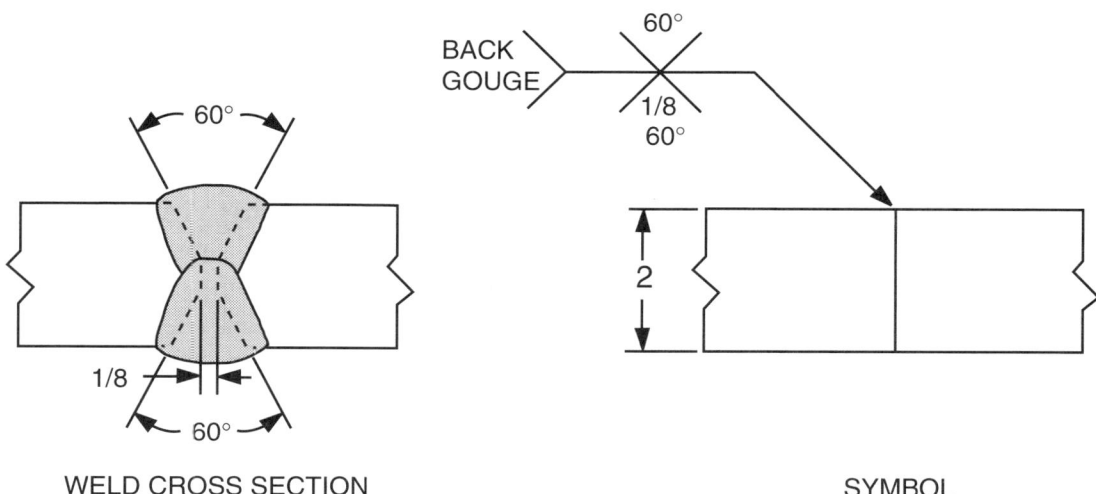

SYMMETRICAL GROOVE WELDS WITH BACK GOUGING

No dimension to the right of the fillet weld symbol indicates that the fillet weld is to be continuous for the entire length of the joint. A specific length of fillet weld is denoted by that single dimension appearing to the right of the weld symbol.

A common welding practice is to use intermittent fillet welds instead of a continuous fillet weld to reduce distortion and the amount of time required for welding. The dimensions for intermittent fillet welds are shown as two numbers separated by a hyphen. The first number is the length of each individual weld segment and the second number refers to the center-to-center spacing of these weld segments. The spacing from one segment to the next is referred to as the pitch. The pitch is measured as the center-to-center distance of each adjacent length of fillet weld.

Intermittent fillet welds may be applied to both sides of a joint in one of two ways. If the individual segments are directly opposite each other, it is referred to as chain intermittent welding. Staggered intermittent welding is when the segments on either side of the joint coincide with spaces between individual segments on the other side of the joint. In both types of intermittent welds, the pitch distance refers to the center-to-center spacing on that side of the joint only.

Plug and Slot Weld Detailing

The symbolization of the next types of welds involves several different features because of the uniqueness of their configurations. They are the plug and slot welds. As discussed in Chapter 3 of this text, they both are welds joining overlapping members by filling a hole in the top member to connect it to the backing member. The weld symbol for both is simply a rectangular box.

Dimensions for plug welds include: plug weld size, depth of filling, pitch distances between adjacent plugs, and groove angle for tapered plug holes. The plug weld size dimension appears to the *left* of the weld symbol. If the hole is only intended to be partially filled, the required depth of filling is indicated *within* the plug weld symbol. Pitch distances are shown to the *right* of the plug weld symbol. If the hole is to be tapered to provide better root access, the angular dimension appears just outside (above or below) the weld symbol.

In general, the same rules apply to the welding symbols for slot welds. <u>An exception, however, is that the length, width, spacing, included angle of countersink, orientation and location of slot welds cannot be shown on the welding symbol.</u> This information shall be shown on the drawing or by a detail referenced on the welding symbol.

Spot and Seam Weld Detailing

Spot and seam welds can also be described using welding symbols. The size of a spot or seam weld is shown as a dimension to the left of the weld symbol. This size refers to the diameter of the spot or width of the seam. Another way in which the degree of welding can be described is by specifying the required shear strength of the resulting spot weld.

The pitch distance of adjacent spot welds is shown in the same manner as for plug and slot welds. The required number of spots is shown by the number enclosed in parentheses just outside the weld symbol.

As might be expected, the dimensioning of seam welds is similar to that of spot welds.

Back and Backing Weld Detailing

Two other types of welds requiring attention are the back and backing welds. While both are represented by the same weld symbol, they differ in that the back weld is deposited after one side has been welded and the backing weld is deposited before depositing the opposite side. Some treatment, such as backgouging, may be required before application of a back weld and after the deposition of a backing weld. There are two ways

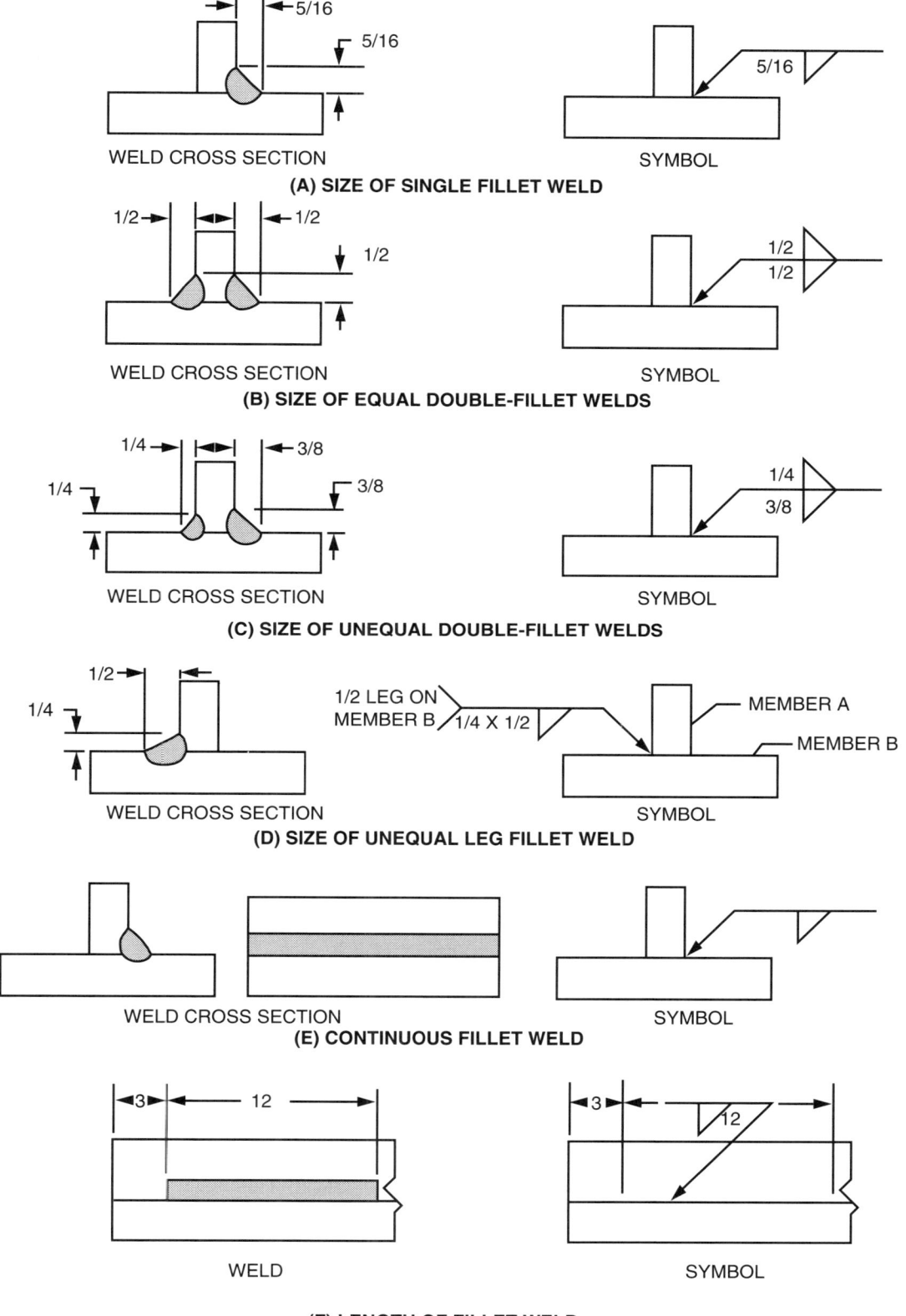

SPECIFICATION OF SIZE AND LENGTH OF FILLET WELDS

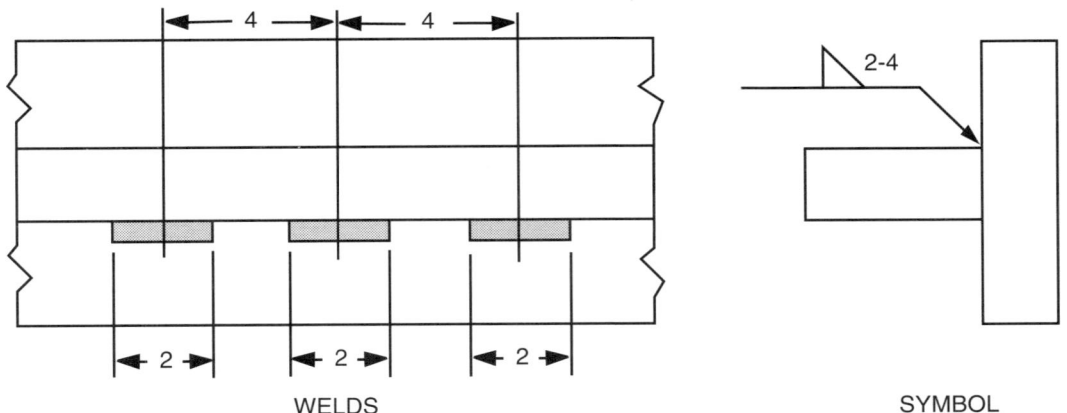

(A) LENGTH AND PITCH OF INTERMITTENT WELDS

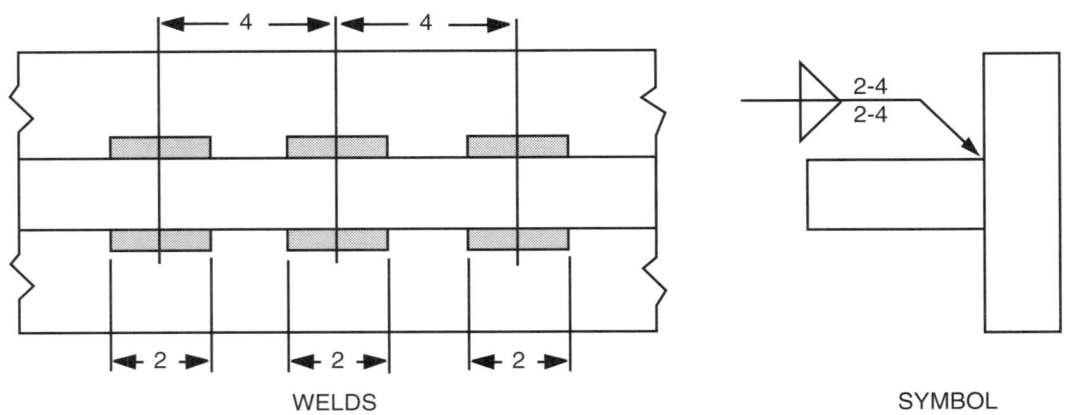

(B) LENGTH AND PITCH OF CHAIN INTERMITTENT WELDS

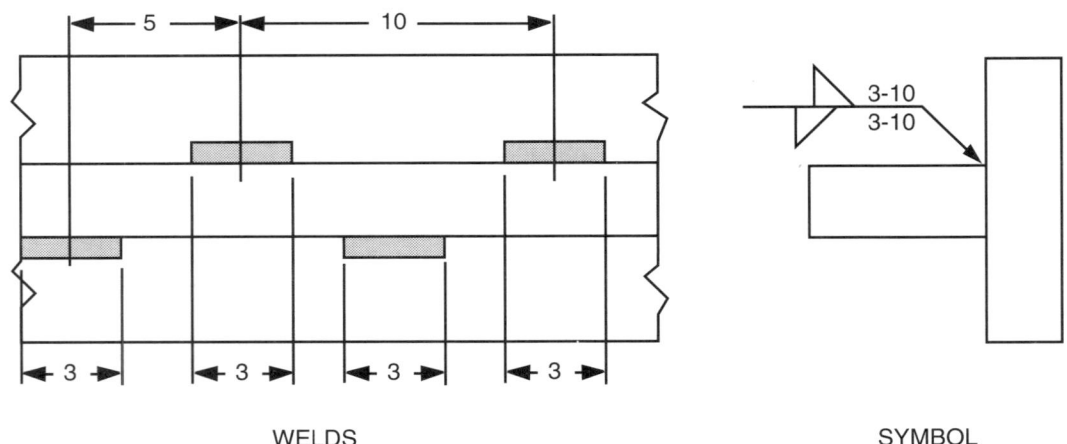

(C) LENGTH AND PITCH OF STAGGERED INTERMITTENT WELDS

APPLICATIONS OF INTERMITTENT FILLET WELD SYMBOLS

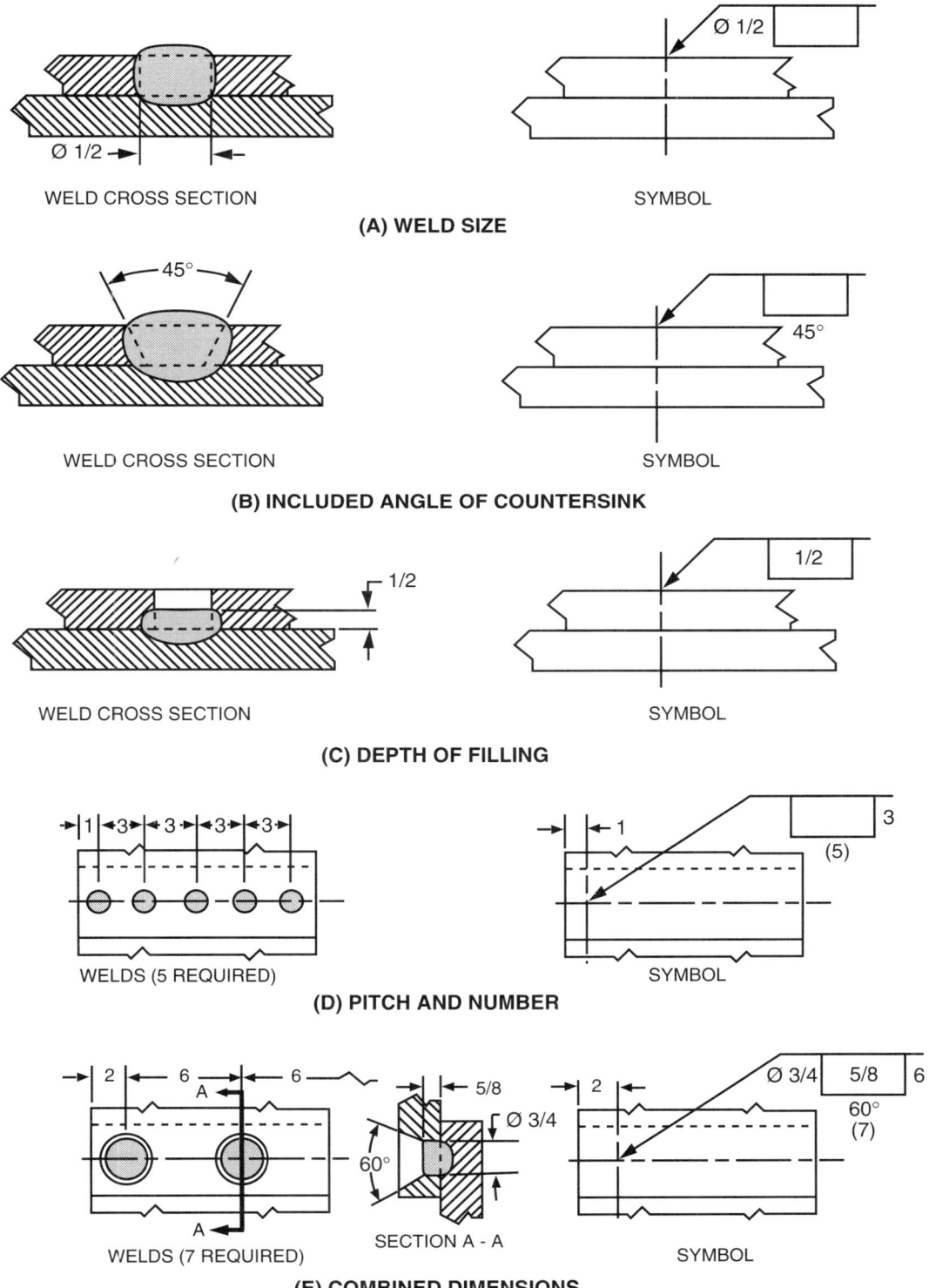

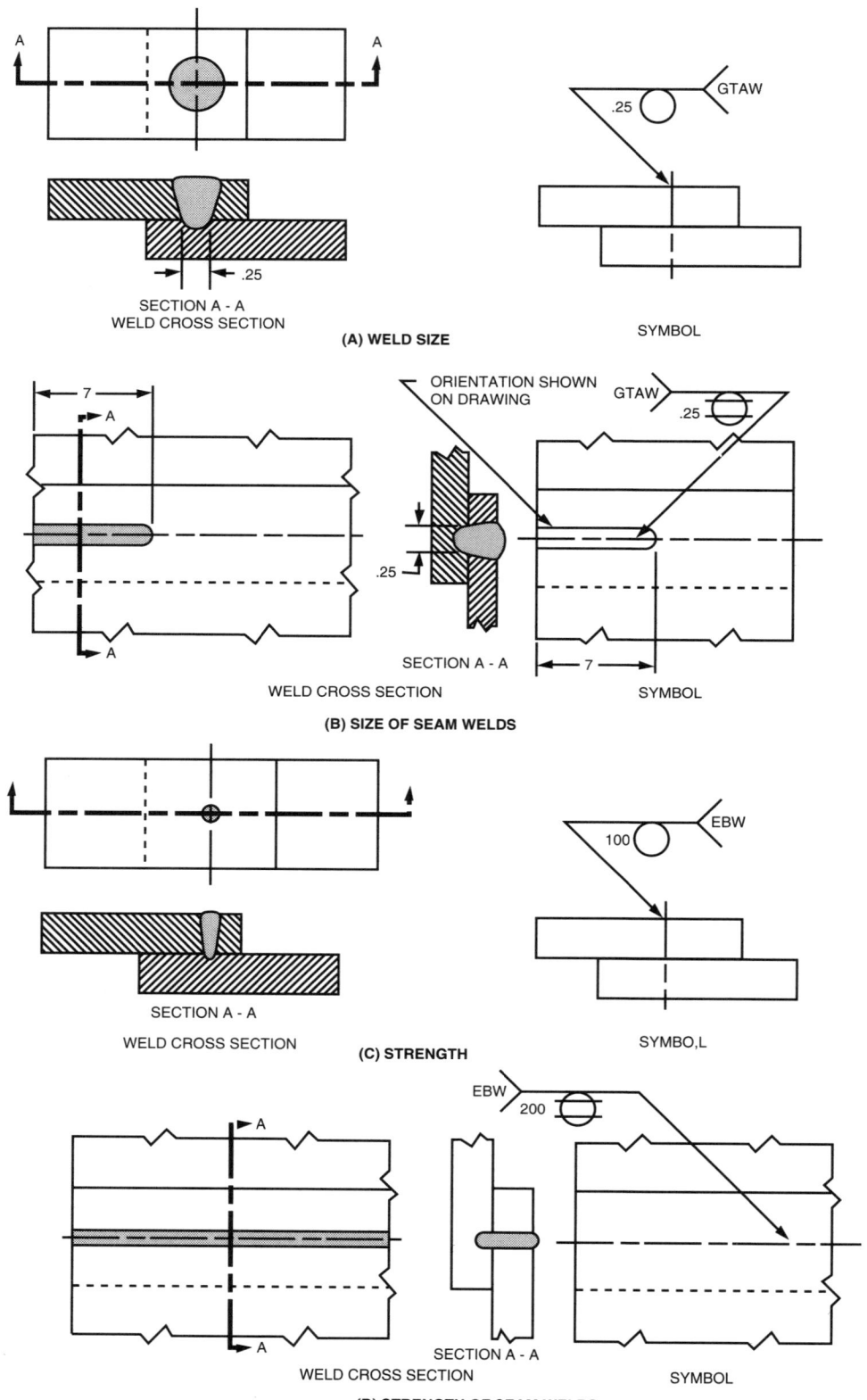

APPLICATIONS OF INFORMATION TO SPOT AND SEAM WELD SYMBOLS

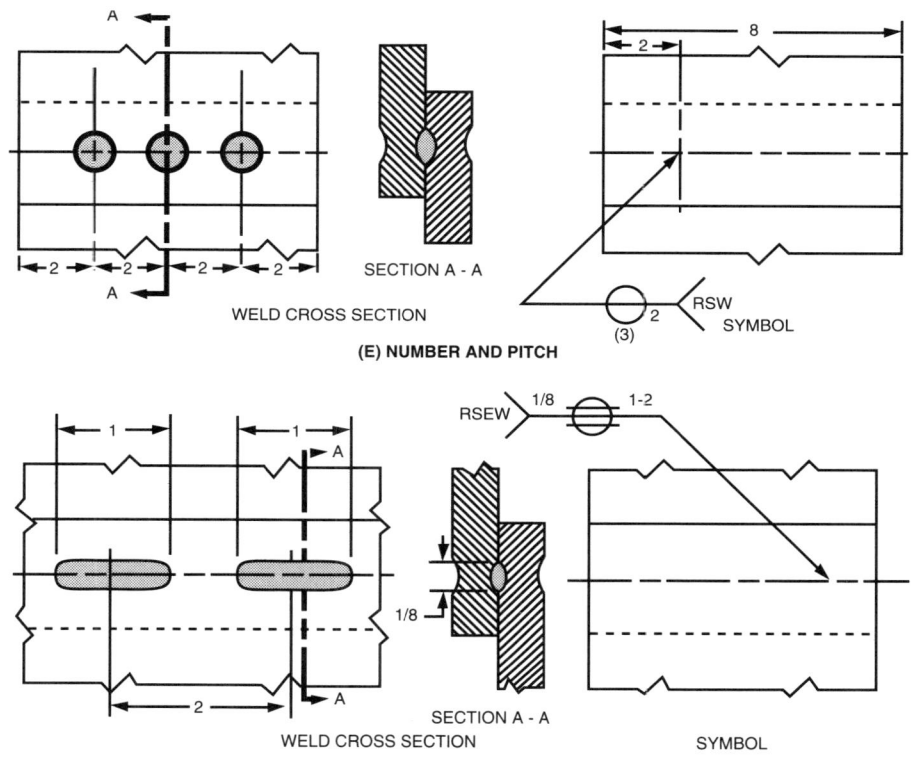

APPLICATIONS OF INFORMATION TO SPOT AND SEAM WELD SYMBOLS

to describe the sequencing of these welds. They can be differentiated by using a symbol with a note in the tail or simply use multiple reference lines to show a sequence of operations.

Surfacing Weld Detailing

The detailing of surface welds is quite simple. The primary information required is the thickness of this surfacing and the weld filler metal. The welding symbol must then indicate the region of the part requiring the surfacing treatment. The surfacing symbol only appears as an arrow side weld.

Stud Weld Detailing

The newest weld symbol is that for stud welds. It appears as a circle with an enclosed "X" and is only shown as an arrow side weld, as was the case for the surfacing weld. A dimension to the left of the stud weld symbol refers to its required size. The number of studs required can be indicated by a number enclosed in parentheses just outside the stud weld symbol. To indicate the spacing of adjacent studs, a number can be placed to the right of the stud weld symbol.

Flange Weld Detailing

The final weld symbols to be discussed are those for the various types of flange welds. Their dimensioning shows not only the size of the weld but also details as to how the flange is to be formed. The two dimensions appearing with a "+" sign between them refers to the forming of the flange. The first number indicates the inside radius of the formed flange while the second dimension indicates the length of the formed flange. The final dimension is an indication of the required weld size.

Supplementary Symbols

The basic weld symbols can be supplemented with additional symbols to detail other important information. This group of symbols is referred to

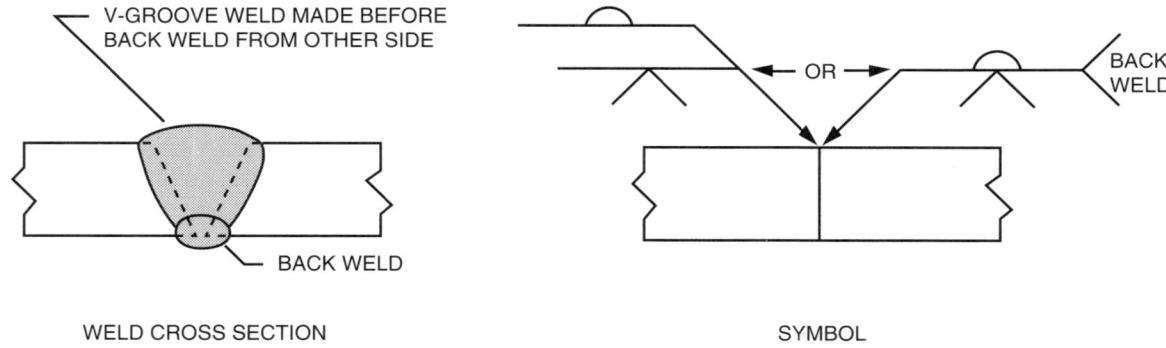

(A) APPLICATION OF BACK WELD SYMBOL

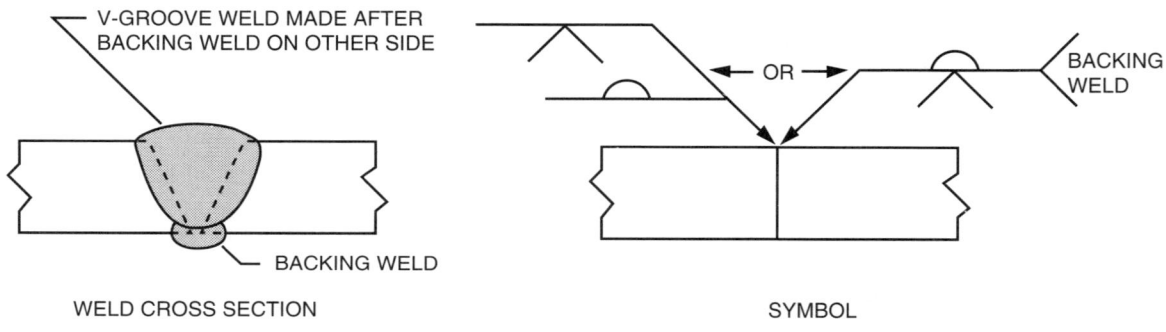

(B) APPLICATION OF BACKING WELD SYMBOL

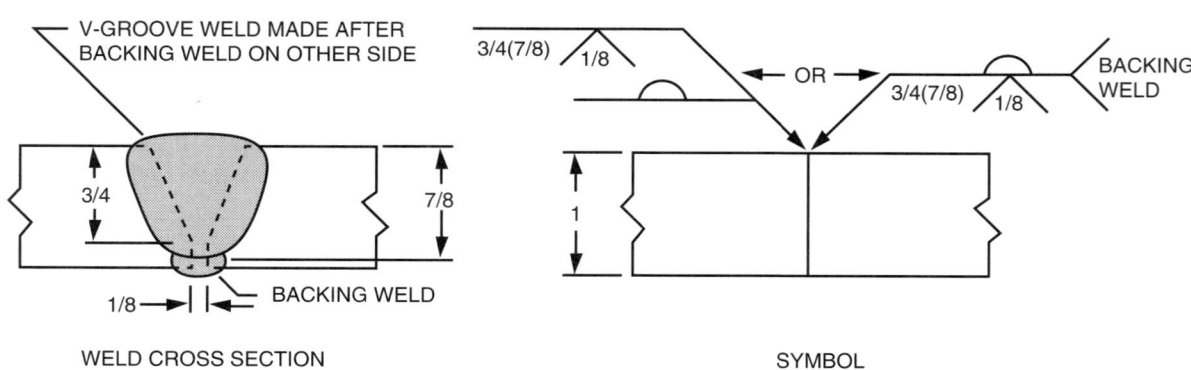

(C) APPLICATION OF BACKING WELD WITH ROOT OPENING SPECIFIED

APPLICATIONS OF BACK OR BACKING WELD SYMBOL

5-13

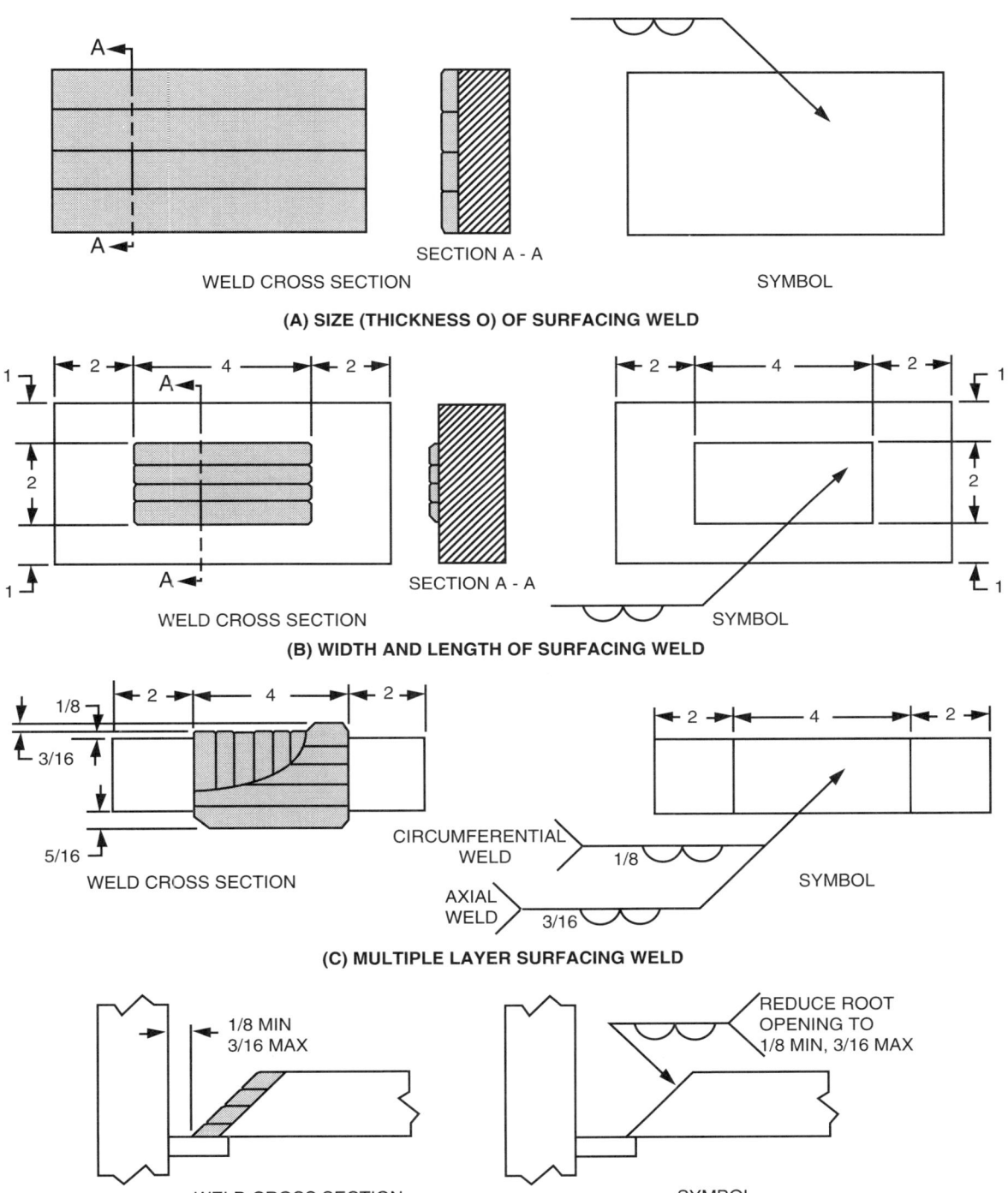

APPLICATIONS OF SURFACING WELD SYMBOL

5-14

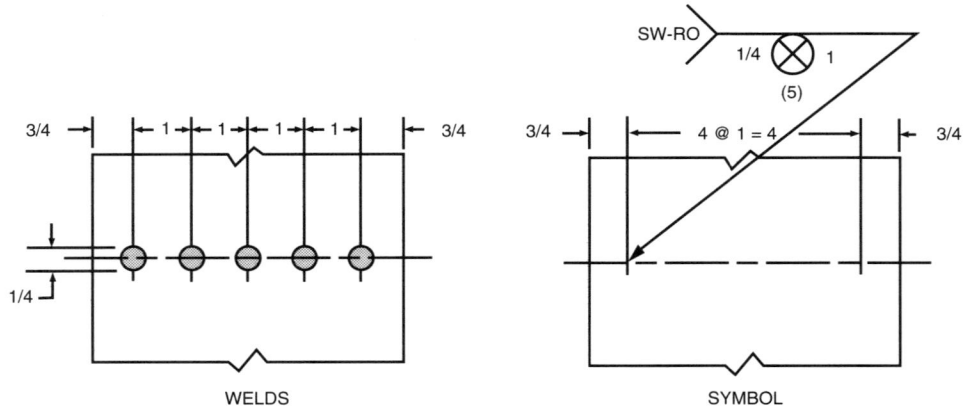

(A) STUD WELD SYMBOL WITH COMBINED DIMENSIONS

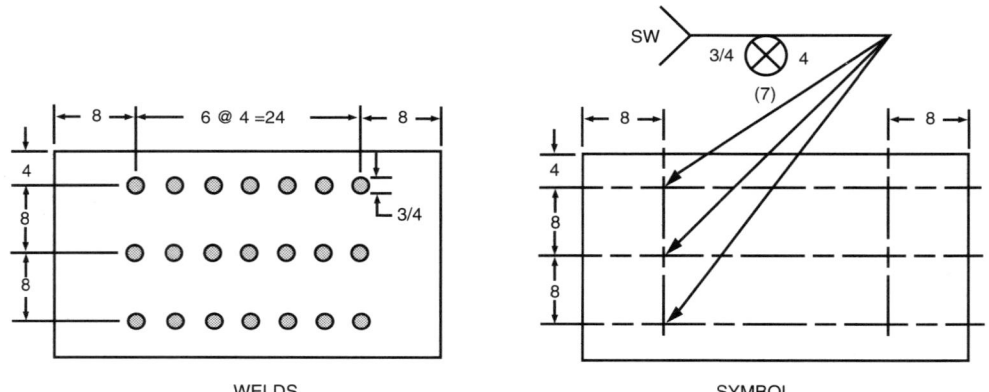

(B) STUD WELD SYMBOL FOR MULTIPLE ROWS

APPLICATIONS OF STUD WELD SYMBOLS

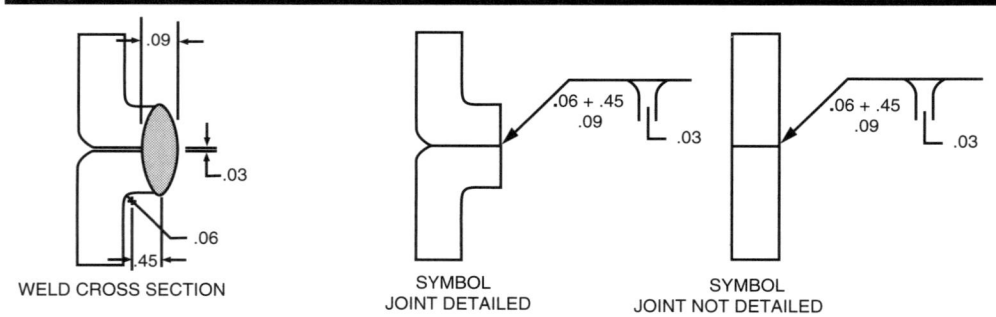

(A) ARROW-SIDE EDGE-FLANGE WELD SYMBOL

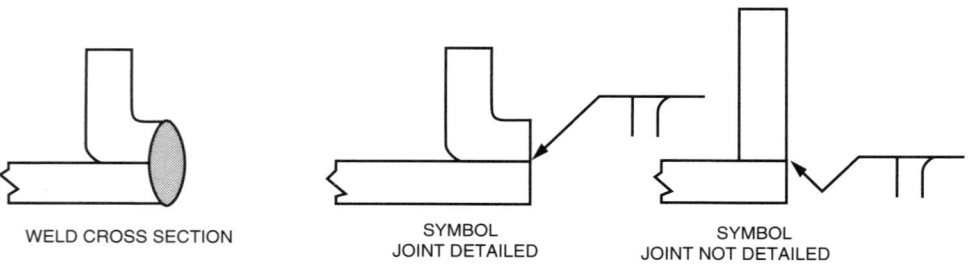

(B) ARROW-SIDE EDGE-FLANGE WELD SYMBOL

APPLICATIONS OF FLANGE WELD SYMBOLS

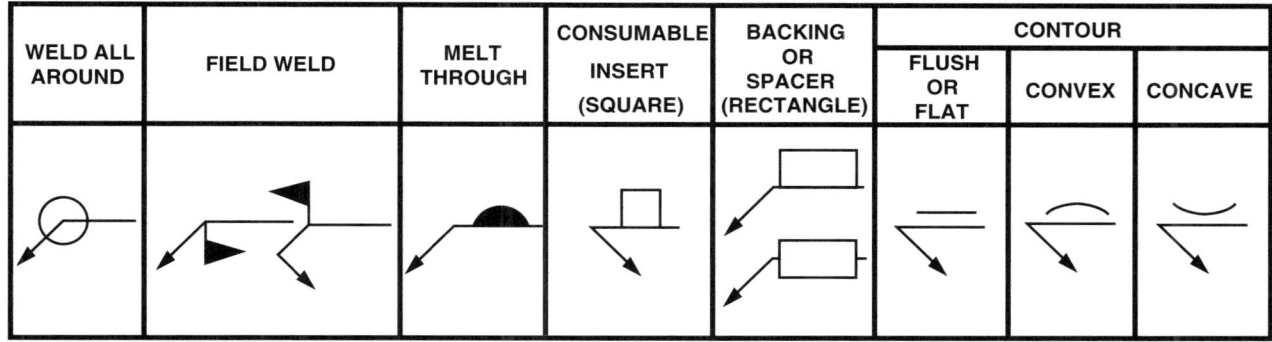

SUPPLEMENTARY SYMBOLS

as supplementary symbols, because they can be combined with many of the basic weld symbols.

The first of these symbols is the weld all around symbol. This symbol, consisting of a circle around the junction of the arrow and reference line, describes a weld which is to be continuous around joint, even though there may be abrupt changes in direction.

Another commonly used supplementary symbol is the field weld symbol. This symbol, shown as a flag at the junction of the arrow and reference line, defines welding in the field rather than in the shop. The field weld flag is placed at a right angle to, and on either side of, the reference line at the junction with the arrow.

The next supplementary symbol commonly used is the melt-thru symbol. It can be used to describe weld penetration beyond the back surface of the joint. The melt-thru symbol itself appears as a darkened-in back/backing weld symbol. The amount of melt-thru can be detailed by including a dimension to the left of the symbol.

Another way to show the backside treatment of a weld is through the use of a symbol for backing or spacer material. If a weld requires some type of backing material or a spacer within the joint, it can be symbolized by placing a rectangular box opposite the groove weld symbol.

While this symbol appears similar to the plug weld symbol, It can be differentiated by the fact that the backing material symbol will always appear in conjunction with some groove weld symbol. Details such as the material type and size can be noted in the tail.

If some material is to be placed at the joint root for a double groove weld, it is referred to as a spacer. To depict its use on a welding symbol, an open rectangular box is placed within the groove weld symbol at the diagrammatic joint root. Use of this symbol is identical to the backing material symbol.

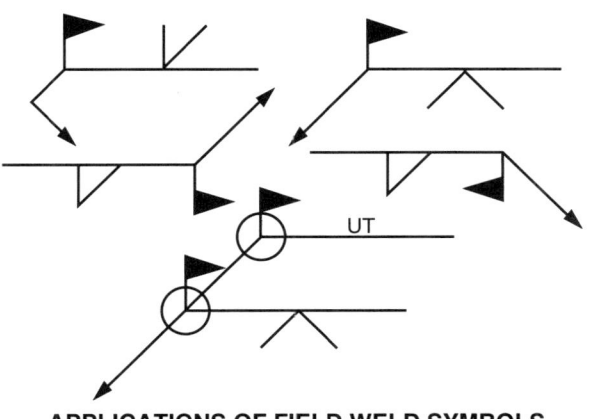

APPLICATIONS OF FIELD WELD SYMBOLS

5-16

APPLICATION OF WELD-ALL-AROUND SYMBOL

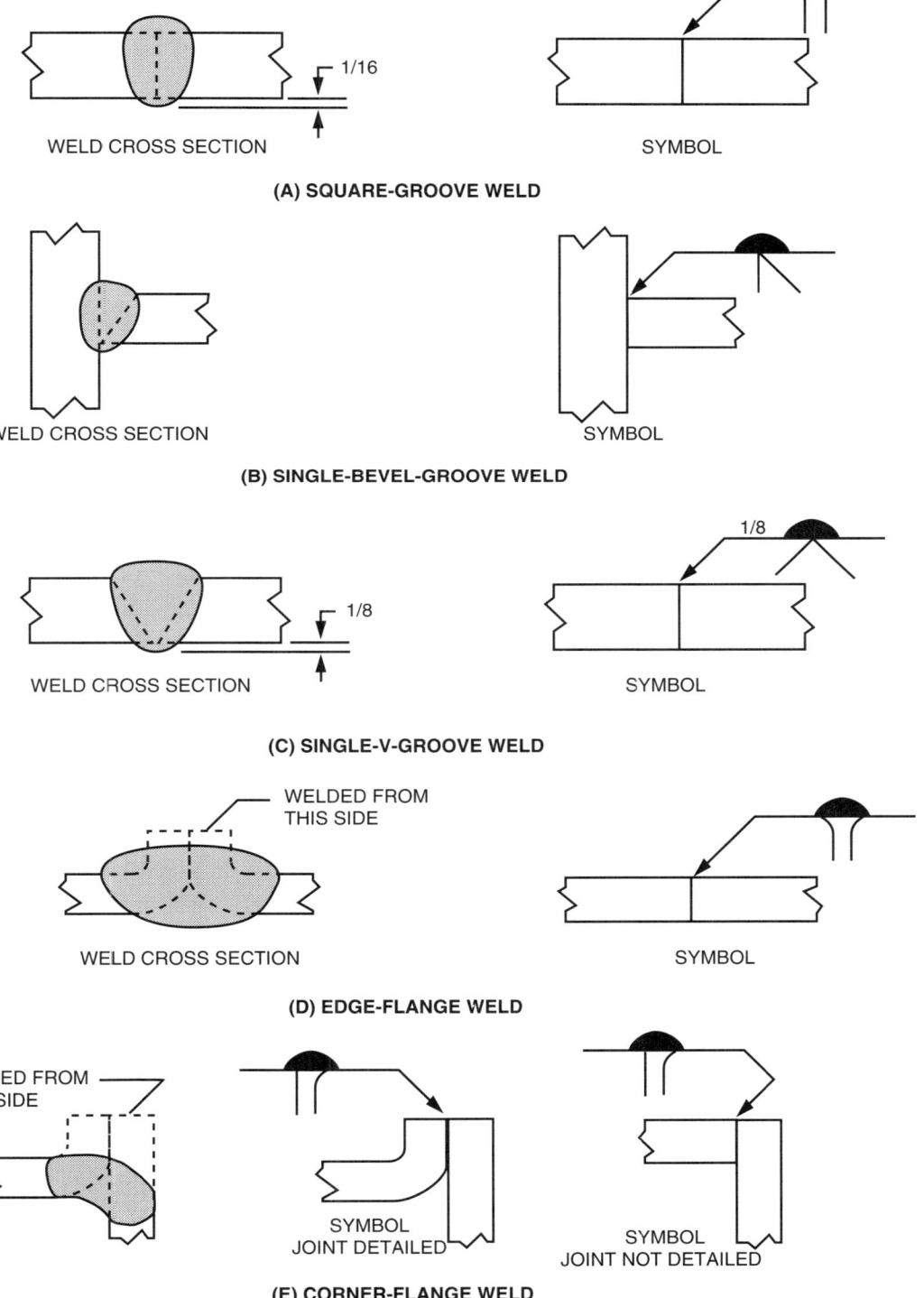

APPLICATIONS OF MELT-THROUGH SYMBOL

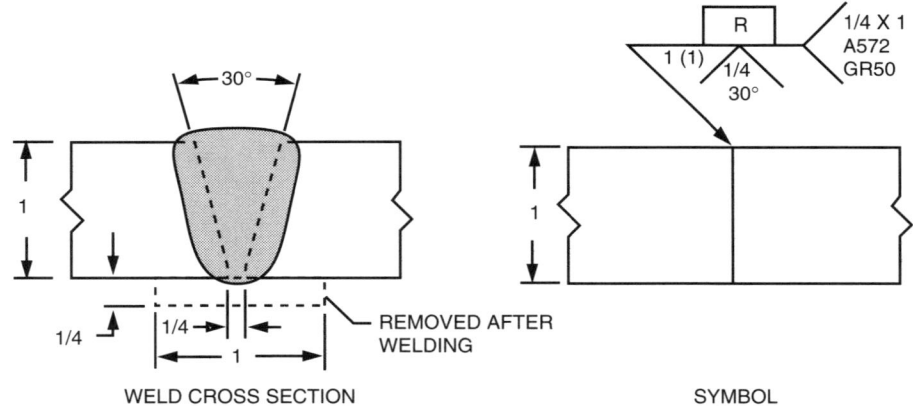

(A) SINGLE-V-GROOVE WELD WITH BACKING

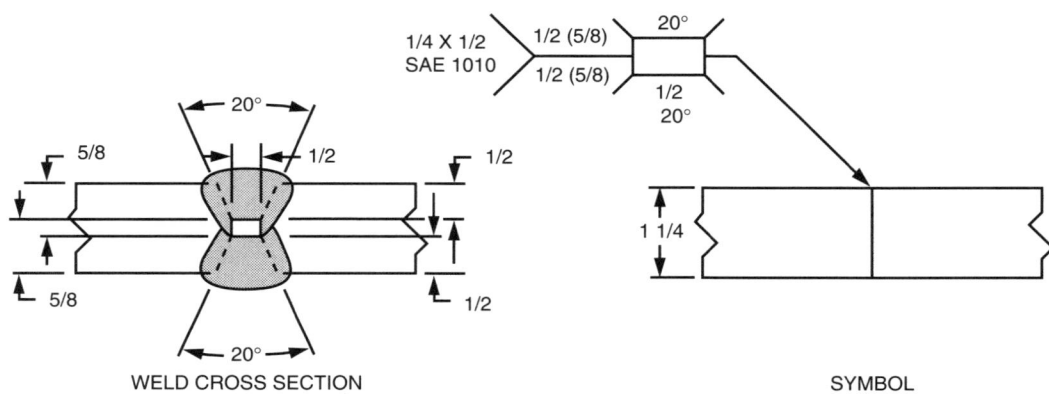

(B) DOUBLE-V-GROOVE WELD WITH SPACER

JOINTS WITH BACKING OR SPACERS

Another type of backing which is commonly used with the gas tungsten arc process is referred to as the consumable insert. The 1986 edition of AWS A2.4 introduced a symbol specifically for this type of backing. It appears much like the backing material symbol except that it is square instead of rectangular.

The last group of supplementary symbols to be discussed are those which describe the desired shape of the completed weld. There are contour symbols for various configurations, including: flush, convex and concave. The symbols for these contours correspond to the actual configurations desired.

The letters outside the contour symbols indicate the method of mechanical finishing to produce the desired contour. The letter designations for the various methods are shown below:

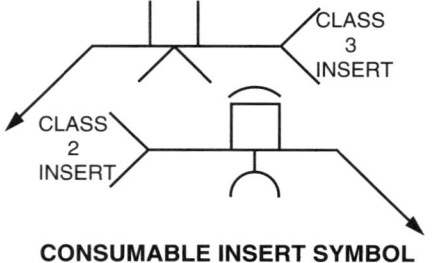

CONSUMABLE INSERT SYMBOL

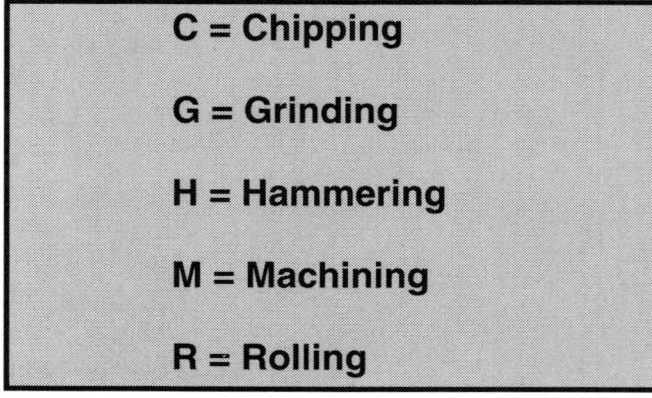

C = Chipping

G = Grinding

H = Hammering

M = Machining

R = Rolling

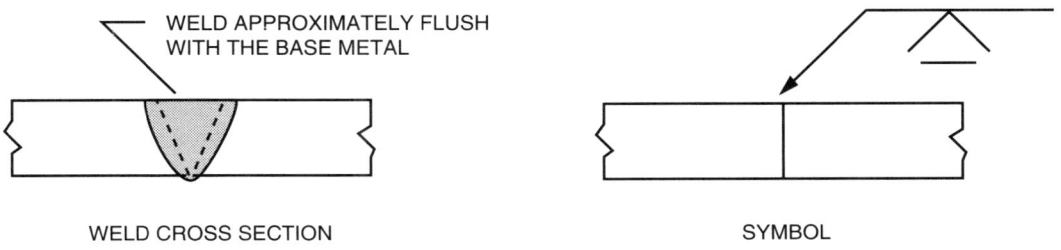

(A) ARROW-SIDE FLUSH CONTOUR SYMBOL

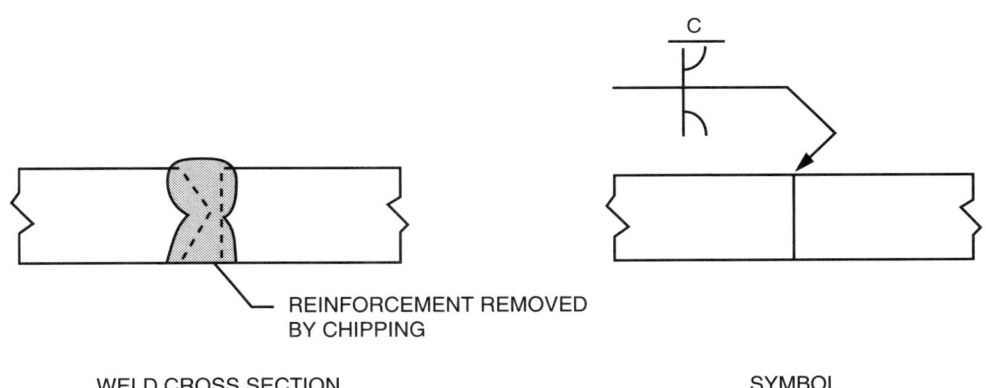

(B) OTHER-SIDE FLUSH CONTOUR SYMBOL

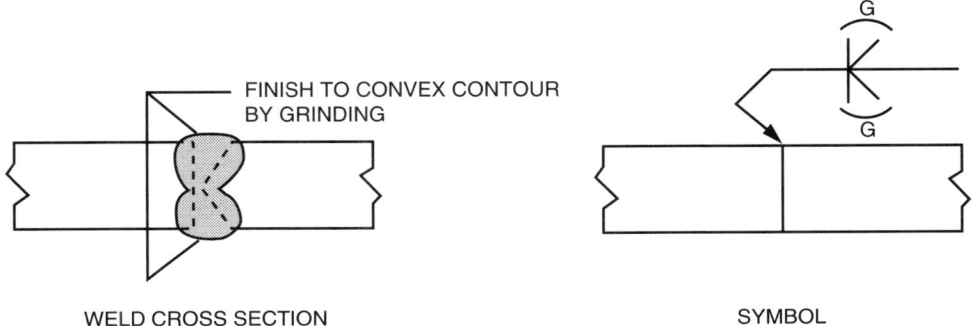

(B) BOTH SIDES CONVEX CONTOUR SYMBOL

APPLICATIONS OF FLUSH AND CONVEX CONTOUR SYMBOLS

The Tail of the Symbol

The final element of the welding symbol is referred to as the tail. While not considered to be an essential element of a welding symbol, it can be used effectively to convey other important information which cannot be conveniently communicated elsewhere on the welding symbol. When used, the tail is placed on the end of the reference line opposite the arrow. Some of the typical information which could be included in the tail are: procedure number, process type, specification number, filler metal type, need for backgouging, reference to other drawing details, need for NDE, etc.

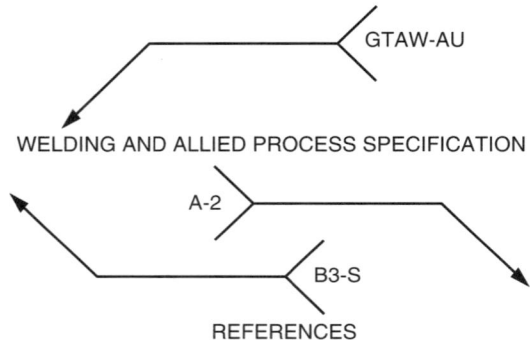

SUPPLEMENTARY DATA SHOWN IN THE TAIL OF THE WELDING SYMBOL

Use of Multiple Reference Lines

All of the discussion thus far has dealt with the use of various weld symbols along with other elements to create a welding symbol capable of describing the requirements for welding some particular weld joint. It sometimes becomes important to convey more detailed explanations of exactly how a weld is to be performed. For one thing, it is often convenient to describe the order, or sequence, of the entire welding operation. This becomes more important when the weld joint in question requires measures to prevent excessive distortion or reduce the possibility of cracking due to high restraint.

One way to describe this sequence of operations is to combine several individual reference lines on the same arrow. Each reference line could contain information to be applied at a certain step in the welding operation. The convention is that the order of operations depends on the relative location of each reference line with respect to the arrow. That is, the first operation is described by the reference line closest to the arrow. Reference lines for subsequent operations will then appear in order moving away from the arrow such that the last operation is described by that reference line furthest from the arrow.

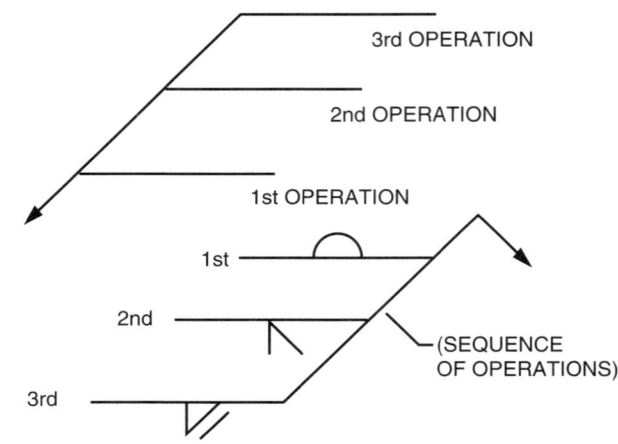

MULTIPLE REFERENCE LINES

Brazing Symbols

The use of welding symbols can also be applied to various brazing joints as well, with some minor changes.

When a brazing symbol is used, there are certain dimensions which should be specified to fully describe the important aspects of the braze joint. A dimension within the square-groove symbol describes the amount of clearance between the two members when fitup. The dimension appearing to the right of the braze symbol refers to the amount of overlap, and the dimension appearing to the left of the braze symbol indicates the size of the reinforcing fillet on the outside of the joint.

While the symbols for some of the braze joints are identical to those used for welding, the scarf groove is a joint design specifically for use with brazing. With this type of joint, the angle of the scarf cut is shown as an angular dimension to the right of the braze symbol.

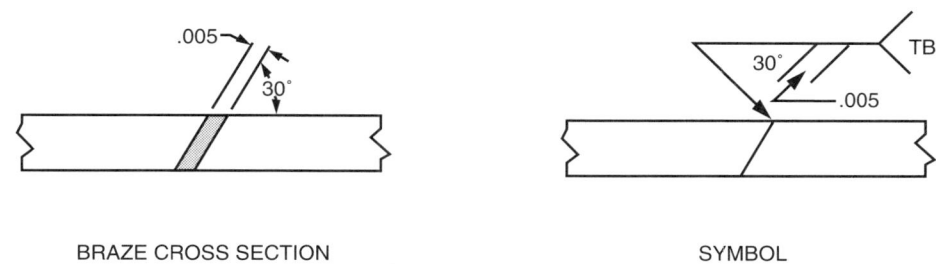

BRAZE CROSS SECTION SYMBOL

CL - CLEARANCE
L - LENGTH OF OVERLAP
S - FILLET SIZE

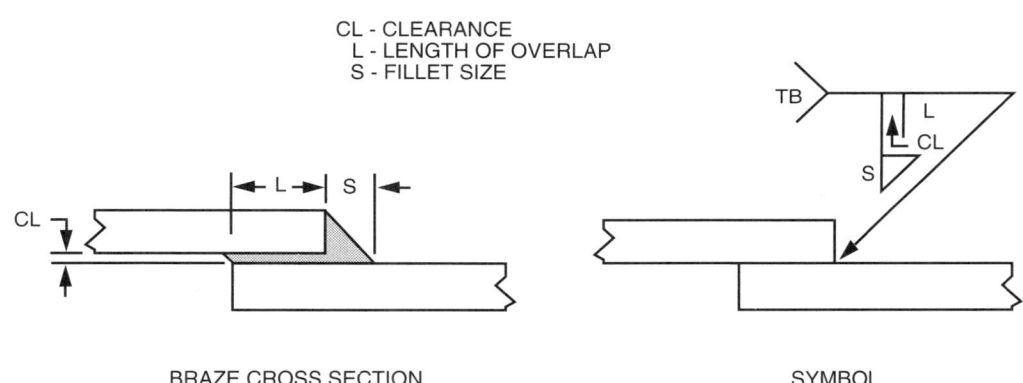

BRAZE CROSS SECTION SYMBOL

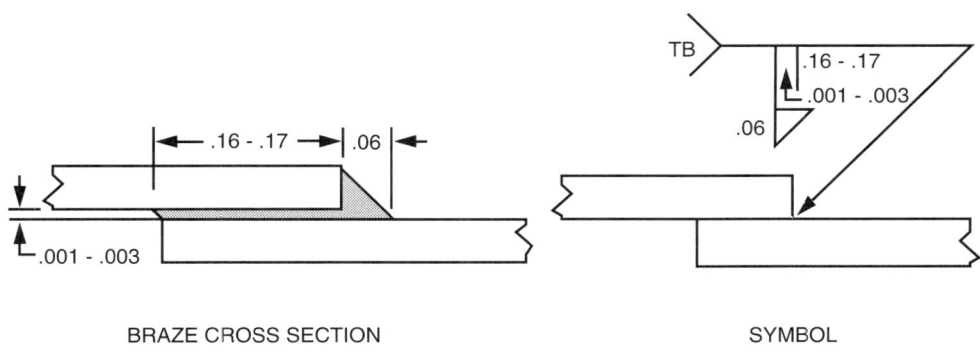

BRAZE CROSS SECTION SYMBOL

APPLICATIONS OF BRAZING SYMBOLS

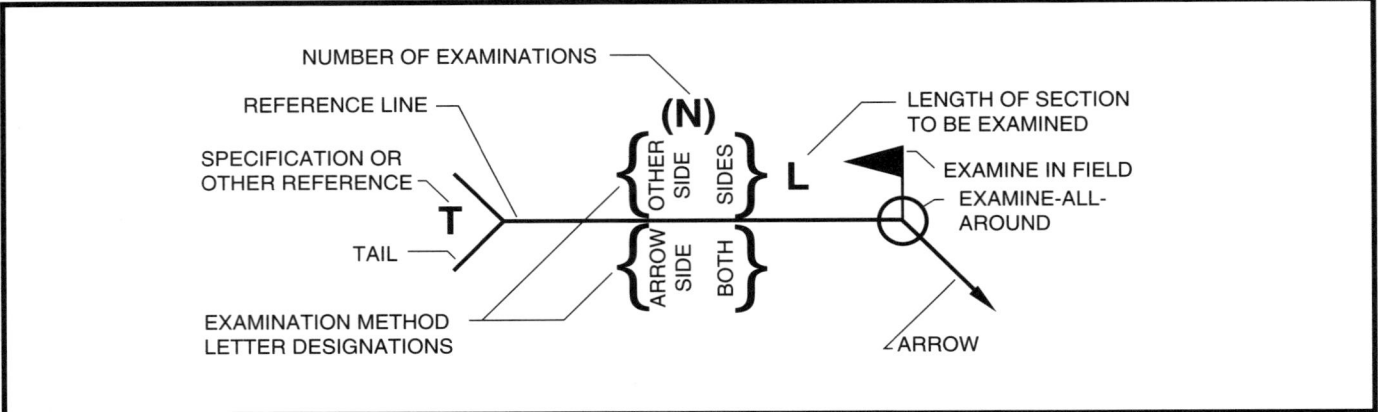

Nondestructive Examination Symbols

The preceding section describes in certain detail the methodology for the application of symbols to welding and brazing joints to detail how members are to be joined. Once joined, it may be necessary to inspect those joints to determine if the applicable quality requirements have been satisfied. As will be discussed later in Chapter 11, there are numerous nondestructive examinations which can be performed to monitor the apparent weld quality. If required, those tests can be specified through the use of nondestructive examination symbols which are constructed in much the same manner as the symbols described earlier.

As was the case for the welding symbol, information below the reference line refers to a testing operation performed on the arrow side of the joint and information above the line describes the treatment of the other side. Instead of weld symbols, there are basic testing symbols which are simply letter designations for the various testing processes. These are shown in the text box at the bottom of the page.

Summary

The welding inspector spends a great deal of time communicating with others involved in the welded fabrication of various structures and components. The use of welding and non-destructive examination symbols is an important part of that communication process, because this is the "shorthand" of welding and inspection used to convey information from the designer to those involved in the production and inspection of that product. So, the welding inspector is expected to understand the many features of these symbols so that weld and inspection requirements can be determined.

Although relatively straightforward, welding symbols can be confusing. Therefore, the welding inspector must learn their meanings. To fully understand the meaning of welding and testing symbols, one must know both the basic elements of the symbols as well as the significance of their relative locations with respect to the reference line. The important thing to remember is that even the most complicated symbol can be interpreted if the meanings of individual parts of the symbol are understood so that a combined determination can be made.

Type of Test	Symbol
Acoustic Emission	AET
Eddy Current	ET
Leak	LT
Magnetic Particle	MT
Neutron Radiographic	NRT
Penetrant	PT
Proof	PRT
Radiographic	RT
Ultrasonic	UT
Visual	VT

REVIEW - CHAPTER 5
WELDING AND NONDESTRUCTIVE EXAMINATION SYMBOLS

Q5-1 The primary element of any welding symbol is referred to as the:
a. tail
b. arrow
c. reference line
d. arrow side
e. weld symbol

Q5-2 Information appearing above the reference line refers to the:
a. near side
b. arrow side
c. far side
d. other side
e. none of the above

Q5-3 The graphical representation of the type of weld is called the:
a. tail
b. welding symbol
c. weld symbol
d. arrow
e. none of the above

Q5-4 Which of the symbols below represents the weld shown?

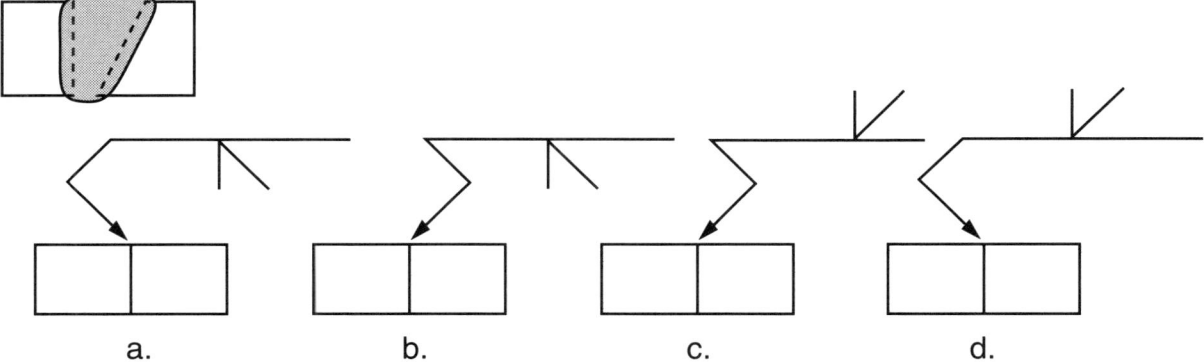

a.   b.   c.   d.

e. none of the above

Q5-5 When a weld symbol is centered on the reference line, this indicates:
   a. that the welder can put the weld on either side.
   b. that there is no side significance.
   c. that the designer doesn't know where the weld should go.
   d. that the welder should weld in whatever position the weld is in.
   e. none of the above

Q5-6 The symbol below depicts what type of weld?

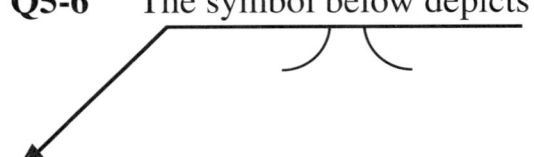

   a. flare-V-groove
   b. flare-bevel-groove
   c. edge-flange
   d. corner-flange
   e. none of the above

Q5-7 In the symbol below, the 1/8 dimension refers to what?

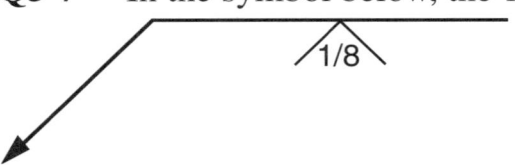

   a. groove angle
   b. root face
   c. depth of preparation
   d. weld size
   e. root opening

Q5-8 In the symbol below, the 3/4 dimension refers to what?

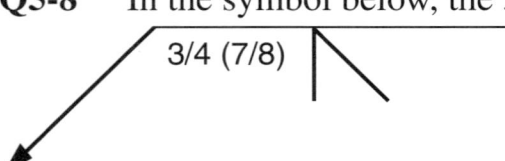

   a. weld size
   b. effective throat
   c. depth of preparation
   d. root opening
   e. none of the above

Q5-9 If applied to a 1 inch thick weld, the symbol below represents what type of weld?

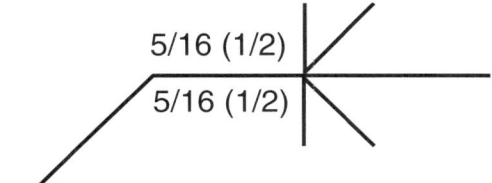

a. full penetration double-bevel-groove weld
b. full penetration double-V-groove weld
c. partial penetration double-bevel-groove weld
d. partial penetration double-V-groove weld
e. none of the above

Q5-10 Dimensions appearing to the left of the weld symbol generally refer to the:
a. weld length
b. weld size/depth of preparation
c. root opening
d. radius
e. none of the above

Q5-11 A triangular-shaped weld symbol represents what type of weld?
a. bevel-groove
b. flare-groove
c. flange-groove
d. V-groove
e. none of the above

Q5-12 The symbol below represents what type of weld?

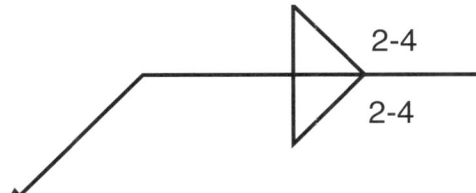

a. staggered intermittent fillet weld
b. chain intermittent fillet weld
c. segmented fillet weld
d. intermittent fillet weld
e. none of the above

Q5-13 Dimensions appearing to the right of the weld symbol generally refer to the:
  a. weld size
  b. root opening
  c. depth of preparation
  d. weld length/pitch
  e. none of the above

Q5-14 A weld symbolized by a rectangular box that contains a dimension represents a:
  a. plug weld
  b. slot weld
  c. plug weld in beveled hole
  d. partially filled plug weld
  e. plug weld in hole having dimension shown

Q5-15 The required spot weld size can be shown as:
  a. a dimension to the right of the symbol
  b. a dimension of the required nugget diameter
  c. a value for the required shear strength
  d. a and b above
  e. b and c above

Q5-16 A number appearing to the right of the spot weld symbol refers to:
  a. spot weld size
  b. spot weld length
  c. number of spots required
  d. pitch distance between adjacent spots
  e. none of the above

Q5-17 In the symbol below, the "A" dimension represents:

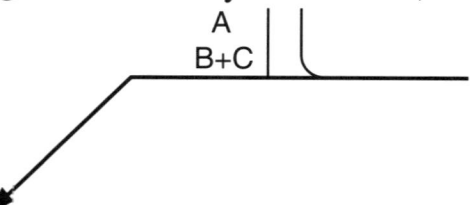

  a. weld size
  b. flange radius
  c. flange length
  d. depth of penetration
  e. none of the above

Q5-18 In the symbol below, the symbol shown on the other side represents:

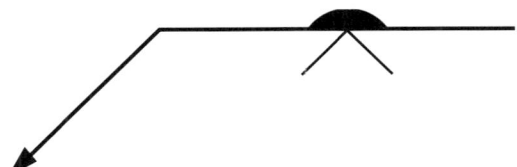

a. back weld
b. backing weld
c. melt-thru weld
d. a and b above
e. b and c above

Q5-19 The symbol below shows the use of what type of weld?

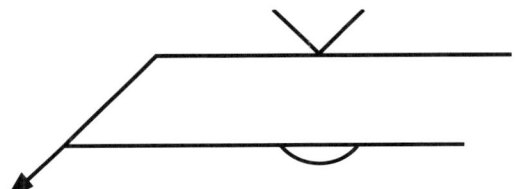

a. single-bevel-groove weld
b. single-V-groove weld
c. backing weld
d. back weld
e. b and c above

Q5-20 The symbol below shows what type of groove configuration?

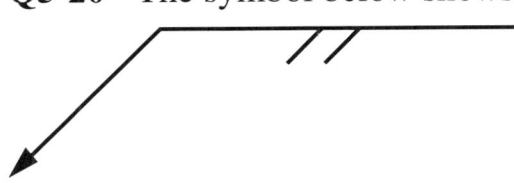

a. square groove
b. skewed groove
c. sloped groove
d. scarf
e. none of the above

Q5-21 The part of the welding symbol which can be used to convey any additional information which cannot be shown otherwise is referred to as:
a. the weld symbol
b. the arrow
c. the reference line
d. the tail
e. none of the above

Q5-22 The symbol below shows what type of weld?

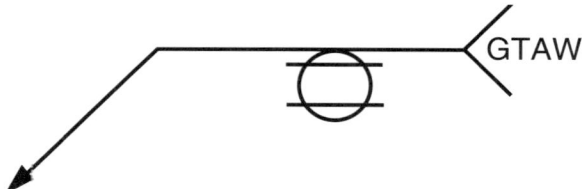

a. gas metal arc spot weld
b. resistance spot weld
c. gas tungsten arc seam weld
d. resistance seam weld
e. none of the above

Q5-23 What nondestructive examination method is to be applied to the arrow side?

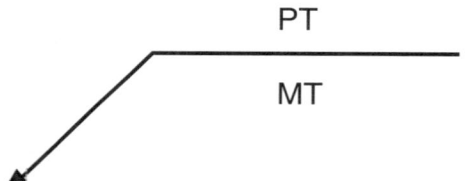

a. magnetic particle testing
b. eddy current testing
c. radiographic testing
d. penetrant testing
e. none of the above

Q5-24 A number in parentheses just outside a test symbol represents:
a. the length of weld to be tested
b. the extent of testing
c. the number of tests to perform
d. the type of test to perform
e. none of the above

Q5-25 A number to the right of a nondestructive examination symbol refers to the:
   a. number of tests to perform
   b. the length of weld to be tested
   c. the applicable quality standard
   d. the test procedure to use
   e. none of the above

Q5-26 Which of the symbols represent the weld shown below?

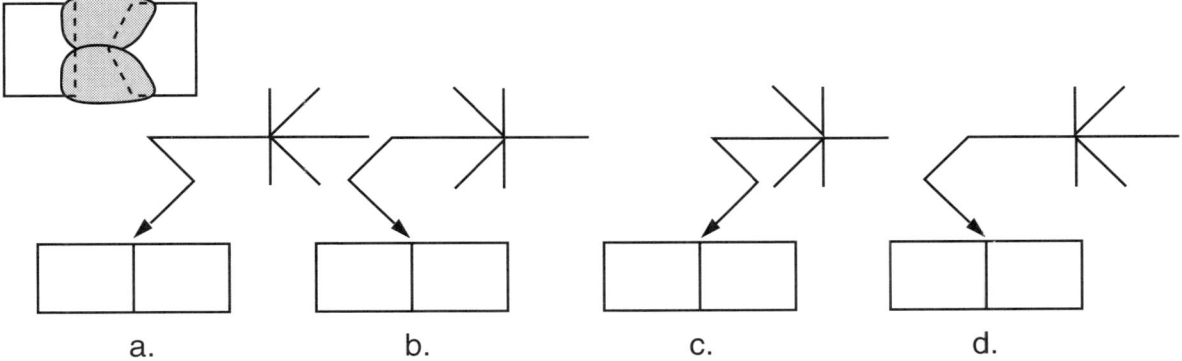

   e. none of the above

Q5-27 Which of the symbols represent the weld shown below?

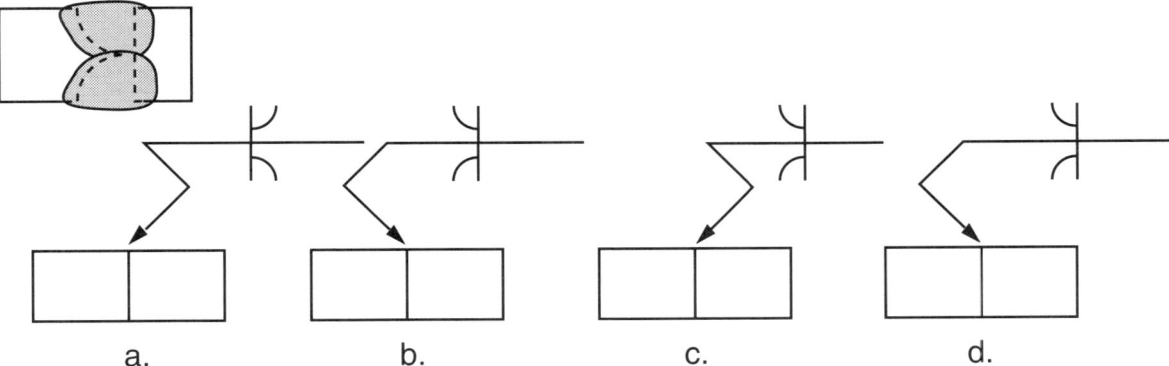

   e. none of the above

Q5-28 Which of the symbols represent the weld shown below?

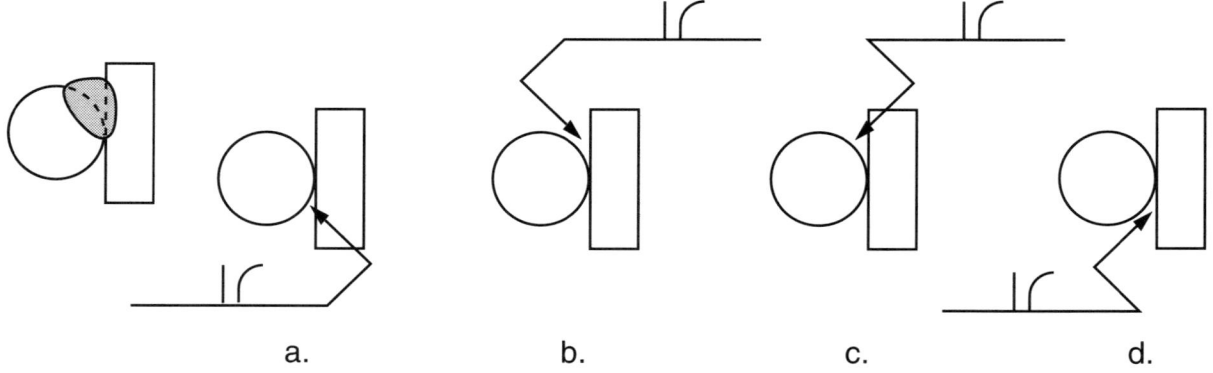

a. b. c. d.

e. none of the above

Q5-29 Which of the symbols represent the weld shown below?

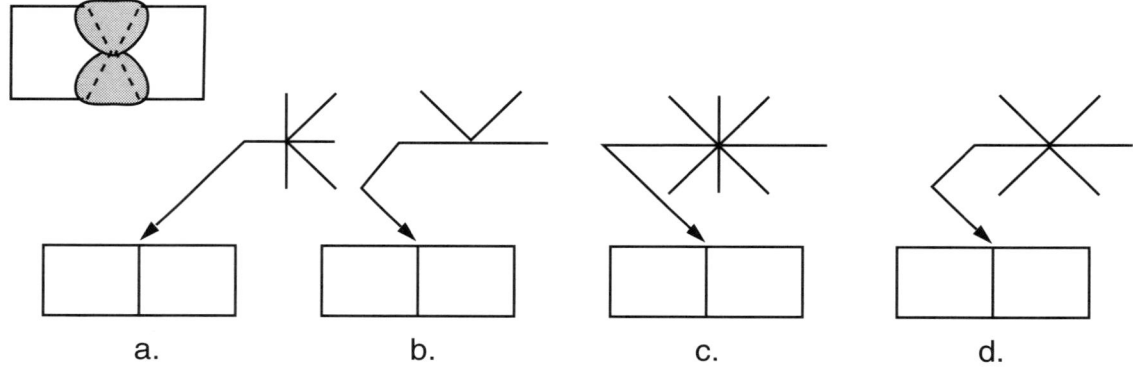

a. b. c. d.

e. none of the above

Q5-30 Which of the symbols represent the weld shown below?

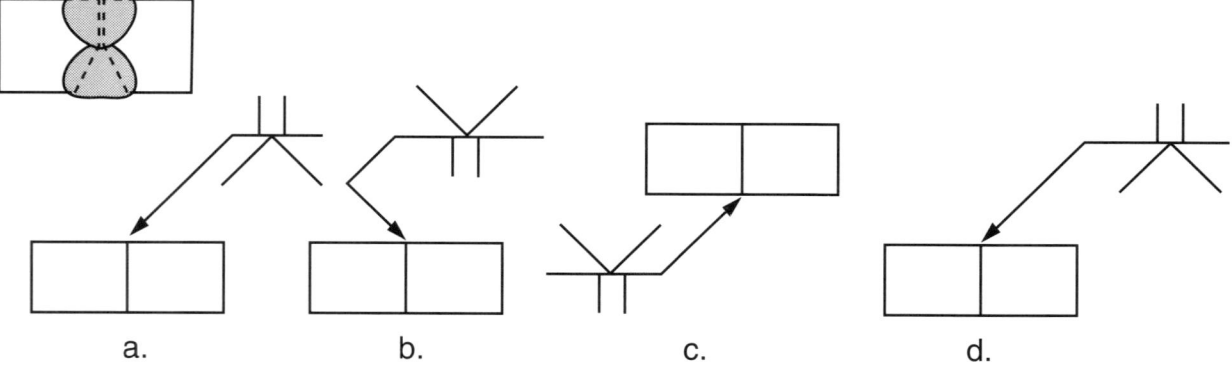

a. b. c. d.

e. none of the above

Q5-31 Which of the symbols represent the weld shown below?

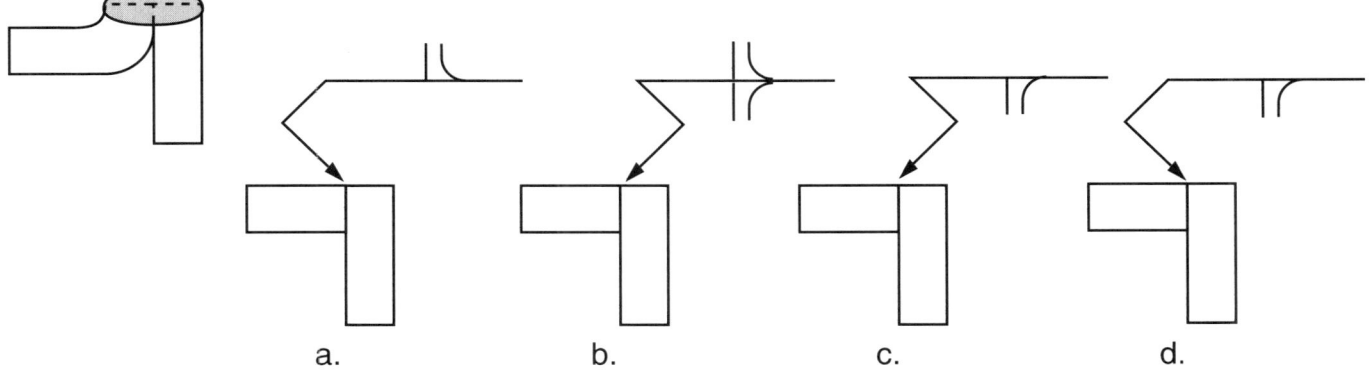

e. none of the above

Q5-32 Which of the symbols represent the weld shown below?

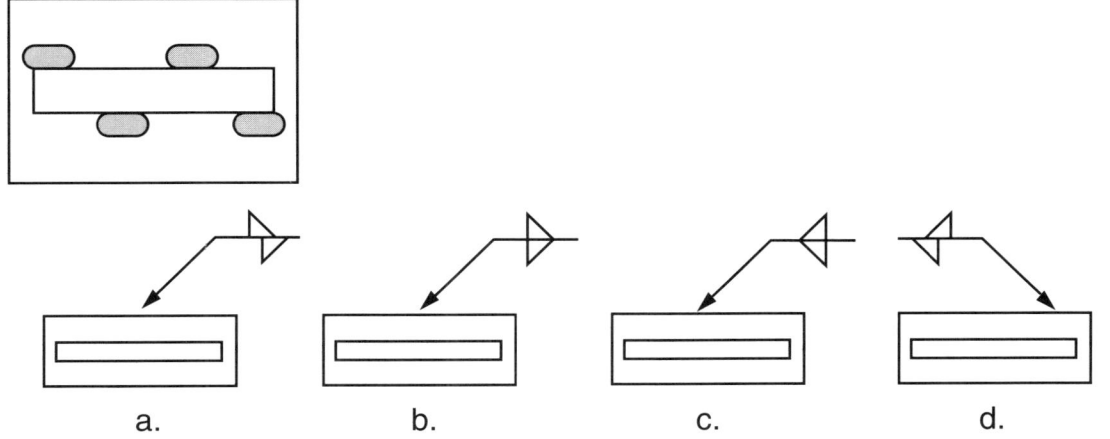

e. none of the above

Q5-33 Which symbol represents the welds shown?

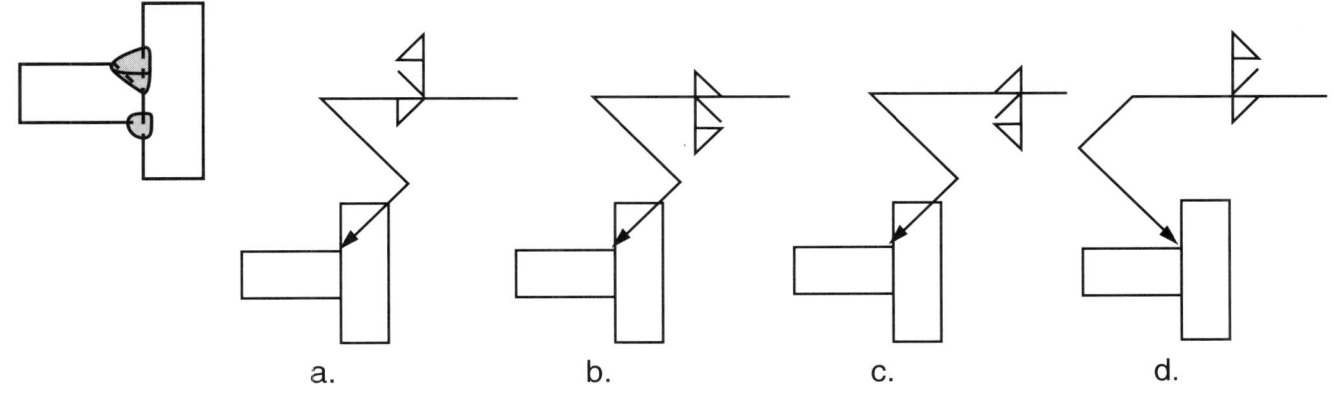

e. none of the above

Q5-34 Which of the welds is represented by the symbol below?

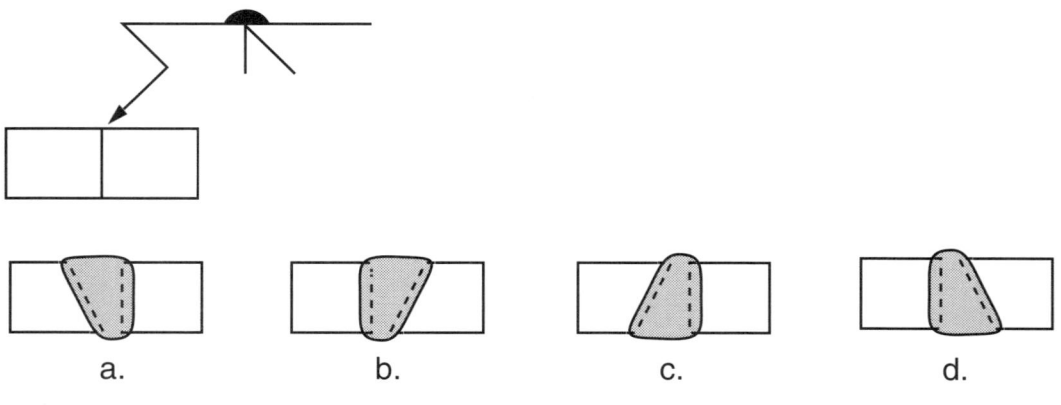

e. none of the above

Q5-35 Which of the welds is represented by the symbol below?

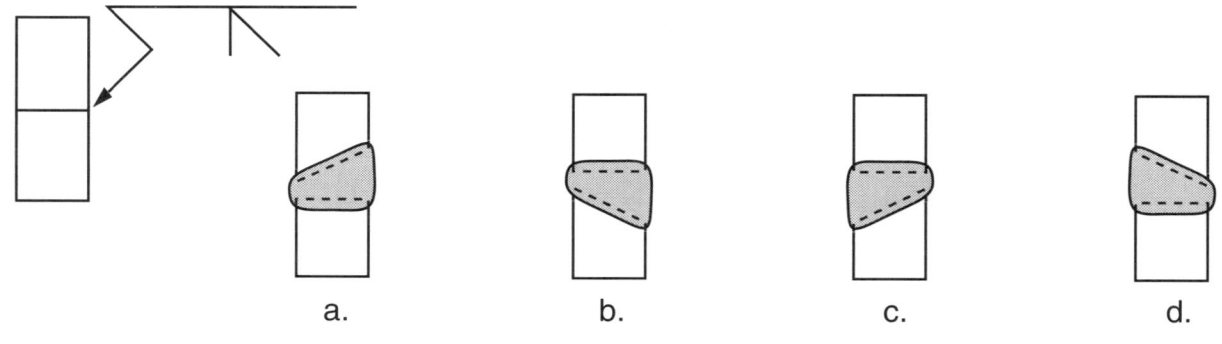

e. none of the above

Q5-36 Which of the welds is represented by the symbol below?

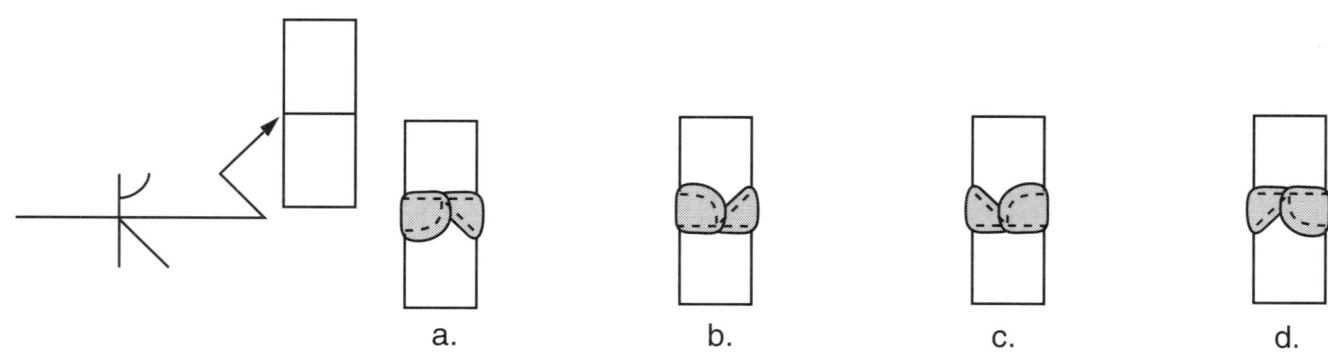

e. none of the above

Q5-37 Which of the welds is represented by the symbol below?

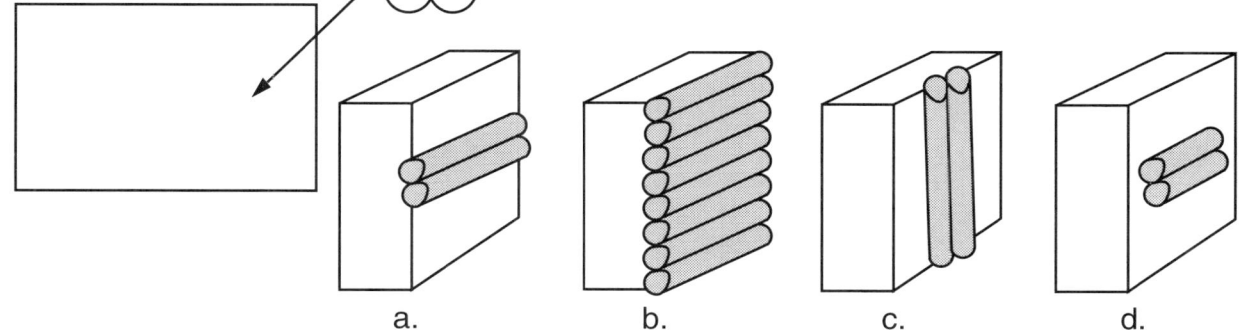

e. none of the above

Q5-38 Which of the welds is represented by the symbol below?

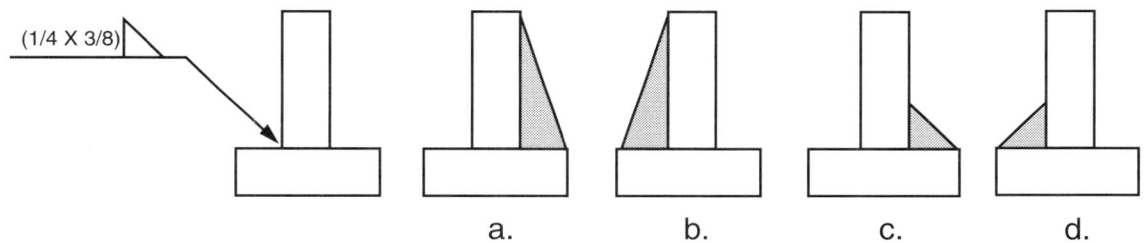

e. insufficient reference or detailing

Q5-39 Which of the welds is represented by the symbol below?

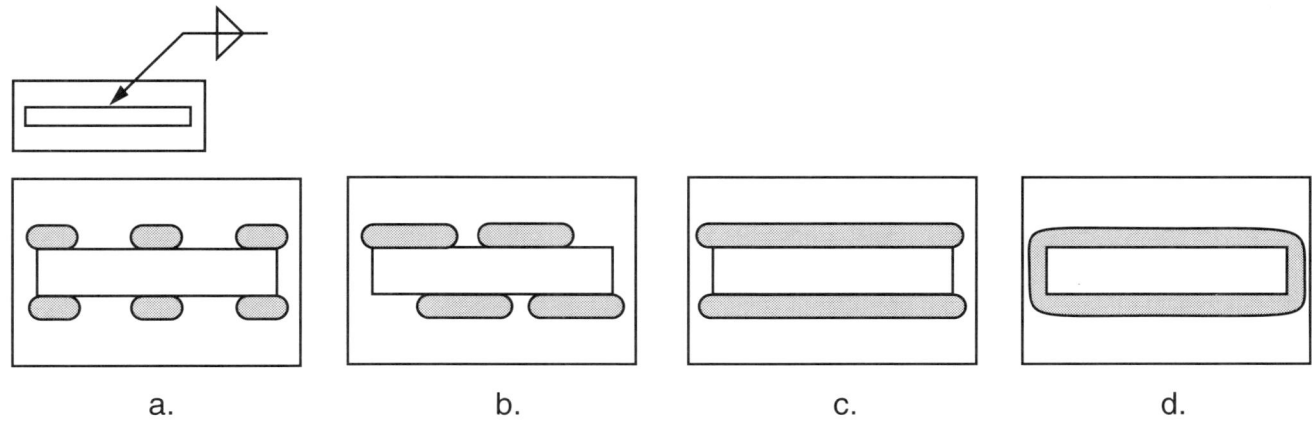

e. none of the above

Q5-40 Which of the welds is represented by the symbol below?

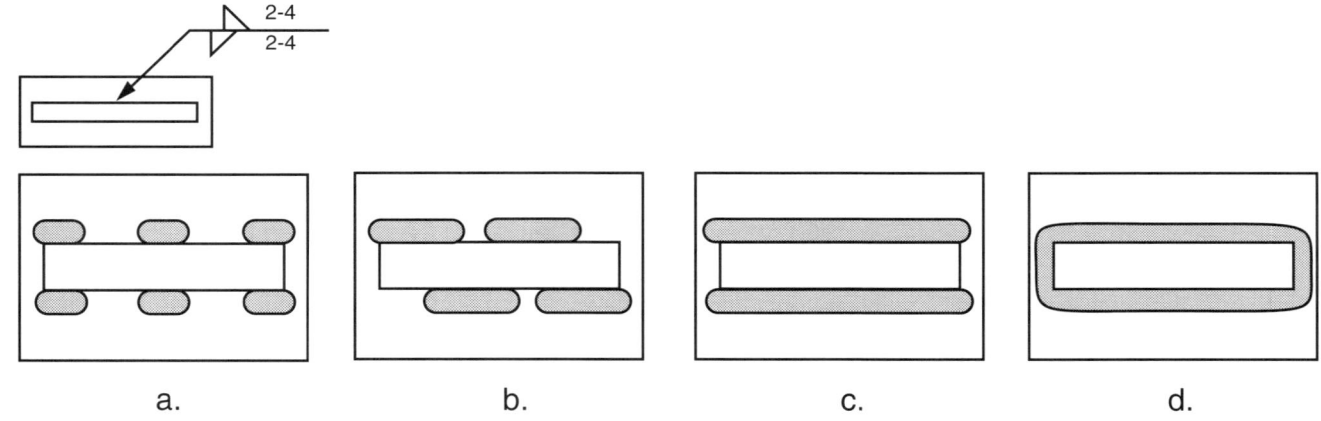

e. none of the above

Q5-41 Which of the welds is represented by the symbol below?

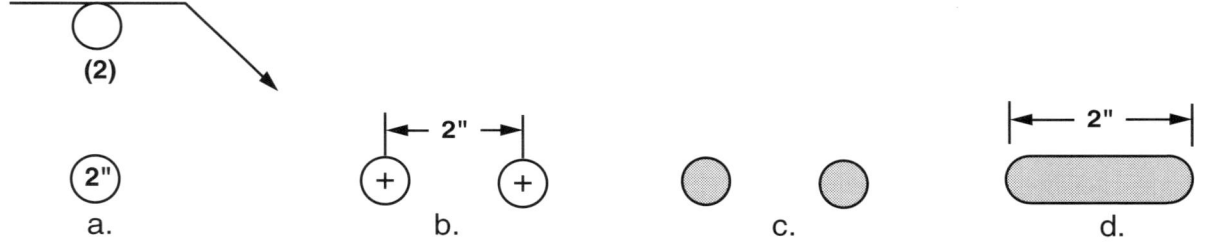

e. none of the above

Questions **Q5-42** through **Q5-46** refers to Figure 1 below:

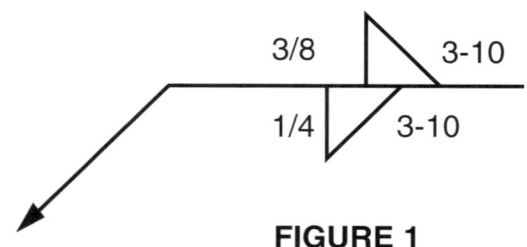

FIGURE 1

Q5-42 What is the weld length?
    a. 1/4"
    b. 3/8"
    c. 3"
    d. 10"
    e. none of the above

Q5-43 What is the pitch distance?
   a. 1/4"
   b. 3/8"
   c. 3"
   d. 10"
   e. none of the above

Q5-44 What is the size of the arrow side weld?
   a. 1/4"
   b. 3/8"
   c. 3"
   d. 10"
   e. none of the above

Q5-45 What is the size of the other side weld?
   a. 1/4"
   b. 3/8"
   c. 3"
   d. 10"
   e. none of the above

Q5-46 What does the symbol represent?
   a. fillet welds on both sides
   b. intermittent fillet welds
   c. chain intermittent fillet welds
   d. staggered intermittent fillet welds
   e. none of the above

Questions **Q5-47** through **Q5-51** refer to Figure 2 below:

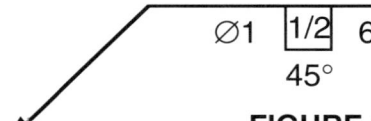

FIGURE 2

Q5-47 What is the pitch distance?
- a. 1"
- b. 1/2"
- c. 45"
- d. 6"
- e. none of the above

Q5-48 What is the angle of the countersink?
- a. 1°
- b. 1/2°
- c. 45°
- d. 6°
- e. none of the above

Q5-49 What is the depth of filling?
- a. 1"
- b. 1/2"
- c. 45"
- d. 6"
- e. none of the above

Q5-50 What is the weld size?
- a. 1"
- b. 1/2"
- c. 45"
- d. 6"
- e. none of the above

Q5-51 What weld is represented by the symbol?
- a. arrow side slot weld
- b. other side slot weld
- c. arrow side plug weld
- d. other side plug weld
- e. a or c above

Questions **Q5-52** through **Q5-56** refer to Figure 3 below:

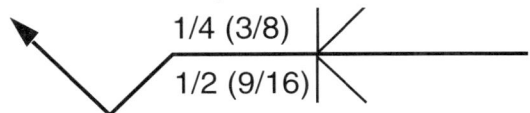

Q5-52 What is the arrow side depth of preparation?
- a. 1/4"
- b. 3/8"
- c. 1/2"
- d. 9/16"
- e. 15/16"

Q5-53 What is the other side depth of preparation?
- a. 1/4"
- b. 3/8"
- c. 1/2"
- d. 9/16"
- e. 15/16"

Q5-54 What is the other side weld size?
- a. 1/4"
- b. 3/8"
- c. 1/2"
- d. 9/16"
- e. 15/16"

Q5-55 What is the arrow side weld size?
- a. 1/4"
- b. 3/8"
- c. 1/2"
- d. 9/16"
- e. 15/16"

Q5-56 What is the total weld size?
- a. 1/4"
- b. 3/8"
- c. 1/2"
- d. 9/16"
- e. 15/16"

CHAPTER 6: WELDABILITY, WELDING METALLURGY, AND WELDING CHEMISTRY

Introduction

A weld joint is the functional unit basic to any welded structure, and the key base metal property is weldability. The better the weldability of a metal, the easier it can be fabricated into a suitably designed structure that will perform satisfactorily. Weldability is determined by several factors, including:

Welding metallurgy
Welding chemistry
Joint surface condition
Joint geometry

Welding metallurgy deals with the changes that occur in a metal when acted upon by various mechanical and thermal effects of a welding process. It considers the basic atomic structure of metals and how that structure can be affected by forces and temperature. Metals' atomic structures have a pronounced effect on their resulting mechanical properties. Consequently, we will be able to establish relationships between the metallurgical treatments given metals (preheat, postheat, stress relief, etc.) and the mechanical properties which result.

Welding chemistry deals with the chemical interaction between the base metals, filler metals and other chemicals present at the joint during the welding process. The ability of the base metal and filler metal to fuse together without adverse chemical effects is an important factor in determining weldability.

The final factors affecting weldability are referred to as joint surface condition and joint geometry. The joint surface condition factor includes the effects of different amounts of surface roughness and cleanliness of the joint. The shape of the joint edge geometry will also have an effect on a metal's (or perhaps a joint's) weldability. The amount of joint restraint can also have an effect on weldability.

Although weldability problems are supposed to be solved by engineers, it may be that the inspector will be the first person to sense that a weldability problem exists. Welders do not generally make poor welds on purpose; instead, they usually struggle to make good welds, and to correct faults. Whenever a rejection is encountered, its cause should be sought out and corrections made. Defects that are not welder-oriented, or are repetitive, should be noted and reported for corrective action.

Why, then, is there a need to study weldability, when the welding inspector's job entails no responsibility for the ease or difficulty of making the welds being inspected? Is not weldability something the engineer should study, to improve the chances that the weld design will be worth inspecting?

The answer is: Sometimes the best choice the engineer can make is a compromise that will still leave the welding inspector examining welds made in steels of poor weldability. The welding inspector will then be vulnerable to overlooking unexpected discontinuities if not prepared to give special attention to those welds that are extra prone to cracking, to welds that may crack the base metal beneath the surface, and to welds that may significantly reduce the corrosion resistance of the base metal's heat affected zone (HAZ). It is important the welding inspector have knowledge of these types of weldability problems to better do his/her inspection function. Understanding the basics also helps when discussing problems with engineers.

Engineers should know the weldability of the metals they select; many problem alloys are listed in technical references on welding metallurgy. The topic of weldability for various steels is covered in great detail in the book, *Weldability of Steels*, published by the Welding Research Council and available through AWS. The AWS "Welding Journal" is also an excellent source for this type of information, since it includes many research papers on the subject of weldability.

An example of a code reference to the weldability of a metal is found in AWS D1.1, *Structural Welding Code- Steel*. The particular reference has to do with the welding of certain high strength, low alloy, quenched and tempered steels (ASTM A514 and ASTM A517). The precaution is clearly stated in several paragraphs in Sections 8, 9 and 10: "Final weld inspection of these materials must not occur until at least 48 hours after welding is completed and the base metal has cooled to room temperature." The reason for this is that these steels are susceptible to underbead cracking, a delayed cracking problem, which premature inspection will not discover.

Welding Metallurgy

Metallurgy is the science which deals with the internal structure of metals and the relationship between those structures and the properties exhibited by metals. When referring to welding metallurgy, the real concerns are about the various changes that occur in metals when joined by welding. From a welding standpoint, there are several topics with which welding engineers are concerned:

Solids and liquids properties
Diffusion
Solid solubility
Melting and freezing
Thermal expansion
Thermal treatments (e.g. stress relief)

Solids Versus Liquids

Since the topic of interest here is welding, the following discussion will be related to the solid (metallic) state, and the liquid (molten) state. The primary difference between these two states of matter in metals is the amount of energy contained in each. The liquid metal contains a much higher level of internal energy than the solid metal. Structurally, the major difference is that for solid metals, the atoms are in a fixed position relative to each other, while in a liquid, the atoms are free to move in relation to each other.

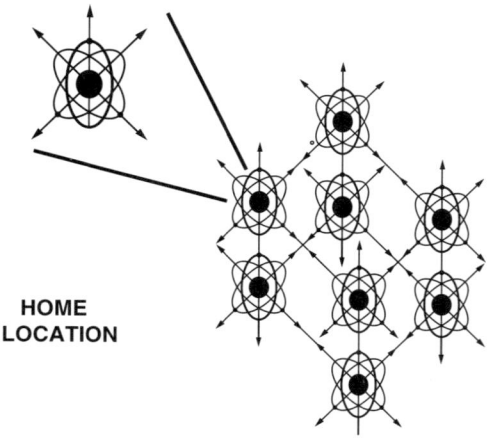

In the solid state, each atom has a specific 'home' position with respect to all the other atoms adjacent to it. These atoms are aligned row upon row, layer upon layer, in a three-dimensional, symmetrical, crystalline structure, or pattern. These 'fixed' atoms are held in place by the attraction and repelling forces of their neighboring atoms. Each atom nucleus has a cloud of encircling electrons to complete its atomic structure. This overall atomic configuration gives metals in the solid state their metallic luster and determines their physical, mechanical, chemical and electrical properties.

While occupying these 'home' positions in the solid state, the atoms are vibrating or oscillating about their home position. The amount of atom movement is directly proportional to the metal's temperature. As the metal temperature increases, the movement of the atoms becomes larger and faster. Eventually, as temperature continues to increase, the internal energy increases to the point where the individual atoms break away from one another and move freely in the liquid state. This 'breakaway' occurs at the melting point of the metal.

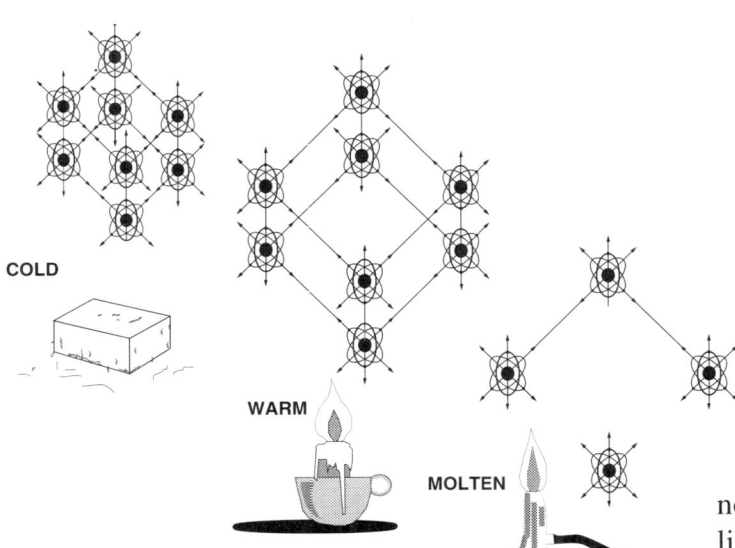

Diffusion

We have previously noted that atoms in the liquid state can move about easily with respect to each other; however, under certain conditions, even atoms in the solid state can change positions. In fact, any atom may 'wander' away, step by step, from its home position. These changes of atom position in the solid state are called diffusion.

An example of diffusion is shown if smooth, flat bars of lead and gold are clamped tightly together. If they are clamped together at room temperature for several days, the two sheets of metal remain attached when the clamps are removed. This attachment is due to the fact that atoms of lead and gold have each migrated, or diffused, into the other metal, forming a very weak metallurgical bond. This bond is quite weak, and the two metals can be broken apart by a sharp blow at their joint line. If the metals' temperatures are increased, the amount of diffusion increases, and at a temperature above the melting point of both, complete mixing occurs.

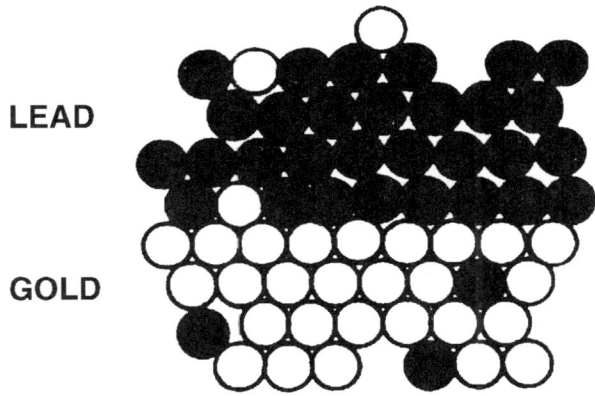

Another example of diffusion occurs when hydrogen, a gas, is allowed in the vicinity of molten metal, such as a weld. The most common source of hydrogen is moisture (H_2O), or contamination on the surfaces of the parts to be welded. Many of the contaminants normally found on metals are organic compounds, like oil, grease, etc., and they contain hydrogen in their chemical makeup. The heat of welding will break down the water or organic contaminants into individual atoms, which includes the hydrogen atom (H^+). The hydrogen atoms are quite small, and they can easily diffuse into the base metal structure. As they enter the base metal, the hydrogen atoms often re-combine into the hydrogen molecule (H_2), a combination of two atoms of hydrogen, which is much larger than a single atom of hydrogen. The larger molecules often become trapped in the metal at discontinuities such as grain boundaries or inclusions. These hydrogen molecules, because of their larger size, can cause high stresses in the internal structure of the metal, and for metals of low ductility, can cause cracking. Hydrogen cracking is often referred to as 'underbead' or 'delayed' cracking.

The primary cure for hydrogen cracking problems is to simply eliminate the source of hydrogen; the first step is to thoroughly clean all surfaces to be welded. Another approach is to specify the 'low hydrogen' electrodes for use on carbon and low alloy steels. These low hydrogen electrodes are specially formulated to keep their hydrogen content quite low, but they do require special handling to avoid moisture pickup after opening the sealed shipping containers. Preheating the base metals is also effective in eliminating hydrogen pickup because hydrogen will diffuse out of most metals at temperatures of 200 - 450 degrees F. The methods noted above all aid in reducing the possibility of hydrogen cracking in those metals which are susceptible.

Solid Solubility

Most of us are familiar with the normal solubility of solids into liquids. Adding a spoonful of salt to a glass of water and stirring will dissolve the salt. However, most of us are not familiar with one solid dissolving into another solid. In the example given above with the lead and gold, the two metals were diffusing through solid solution into each other. And, returning to our example of salt and water, if additional salt is added, we find that some of it will not dissolve regardless of how much we stir. What has happened is that for that amount of liquid, and its temperature, we reach a 'critical solubility limit'. No amount of stirring will dissolve any more salt. In order to dissolve more salt, the liquid volume would either have to be increased, or its temperature raised. Thus we see that for a solid dissolving into a liquid, there is a critical solubility limit depending on liquid volume and temperature. Metals behave similarly, except through diffusion, and they 'dissolve' into each other when both are solid.

But just like the salt and water, there are solubility limits of one solid dissolving into another and the critical solubility limit is dependent on temperature. The higher the temperature of a metal, the more dissolving of a second element will occur. Thus, we can get metals combining even when both are solid. Of course, as a metal's temperature is raised, the amount of diffusion and solubility increases.

An example of a solid dissolving into another solid is a method we use for increasing the surface hardness of a steel. If the steel is packed into a bed of carbon particles, and then heated to a temperature of about 1600 - 1700 degrees F, which is well below the melting point of both the carbon and the steel, some of the carbon will diffuse (dissolve) into the surfaces of the steel. This added carbon in the steel's surface makes the surface much harder, and is useful for resisting wear and abrasion. This technique is commonly called 'carburizing'. The surface of steels can also be made hard by exposing the steel to an ammonia environment at similar temperatures to carburizing. The ammonia (NH_3) breaks down into its individual components of nitrogen and hydrogen, and the nitrogen atoms enter the surface. This technique is called 'nitriding'. Both of these surface hardening techniques demonstrate the diffusion and solid solubility of metals. Knowledge of diffusion and solid solubility will aid the welding inspector in understanding the importance of cleanliness in welding, and the need for proper shielding during all welding operations.

Melting and Freezing

When a metal is heated to the liquid state, atoms move about very energetically with complete mixing. The mixing action is created by convection and conduction, which occurs due to the flow of heat from hot areas to colder areas. In welding, the mixing is also stimulated by magnetic forces, and by arc pressure, or pressure from a gas flame and movement of a welding electrode. The welder can observe this mixing as the metal swirls beneath the heat source. The result is that the atoms from the base metal mix with the atoms from the metal being added.

Freezing of liquid metal cannot occur until the heat source is removed and the atoms lose enough energy to 'settle down' into a metallic crystalline structure. The liquid's energy must be dissipated, in the form of heat loss. Heat is quickly given to surrounding atoms already in the solid state. When this happens, the surrounding metal warms, and the liquid cools enough to solidify.

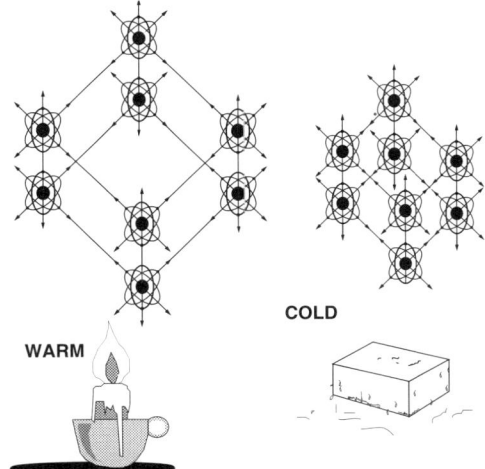

The initial configuration of the crystalline structure as the weld solidifies is determined by the crystalline structure of the metal already existing in the solid state. Growth continues in the same pattern if that is the natural pattern for the mix of atoms present. Atoms of a dissimilar weld composition may take their own preferred alignment shortly after the solidification has begun. Some atoms in a mixture may tend to segregate when solidifying, but in general, atoms mixed together as a liquid will form a homogeneous (consistent) weld. Segregation refers to the separation of like-atoms or phases resulting in a non-homogeneous structure.

Thermal Expansion

Metals expand and contract on heating and cooling because of the effects of energy on the atomic oscillations. Heating the metal puts more energy into the oscillations, which tends to move the atoms further apart. As a result, the metal expands. When the metal cools, this process is reversed and the metal contracts, or shrinks.

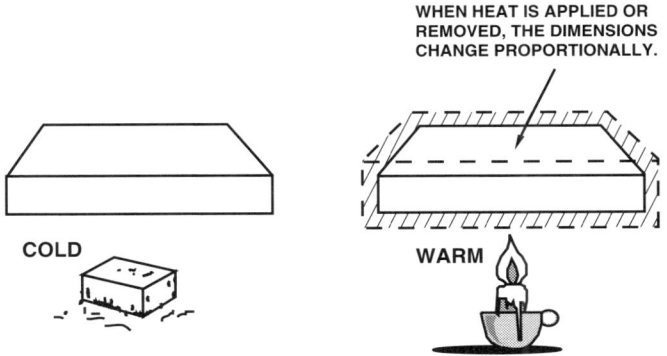

THERMAL EXPANSION

When heat is applied or removed uniformly from a piece of metal, the dimensions change, but stress is not induced from within because the expansion or contraction is uniform throughout. However, when the application or removal of heat is not uniform, as occurs in welding, stress is induced into the part, and some distortion may result because the hotter parts of the metal expand more than the portions at lower temperatures.

The figures on the next page illustrate the dimensional changes that occur in a straight bar (a) that is to be heated just on one side. In (b), the arc has been struck and the plate begins to heat under the influence of the arc. It is quite apparent the heating is very localized in a small spot. In (c), the heated portion expands, and because it is partially restrained by the portion of the bar that is not heated, the bar tends to bow slightly upward as shown. Since the hot part is weaker (the strength of a metal decreases as its temperature is raised) and, in fact, a small portion has actually melted, the hotter portion does not succeed in forcing the bar to bend very much. (The sketch exaggerates the amount of bending for illustration.)

In (d), the arc has been removed, and the molten portion begins to cool as the heat is conducted into the cooler portions of the bar. Heat always flows from the hot to cold sections. As cooling continues, in (e), and solidification occurs, the contraction on cooling causes the bar to bend in the opposite direction and results in a concave shape when reaching room temperature. Just from the simple act of melting a portion of the bar non-uniformly, and allowing it to cool, distortion has occurred and the bar now contains a low level of residual (remaining) stress. If the bar were restrained during the heating and cooling cycle, the residual stress would be much higher. Residual stresses can be great enough to cause cracking on cooling, or cracking in service from fatigue or corrosion stress cracking mechanisms. It is often necessary to remove these residual stresses by a post weld thermal heat treatment referred to as 'stress relief'.

TYPICAL STRESS RELIEF HEAT TREATMENTS FOR WELDMENTS		
MATERIAL	Soaking °C	Temp. °F
Carbon Steel	595-680	1100-1250
Carbon - 12% Mo steel	595-720	1100-1325
1% Cr -1/2% Mo steel	620-730	1150-1350
2-1/4% Cr - 1% Mo steel	705-770	1300-1425
5% Cr - 1/2% Mo steel (502)	705-770	1300-1425
7% Cr - 1/2% Mo steel	705-760	1300-1400
9% Cr - 1% Mo steel	705-760	1300-1400
12% Cr (Type 410) steel	760-815	1400-1500
16% Cr (Type 430) steel	760-815	1400-1500
Low-alloy Cr-Ni-Mo steels	595-680	1100-1250
2% to 5% Ni steels	595-560	1100-1200
9% Ni steels	550-585	1025-1080
Quenched & tempered steels	540-550	1000-1025

Stress Relief

The basis for stress relief is the fact that as metals are raised to higher temperatures, their mechanical strength is decreased. Typically, a common carbon steel alloy has a tensile strength of 70-80 KSI (1 KSI = 1 thousand pounds per square inch), and a yield of 30-40 KSI at room temperature. Residual stresses after welding can approach the level of the material's yield strength. When the metal is raised to its stress relieving temperature of 1100 or 1150 degrees F, the mechanical properties are greatly reduced, and the yield strength at the higher temperature typically drops to a much lower level, about 3-7 KSI. At this higher temperature, the residual stresses are much higher than the hot yield strength, and internal microstructural yielding occurs, relieving and reducing the residual stress to about the new, lower level of yield strength.

It is imperative when stress relieving a structure, that the component is heated uniformly at a prescribed rate, typically 400 degrees F/inch of thickness/hour, held for the appropriate time at its stress relieving temperature, typically one hour per inch of thickness, and then cooled uniformly, typically at a rate of 500 degrees F/inch of thickness/hour. This controlled heat-up, hold, and cool-down assures very even heating and cooling, eliminating the thermal gradients in the part and affecting a complete and thorough stress relief.

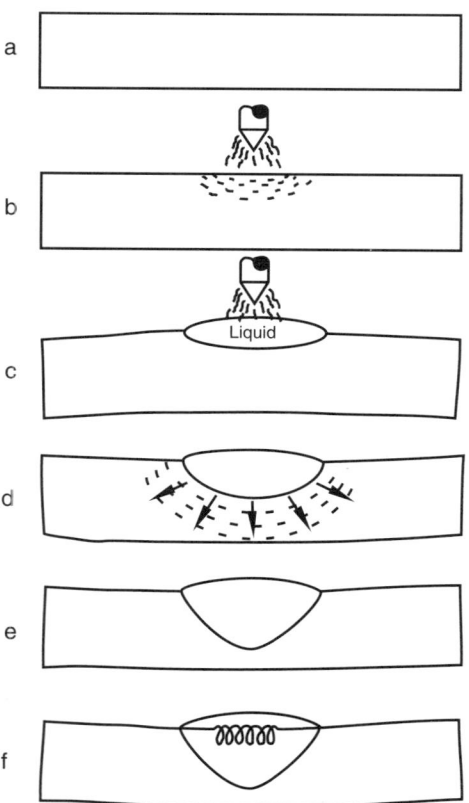

Shrinkage in a weld caused by internal restraint. Also shown is the nature of residual stress.

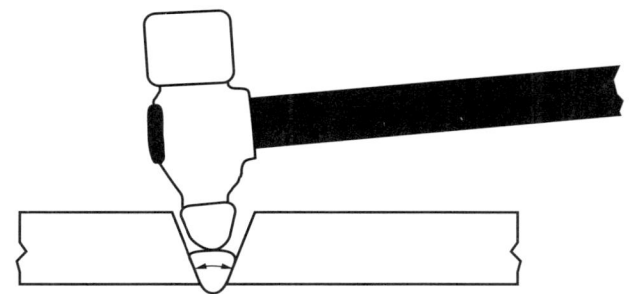

Peening the middle weld layer distributes and balances distortion.

Many of the fabrication codes require stress relief and specify the heating rates, holding temperatures and cooling rates. Stress relief can also be done using peening techniques, and vibratory stress relief. Peening is the mechanical distortion of the weld bead through mechanical means, usually when the metal is still quite hot. Peening should not be done on the root pass nor final pass of a weld, but only on the intermediate layers. Peening the root pass when it is hot can result in cracking of the weld since its size cannot withstand the mechanical distortion, and peening of the final pass can interfere with later visual inspection or other nondestructive testing. The vibratory technique imparts a high frequency vibrational energy into the part. The table on the previous page shows typical thermal stress relief treatments for several alloys.

One technique that may be used to reduce the need for postweld stress relief is 'preheat'. Preheating a part prior to welding slows the cooling rate, and may eliminate the need for post weld stress relief. Preheating reduces the thermal gradients within a weldment and slows the cooling rate, resulting in a more ductile structure with lower residual stress. Preheat is very effective in reducing or eliminating hot cracking of many alloys. Preheat also aids in removing moisture from the part, helps remove hydrogen, and retards the formation of martensite, a hard brittle phase that forms on rapid cooling.

Preheating can be done by several methods, including gas fired or electric furnaces, electric resistance heating elements, gas torches, or high energy quartz lamps. Preheating should be done on an area much larger than just the weld zone; if done on too small an area, insufficient heat energy is put into the part, and it cools too quickly. The AWS D1.1, *Structural Welding Code-Steel*, requires that the preheat temperature be checked to assure that all parts on which the weld metal is being deposited are above the minimum specified temperature for a distance equal to the thickness of the part being welded, but not less than 3 inches, in all directions, from the area being welded.

Welds Under the Microscope

When properly prepared, and examined under a metallurgical microscope, base metals and welds have a revealing structure that correlates with their mechanical properties. A brief review of the appearance of carbon steels as seen by the metallographer will expand your appreciation of the help that a metallographic study can bring to any welding discussion.

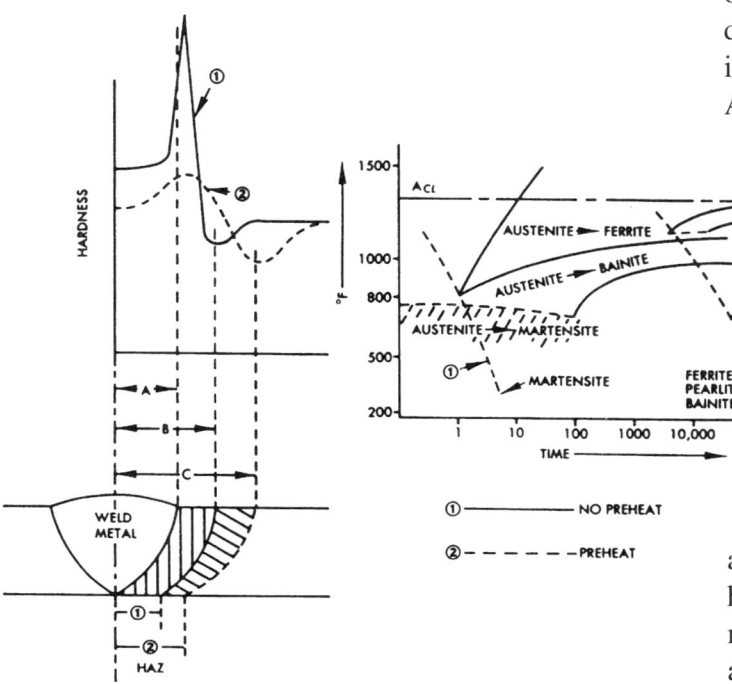

Variation of cooling conditions in a weld as a function of preheat; diagram includes some of the property changes brought about by these conditions. Also shown is a CCT diagram for a steel, to illustrate why the cooling conditions promote the property changes.

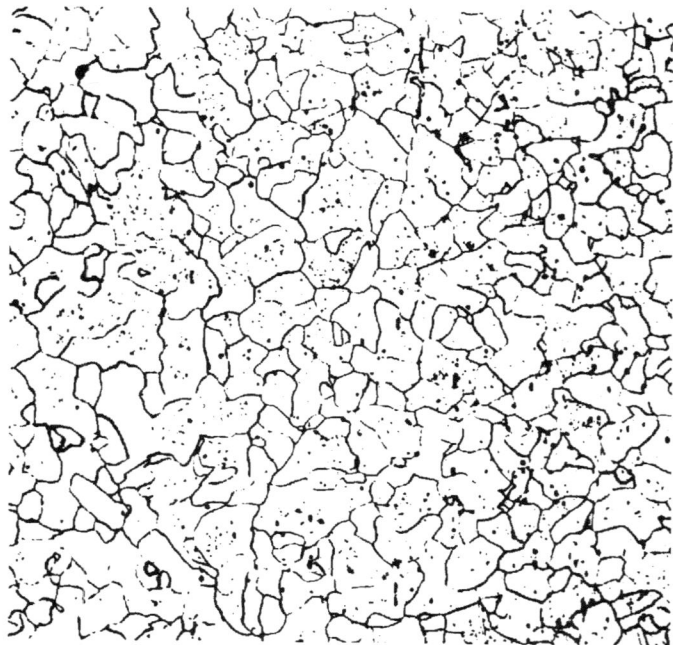

Photomicrograph of commercially pure iron, nominally called ferrite. No carbides are present, the acid etch reveals grain boundaries. The dark globules are nonmetallic inclusions.

The metallograph, a microscope designed for examining metals, shows that all metals are composed of grains, or crystals. In very low carbon iron (ingot iron), the grains are all one phase, called ferrite. Ferrite is composed of iron atoms in a body centered cubic structure (BCC), called a unit cell. A sketch of the BCC structure is shown, and you can see it is a cubic shaped structure formed by an iron atom at each corner, and one iron atom at the center of the cube. The atoms at the corners are actually shared with the adjacent unit cells.

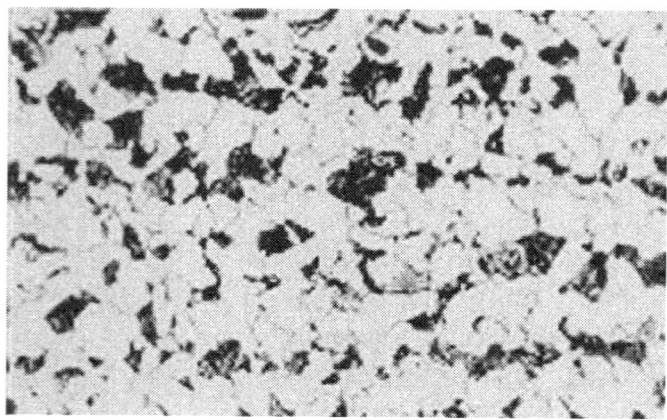

Small grains of Ferrite and Pearlite

A medium carbon steel contains more carbon and on cooling, its structure is formed into ferrite and other phases, depending on its cooling rate. One phase formed in addition to the ferrite is an iron carbide, Fe_3C, called 'cementite'. The cementite forms in several patterns; it can occur at the grain boundaries of the ferrite grains, or in alternate bands with the ferrite grains, called 'pearlite'. Pearlite has slightly higher strength than ferrite and slightly less ductility. Pearlite is referred to as plate-like, with alternate bands of ferrite and cementite. Iron carbides can also form in a feathery form called 'bainite'. Bainite has still more strength than pearlite, with still less ductility.

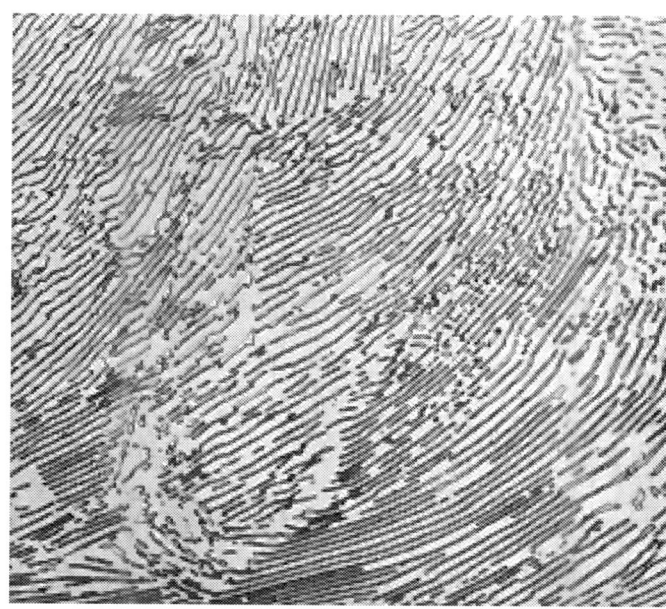

Typical lamellar appearance of pearlite. 1500X Etchant:Picral

For higher carbon alloys, quenching them rapidly forms a phase which is very hard and brittle; this phase is called 'martensite'. Martensite is very strong but has very little ductility. It is necessary to 'temper' the steel after forming martensite to restore ductility and toughness. Tempering is a heat treatment that consists of re-heating the metal to temperatures ranging from 100 - 1300 degrees F. Tempering a quenched steel results in a steel having higher strength than when annealed, with good toughness and ductility. As-quenched martensite is seldom used in components because of its brittleness and lack of toughness; tempering is almost always required. The unit structure of martensite is body centered tetragonal (BCT), and is similar to BCC except that one axis is longer than the other two, forming a rectangular unit cell.

Quenched martensite (500X etched), showing acicular structure

Phase Transformations

The various phases noted above are only a small part of the complexity of the metallurgy of steels. The primary reason steel is such an important part of our civilization and the subject of the majority of welding inspection duties is that when heated and cooled at various rates, steel undergoes changes in its atomic structure which can be predetermined, resulting in parts having very desirable mechanical properties with a wide spectrum of usage. The properties can vary from a very high hardness, brittle structure which is ideal for files and cutting tools, to a very ductile, tough structure suitable for bridge beams. These changes occur primarily by controlling the percentage of carbon and other alloying elements, and the metal's heat treatment.

It was noted above that low carbon steel is BCC at room temperature. When heated to a temperature of 1600-1800 degrees F, the BCC structure changes, or 'transforms' to a structure of face centered cubic (FCC) which consists of a cubic structure formed by having one iron atom at each corner of a cube and an atom on each face of the cube. The atoms on each face and corner are shared with adjacent unit cells. The FCC structure for steel is called 'austenite'. The FCC structure can dissolve more carbon than the BCC structure, and as learned earlier, heating a metal also increases its solid solubility limits. Thus, a heated steel can dissolve more carbon in its structure. We can take advantage of this higher solubility for carbon by increasing the dissolved carbon content by heating into the 'austenitizing range', holding for a short period of time, and then quenching rapidly to form the martensitic structure having very high strength. Tempering completes the heat treatment, and we have significantly increased the mechanical strength while maintaining good ductility and toughness. Toughness is defined as the ability of a metal to withstand an impact load without failing in a brittle manner.

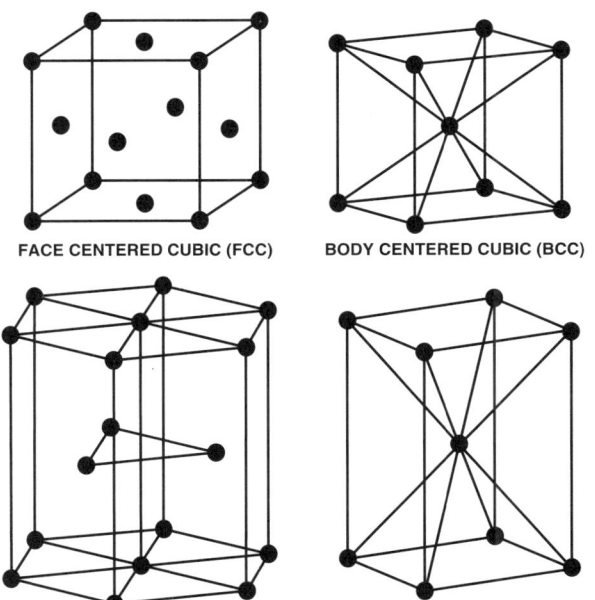

Common crystalline structures of metals and alloys

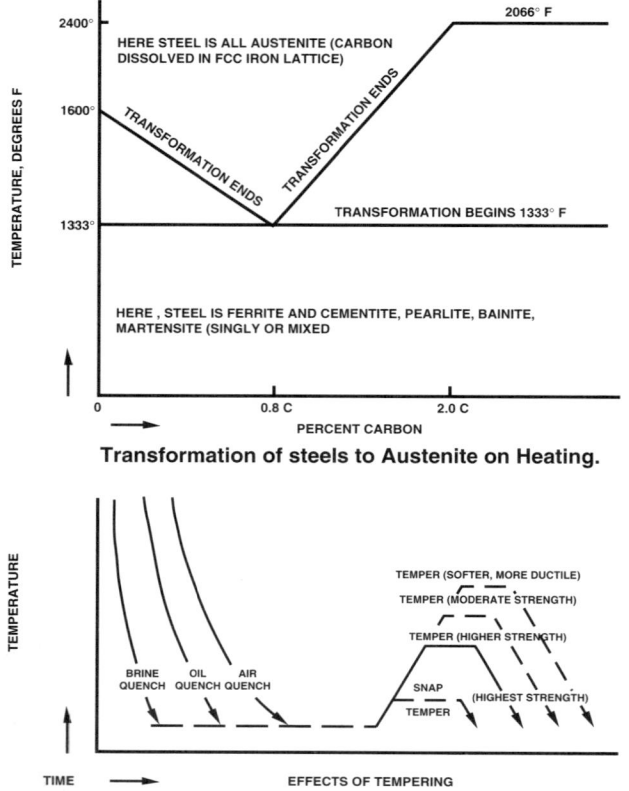
Transformation of steels to Austenite on Heating.

Austenite with Dissolved Carbon

Transformation of the various phases of steel found at room temperature such as ferrite, martensite, pearlite and bainite into austenite is accomplished by heating the metal above its transformation temperature into the austenitizing zone of its phase diagram. The transformation always begins for any steel containing more than a trace amount of carbon at a temperature of 1333 degrees F. In most carbon steel alloys, this transformation occurs over a temperature range from the 1333 degrees F to about 1670 - 2066 degrees F. For a steel containing exactly 0.8% carbon, called a 'eutectoid' composition, the transformation occurs quickly and completely at the 1333 degree F temperature. Transforming the room temperature phases into austenite results in dissolving all the carbon into the austenite structure since the solubility of carbon is much greater due to the increased temperature and the FCC structure being able to dissolve a greater percent of carbon than the BCC structure.

Cooling Rates

On cooling, the transformation reverses; the FCC crystalline structure changes back to BCC. Since the BCC arrangement will not accommodate as much carbon, the excess carbon must form iron carbides (cementite) in one of its several phases. The phase that forms is determined by the speed of cooling. Slow cooling forms ferrite and cementite (pearlite); somewhat faster cooling forms bainite. A very rapid quench forms martensite. The critical cooling rate is governed by the carbon content, and for alloy steels, by their additional chemical composition.

It should be apparent by now why the cooling rate for a specific steel composition determines the need for stress relief heat treatment of a weldment. A cooling rate that forms martensite has moved the iron and carbon atoms into the strained martensitic crystalline structure. Tempering is accomplished by heating the martensite to a temperature between 100 degrees F and 1300 degrees F to soften the steel. At the lower temperatures of tempering, no structural change can be seen on the microscope, but the strength and hardness are reduced, and the ductility and toughness are increased. The changes are more or less proportional to the tempering temperature.

The language of cooling rates and heat treatments is precise. Heating the steel above the transformation temperature to form austenite and cooling it at the slowest rate, as in a furnace cool, is called 'annealing'. Annealing a steel results in placing it in its softest, weakest condition.

Heating to form austenite, and cooling in 'still air' forms new grains of ferrite and pearlite; this heat treatment is called 'normalizing'. Cooling the steel very rapidly, by quenching in water, oil, or a cold blast of air is called 'quenching'. Quenching the steel results in a martensitic structure, as noted previously. As noted earlier, steels quenched to form martensite usually require a 'tempering' heat treatment to lower their hardness and strength, improve ductility and toughness.

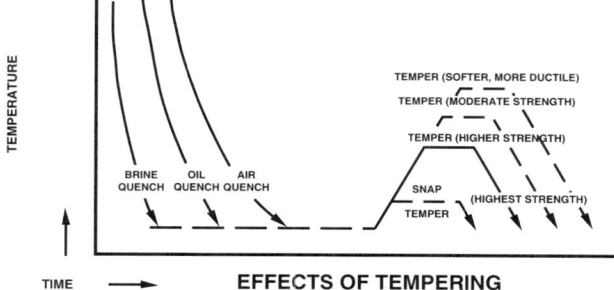

Tempering, as described above, is simply the reheating of the metal, containing martensite, to some appropriate temperature below 1333 degrees F and holding at that temperature for a specified period of time. This permits the as-quenched martensite to change to tempered martensite. The desired compromise between hardness, strength and toughness is obtained by choosing the proper tempering temperature and time cycle. Higher temperatures make the steel softer and weaker, but tougher.

In simple terms, for steels, the faster the cooling rate, the harder and less ductile the resulting structure. While the increased strength is often desirable, the accompanying low ductility will increase the steel's susceptibility to cracking. When welding, the base metal directly adjacent to the molten weld metal will be subjected to the maximum cooling rate because it is heated to a very high temperature and then rapidly quenched due to its contact with the massive, colder base metal. This region, called the heat affected zone (HAZ), will often contain martensite because of the quenching action of the rapid cooling. Consequently, this region is likely to crack unless corrective steps are taken.

One of the more common heat treatments used to reduce the tendency for this high hardness and low ductility in the HAZ is preheat. By preheating the base metal prior to welding to a temperature of 150 - 700 degrees F, the cooling rate will be effectively reduced. This will allow more time for a more ductile microstructure to form rather than martensite, resulting in a more ductile, less crack sensitive weld and HAZ.

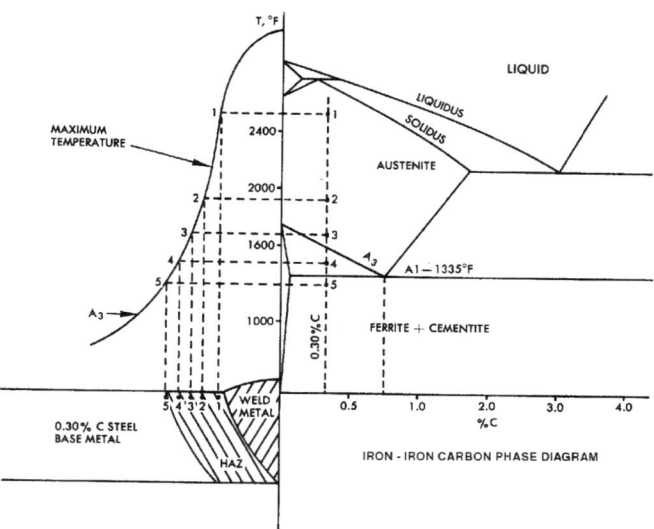

Relation between the peak temperatures experienced by various regions in a weld, and how these correlate with the iron-carbon phase diagram.

Another factor having an affect on the cooling rate of the weld zone is heat input or the amount of energy supplied by the welding arc to heat the base metal. As the heat input increases, the cooling rate decreases. Use of smaller diameter welding electrodes, lower welding currents and faster travel speeds will all tend to decrease the heat input and therefore increase the cooling rate. For arc welding processes, the heat input can be calculated using the formula shown. It is dependent only on the welding current, arc voltage, and travel speed as measured along the longitudinal axis of the weld joint. The heat input formula is:

Heat input = $\dfrac{\text{welding current x welding voltage x 60}}{\text{travel speed}}$

Heat input is expressed in joules/inch, and travel speed in inches/minute. Joules can also be expressed as watt-seconds. Therefore the number '60' in the numerator simply converts the 'minutes' of the travel speed into 'seconds'.

Other Metallurgical Factors

Fatigue: Any welds which are to be inspected will have to be designed to resist with elastic deformation (temporary) all stresses up to the yield point of the steel, except for weldments subject to fatigue. Welds subject to cyclic stresses must be designed for lower applied stresses to avoid fatigue failures, since metals usually fail by fatigue at stresses below their yield point.

The allowable stresses for bridge welds are defined in special tables that impose higher than normal factors of safety to account for fatigue loading. Safety factors normally are 3 or 4; bridges often have safety factors of 5 or even higher. The design engineer translates these higher factors of safety into larger fillet welds, longer welds, heavier beams, etc., so the bridge will be able to handle the applied cyclic stresses. The welding inspector should be aware of these design differences and inspect accordingly.

Another important factor affecting a metal's resistance to fatigue stresses is its surface configuration. Sharp crevices or notches on the metal's surface result in creating a 'stress riser' which can increase the actual applied load to much higher levels of stress. Typically, engineers have determined that applied stresses can be multiplied by factors from 1 to 10, depending on the nature and shape of the 'notch'. A crack, with its very sharp end conditions, can multiply the applied stress by as much as 10. Undercut can multiply the applied stress by a factor of 3-5, and weld ripples can increase the applied stress by a factor of 2-3.

With the information above, it is easy to see why the surface discontinuities such as undercut, overlap, excess convexity, etc., cause stress risers, or notches, that can significantly reduce the component's fatigue strength. For that reason, fatigue-loaded structures may have their weld surfaces ground to remove surface irregularities. An example of this is found in AWS D1.5, where the allowable depth of undercut for bridges is significantly less than for some other structures.

Many fabricated structures are subject to some level of fatigue loading; in fact it is difficult to find a complex structure that has not had some aspect of its design modified to account for fatigue (dynamic) loading.

Internal structure: When alloys are melted in the steel mill, and cast into large ingots, they have a cast structure. These large ingots are formed into the required shapes by a combination of hot working and cold working of the metal. This 'working' changes the internal structure of the metal, but metallurgical discontinuities such as segregation of elements, laminations, seams and laps may remain in the finished product. These discontinuities can become 'planes of weakness' in the finished product. Metals have their highest mechanical properties in the direction of rolling, are somewhat weaker in the direction transverse to the rolling direction, and may often be significantly weaker in their through-thickness direction. The designer and welding engineer must take these factors into account in the design and fabrication of a structure. The inspector should be aware of these conditions which relate to the inspection responsibilities.

Welding Chemistry

The final aspect of weldability to be discussed is welding chemistry. Welding chemistry deals with the chemical interaction between the base metal elements, the weld metal elements, and the other elements introduced into the area of the weld. The base and weld metals must be compatible; however, other elements in the welding environment must also be considered. For example, the air atmosphere, containing oxygen, nitrogen, and other elements, must be excluded from the weld zone to maintain weld quality. And the elements contained in fluxes must be considered for compatibility as well as any shielding or purge gasses used.

Metals react with their environment to different degrees, depending on the metal and the environment to which it is exposed. We are

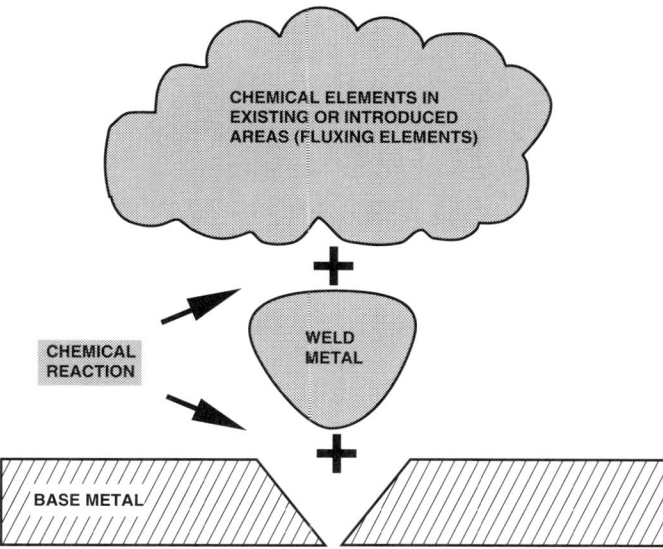

Shielding

When welding occurs, the molten weld puddle must be protected from chemical interactions with the unwanted elements. The primary elements causing difficulties are carbon, oxygen, hydrogen, and nitrogen. These unwanted elements are found in moisture, air, improperly adjusted heating flames, and surface contamination. Since the molten metal must be protected, it is accomplished in part through 'shielding', a protective 'blanket of gas' formed from electrode coverings, welding fluxes, shielding gasses, or purge gasses.

familiar with the 'rusting' of bare steel in moist air; the steel is reacting with the water and the oxygen through a chemical reaction. Many of the stainless steels can also corrode at very high rates if the environment is severe enough. And, as metals corrode, atomic (nascent) hydrogen atoms (H^+) are formed. Earlier discussions about problems with hydrogen emphasize the importance of avoiding corrosion of our welded structures not only because of the wall thinning aspect, but also because of hydrogen formation and diffusion into the steels.

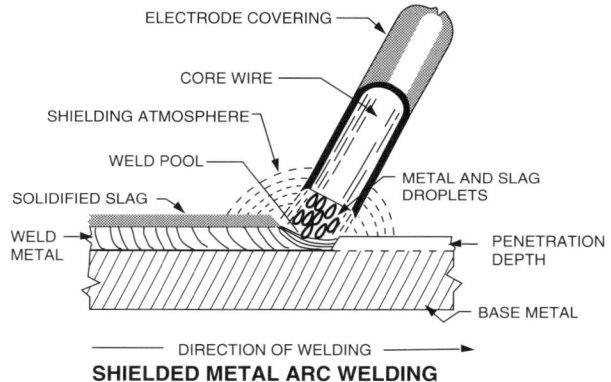

SHIELDED METAL ARC WELDING

In shielded metal arc welding, the shielding comes from the vaporization of the electrode covering; in gas tungsten arc welding, the shielding comes from the inert gas passing through the torch; in oxyacetylene welding, the shielding comes from the neutral welding flame.

In gas metal arc welding and flux cored arc welding, the shielding gas may be carbon dioxide, CO_2, or gas mixtures containing carbon dioxide. Since carbon dioxide decomposes on heating into oxygen atoms (O^+) and carbon monoxide (CO), the oxygen can be absorbed into the weld metal. To alleviate this possibility, deoxidizers are added to the electrode compositions to combine with any oxygen present and prevent it from degrading the weld metal properties.

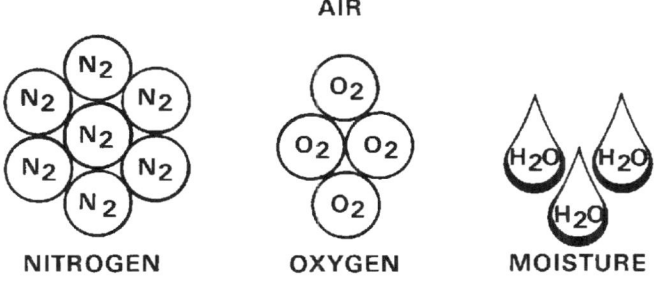

And when molten, metals react with the environments to a much greater degree. Because of this, they must be protected while molten from undesirable elements that may significantly alter their mechanical properties.

Shielding can also take the form of a blanket of flux, which physically excludes the surrounding atmosphere from the weld zone, and forms a protective slag when heated. Submerged arc

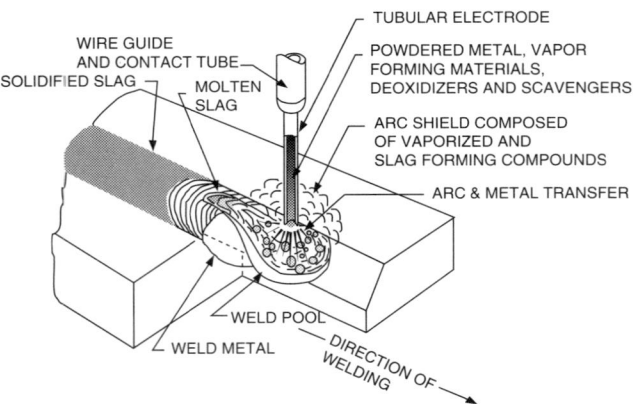

SELF-SHIELDED FLUX CORED ARC WELDING

welding uses a flux covering for shielding and slag formation which protects the weld metal until it cools to room temperature. The protective slag is formed during the heat of welding, and contains deoxidizers added to the flux. Electroslag uses a similar granular flux for shielding.

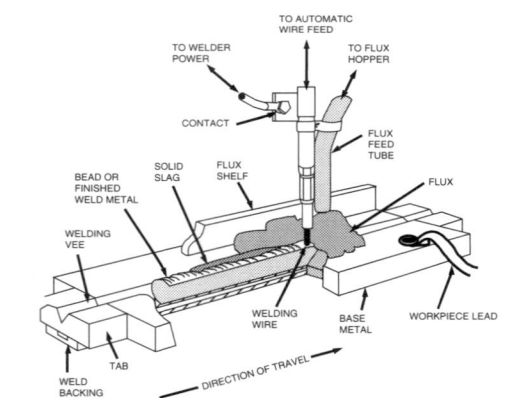

SCHEMATIC VIEW OF SUBMERGED ARC WELDING PROCESS

Weld Metal Composition

The elements composing a weld come from a variety of sources; the base metal adjacent to the weld which is melted, the weld filler metal, fluxes if used, and the shielding agents coming into contact with the weld puddle. Some welds are made without filler metal, and are referred to as 'autogenous' welds. Examples are longitudinal seam welds in tubing, and edge welds and corner welds in sheet metal fabrication. For these welds, the base metal is formed to permit its being melted and supplying the necessary molten puddle to complete the weld. Some groove welding procedures start with an autogenous root pass, and are completed with filler metals.

The welding inspector must recognize the positive and negative chemical aspects of weld metal composition, and their effects on the finished product. For example, if a weld is made without filler metal, the absence of deoxidizers normally contained in filler metals and fluxing agents may result in the presence of porosity or even cracking.

Hydrogen Contamination

Since hydrogen is present in moisture, oil, and other organic compounds, and always forms during metal corrosion, the inspector must be aware of its formation and possible welding problems. The molten puddle must be protected from hydrogen to avoid cracking or reduction of the mechanical properties of the weld. Cleanliness of the joint is very important, as well as proper storage of fluxes and welding electrodes. Low hydrogen electrodes must be kept in heated storage ovens after opening the original containers. Shielding and purging gasses should be purchased with a - 40 degrees F dewpoint to make sure they don't contain moisture. Some special welding procedures may require these gases having a dewpoint of - 80 degrees F. The lower the dewpoint, the drier the gas.

It must be noted the shielded metal arc electrodes with a cellulosic coating such as EXX10 and EXX11 must not be kept in heated storage. They should be kept dry and stored at room temperature. Heating these electrodes destroys the covering chemistry by removing moisture from the covering which is required for satisfactory performance. These two classes of electrodes are usually limited to welding plain carbon steels, with low to medium carbon content.

Welding Chemistry of Specific Base Metals

The chemistry of the base metals and their corresponding filler metals must be matched for best results. Electrodes have been designed for welding almost every metal. We have noted that as the carbon content increases, the weldability decreases. Thus, plain carbon steel base metals with carbon contents of 0.15% to 0.40% are generally considered easily weldable. As the base metals' carbon content increases above about 0.40%, they become more difficult to weld, and may require changes in the welding procedures such as adding preheat.

Also, as the percentage of other elements added to the base metals increase in percentage, weldability usually goes down. Increased chromium, molybdenum, nickel and other elements may require the use of preheat, interpass temperature control, and post weld heat treatments. Preheating the higher carbon base metals, and those with higher element percents noted above, improves their weldability. This improvement occurs due to several aspects; the weld and HAZ cool more slowly when preheated, the formation of martensite is reduced, hydrogen is removed from the base metal, and the effects of any contamination are minimized.

Carbon Equivalent Calculation

The weldability of carbon steels certainly depends on the base metal's chemistry. A formula has been developed to aid in quantifying its weldability, and determining whether preheat is needed or not. Many such formulas have been developed, but they all take into account the primary alloying elements of the base metal, and determine the effect of each based on a corresponding percent of carbon. The elements usually found in these 'carbon equivalent' (CE) formulas include carbon, manganese, chromium, nickel, copper, silicon, and molybdenum. One such formula is listed, and it determines the effect of each element noted regarding martensite formation. Most of the elements promote martensite formation, but to lesser degrees than carbon, and their actual percentage is reduced to a 'carbon equivalent' by dividing by some number.

$$CE = \%C + \frac{(\%Mn+\%Si)}{6} + \frac{(\%Cr + \%Mo)}{5} + \frac{(\%Ni + \%Cu)}{15}$$

When the CE computed by this formula exceeds 0.40, it is recommended that the base metal be preheated in the range of 200 - 400 degrees F. If the CE exceeds 0.60, the preheat range should be increased to 400 - 700 degrees F. Additionally, as the CE rises above 0.40, low hydrogen electrodes are helpful, and post welding heat treatments may be required. Since there are many different CE formulas, the welding inspector must get assistance from the welding engineer in their selection to insure their proper usage.

Welding Chemistry of Stainless Steels

The word 'stainless' is a bit of a misnomer when applied to the classes of metals referred to as stainless steels, since it usually means they resist corrosion. However, in severe corrosive environments, many of the stainless steels corrode at very high rates. The stainless steels are defined as having at least 12% chromium. There are many types of stainless steels, and the welding inspector should recognize this when discussing them and use the proper designation for each type.

The four main classes of stainless steels are ferritic, martensitic, austenitic, and precipitation hardening. The first three categories refer to the stable room temperature phase found in each class. The last one, often called 'PH' stainless steels, refers to the method of hardening them by an 'aging' heat treatment, a precipitation hardening mechanism as opposed to the quenching and tempering mechanism. The stable room temperature phase found in stainless steels depends on the chemistry of the steel, and some stainless steels may contain a combination of the different phases. The more common stainless steels are the austenitic grades, which are identified by the '200' and '300' series grades; 304 and 316 stainless steels are austenitic grades. A 416 steel is a martensitic grade, and 430 is a ferritic grade. One of the common PH stainless steels is a 17-4 PH grade.

As you might expect, the weldability of these grades varies significantly. The austenitic grades are very weldable with today's available filler metal compositions. These grades can be subject to hot short cracking, which occurs when the metal is very hot. This problem is solved by controlling the composition of the base and filler metals to promote the formation of a 'delta ferrite' phase which helps eliminate the hot short cracking problem. Typically, cracking will be avoided by selecting filler metals with a delta ferrite percent of 4-10%. This percentage is often referred to as the 'Ferrite Number' and can be measured with a magnetic gage. The delta ferrite can be measured using the magnetic gage since delta ferrite is BCC and magnetic, while the primary phase, austenite, is FCC and non-magnetic.

The ferritic steels are also considered weldable with the proper filler metals. The martensitic grades are the most difficult to weld, and often require special preheating and post weld heat treatment. Procedures have been developed to weld these materials, and must be followed carefully to avoid cracking problems and maintain the mechanical properties of the base metals. The PH stainless steels are also weldable, but attention must be given to the changes in mechanical properties caused by welding.

One of the common problems found when welding the austenitic grades is referred to as 'carbide precipitation', or 'sensitization'. When heated to the welding temperatures, a portion of the base metal reaches temperatures in the 800 - 1600 degree F range, and within this temperature range, the chromium and carbon present in the metal combine to form chromium carbides. The most severe temperature for their formation is about 1250 degrees F, and this temperature is passed through twice on each welding operation cycle; once on heating to weld and again on cooling to room temperature.

These chromium carbides typically are found along the grain boundaries of the structure. The result of their formation is the reduction of the chromium content within the grain itself adjacent to the grain boundary, called 'chromium depletion', resulting in reducing the chromium content below that desired. The final result of this chromium depletion of the grain is a reduced corrosion resistance of the grain itself due to its reduced chromium content. In certain corrosion environments, the edges of the grains corrode at a high rate, and is called 'intergranular corrosion attack', or IGA.

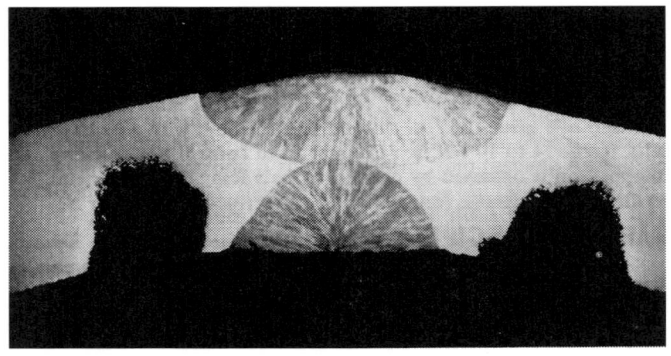

Corrosion by intergranular attack (IGA) caused by sensitization of the HAZ.

Sensitization of austenitic stainless steels during welding can be attacked by several methods. The first method involves re-heat treating the completed structure by heating to 1950 degrees F and quenching rapidly in water. This reheating breaks up the chromium carbides permitting the carbon to be redissolved into the structure. However, this heat treatment can cause severe distortion of welded structures.

A second method is the addition of stabilizers to the base and filler metals. The two most common examples of stabilization are the addition of titanium or columbium (niobium) to the 300 series alloys in amounts equal to 8 or 10 times the

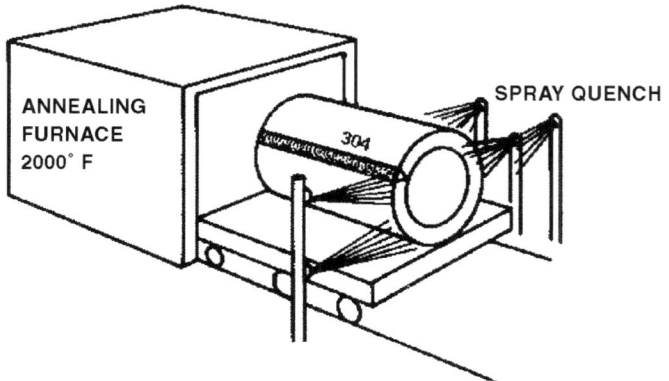

For Stabilization Heat Treatment use 1500° F, no quench needed.

carbon content. These alloying stabilizers preferentially combine with the carbon and reduce the amount of carbon available for chromium carbide formation, maintaining the alloys' chromium content and corrosion resistance. When titanium is added, we have the austenitic stainless alloy 321; when columbium is added, we have the 347 grade.

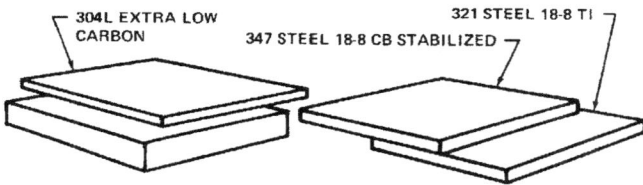

Prevention of corrosion in welded stainless steel.

The third method is the reduction of the carbon content in the base and filler metals. Initially, these low carbon austenitic stainless steels were referred to as 'Extra Low Carbon', or ELC for abbreviation. Today, they are referred to by the letter "L" meaning the carbon content is less than 0.03%. (The standard grades contain up to 0.08% carbon.) By reducing the carbon content in the alloy, less carbon is available to combine with the chromium, and sensitization is reduced during welding. These low carbon grades do have slightly reduced mechanical properties because of their lower carbon content, and this must be considered when selecting these alloys, especially for high temperature use.

Welding Chemistry of Aluminum Alloys

Aluminum alloys have a very tenacious oxide film on their surfaces which forms very rapidly when the bare aluminum is exposed to air, and these oxide films give protection in corrosive environments. These same oxides on the surface do interfere with the joining processes, however. To braze or solder these alloys, fluxes are used to break down the oxide film so the parts can be joined. When welding, alternating current is used which results in breaking down of the oxides by the current reversal of AC welding, and re-formation of the oxide film is avoided by shielding with helium or argon gas. The AC welding method is sometimes referred to as a 'surface cleansing technique'.

The metallurgy of aluminum and its alloys is very complex, especially regarding the great number of alloy types and heat treatments. The proper filler metals for most every grade and heat treat condition can be found in ANSI/AWS A5.10, *Specification for Bare Aluminum and Aluminum Alloy Welding Electrodes and Rods.*

Welding Chemistry of Copper Alloys

Pure copper and many of its alloys cannot by hardened by a quench and temper heat treatment as steel can. These alloys are usually hardened and made stronger by the amount of 'cold work' introduced when forming them into the various shapes. The act of welding softens the cold worked material and must be considered before welding on work-hardened copper alloys. There is a series of copper alloys that are strengthened by 'aging', a treatment similar to the precipitation hardening used on the PH stainless steels. When welding on these alloys, a postweld heat treatment is usually specified to restore the original mechanical properties.

One of the major problems found on welding copper and its alloys is due to their relative low melting point and very high thermal conductivity. Considerable heat must be applied to the metal to overcome its loss through conductivity, and the relatively low melting point often results in the metal melting earlier than expected and flowing out of the weld joint. Most copper alloys are weldable with proper technique and practice.

Welding Chemistry of the Reactive Metals

There are three metals grouped into the class of 'Reactive Metals'. These are titanium, zirconium and tantalum. These metals are so named because they are extremely reactive with other elements, especially when heated to the temperature required to weld. When they react with even traces of other elements, such as oxygen, hydrogen, and nitrogen, they become very embrittled and hard. Extra precaution is required in shielding and purging, and often, tantalum and zirconium are required to be welded in a purging chamber. For titanium, a full shielding technique, or a trailing shield may be sufficient.

It has been stated that because of the great affinity these alloys have for trace contaminants, the 'art of welding' is really the art of proper cleaning and purging. When attention is given to these two aspects, cleaning and purging, high quality welds can be made.

Summary

While the welding inspector is not responsible for selecting the base metals and filler metals, the inspection responsibility will involve many decisions which have a basis on metallurgical principles. This information has been covered to provide the inspector with a background and basic understanding of many of the principles involved. The inspector will benefit from the additional knowledge of the metallurgical considerations applied to the fabrication of structures that explain the behavior of metals during welding.

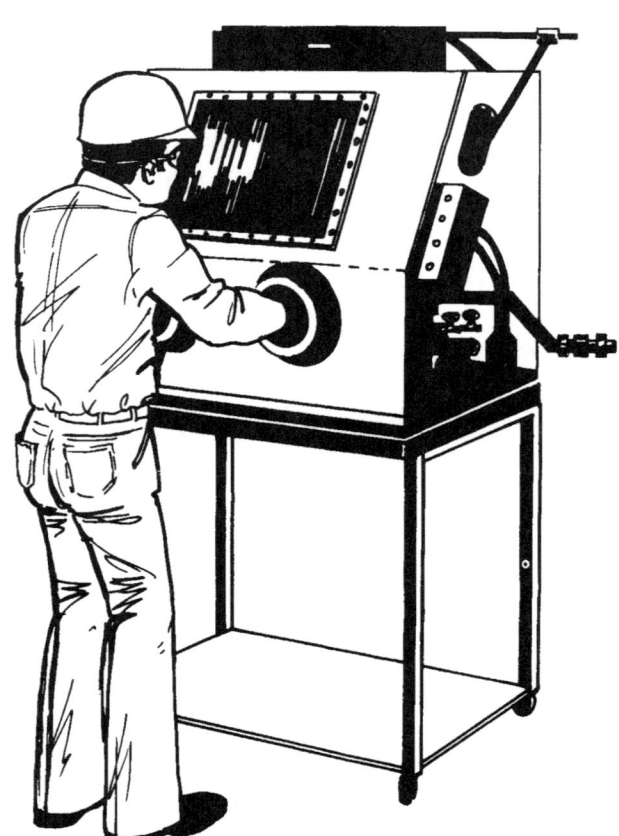

Welding in a Controlled Atmosphere

REVIEW – CHAPTER 6
WELDABILITY, WELDING CHEMISTRY AND WELDING METALLURGY

Q6-1   As a metal is heated:
   a. energy is added to the structure.
   b. the atoms move farther apart.
   c. the atoms vibrate more vigorously.
   d. the metal expands.
   e. all of the above

Q6-2   The state of matter which exhibits the least amount of energy is:
   a. solid
   b. liquid
   c. gas
   d. quasi-liquid
   e. none of the above

Q6-3   A problem occurring in weldments caused by the nonuniform heating produced by the welding operation is:
   a. porosity
   b. incomplete fusion
   c. distortion
   d. slag inclusions
   e. none of the above

Q6-4   All but which of the following will result in the elimination or reduction of residual stresses?
   a. vibratory stress relief
   b. external restraint
   c. thermal stress relief
   d. peening
   e. postweld heat treatment

Q6-5 Steel exists in which of the following crystal arrangements?
a. HCP
b. FCC
c. BCC
d. a and b above
e. b and c above

Q6-6 Rapid cooling of a steel from the austenitic range results in a hard, brittle structure known as:
a. pearlite
b. carbide
c. cementite
d. bainite
e. martensite

Q6-7 Very slow cooling of steel may result in the production of a soft, ductile micro structure which has a lamellar appearance when viewed under high magnification. This structure is referred to as:
a. martensite
b. pearlite
c. bainite
d. ferrite
e. cementite

Q6-8 When rapid cooling produces a martensitic structure, what non-austenitizing heat treatment may be applied to improve the ductility of the steel?
a. quenching
b. tempering
c. annealing
d. normalizing
e. none of the above

Q6-9 The use of preheat will tend to:
a. result in a wider heat affected zone
b. produce a lower heat affected zone hardness
c. slow down the cooling rate
d. reduce the tendency of producing martensite in the heat affected zone
e. all of the above

Q6-10 Which of the following changes will warrant the addition or increase in the required preheat?
a. decreased carbon equivalent
b. increased carbon equivalent
c. increased base metal thickness
d. a and c above
e. b and c above

Q6-11 What heat treatment is characterized by holding the part at the austenitizing temperature for some time and then slow cooling in the furnace?
a. normalizing
b. quenching
c. annealing
d. tempering
e. stress relief

Q6-12 What heat treatment is characterized by holding the part at the austenitizing temperature for some time and then slow cooling in still air?
a. normalizing
b. quenching
c. annealing
d. tempering
e. stress relief

Q6-13 Increasing the heat input:
a. decreases the cooling rate and increases the likelihood of cracking problems.
b. decreases the cooling rate and decreases the likelihood of cracking problems.
c. increases the cooling rate and increases the likelihood of cracking problems.
d. increases the cooling rate and decreases the likelihood of cracking problems.
e. none of the above

Q6-14 Increasing preheat:
a. decreases the cooling rate and increases the likelihood of cracking problems.
b. decreases the cooling rate and decreases the likelihood of cracking problems.
c. increases the cooling rate and increases the likelihood of cracking problems.
d. increases the cooling rate and decreases the likelihood of cracking problems.
e. none of the above

Q6-15 Increasing the carbon content:
a. decreases the likelihood of cracking problems.
b. increases the likelihood of cracking problems.
c. has nothing to do with the likelihood of cracking problems.
d. all of the above
e. none of the above

Q6-16 Which of the following generally follows quenching?
a. annealing
b. normalizing
c. quenching
d. tempering
e. stress relief

Q6-17 Which of the following can be accomplished using either thermal or mechanical techniques?
a. annealing
b. normalizing
c. quenching
d. tempering
e. stress relief

Q6-18 Which of the following results in the softest structure for steel?
a. annealing
b. normalizing
c. quenching
d. tempering
e. stress relief

Q6-19 For a steel having the following composition: 0.11 carbon, 0.65 manganese, 0.13 chromium, 0.19 nickel, 0.005 copper, and 0.07 molybdenum, what is its carbon equivalent using the following formula?

$$CE = \%C + \frac{\%Mn}{6} + \frac{\%Ni}{15} + \frac{\%Cr}{5} + \frac{\%Cu}{14} + \frac{\%Mo}{4}$$

a. 0.15
b. 0.23
c. 0.28
d. 0.31
e. 0.42

Q6-20 For a steel having the following composition: 0.16 carbon, 0.85 manganese, 0.25 chromium, 0.09 nickel, 0.055 copper, and 0.41 molybdenum, what is its carbon equivalent using the following formula?

$$CE = \%C + \frac{(\%Mn)}{6} + \frac{(\%Ni)}{15} + \frac{(\%Cr)}{5} + \frac{(\%Cu)}{13} + \frac{(\%Mo)}{4}$$

a. 0.23
b. 0.31
c. 0.34
d. 0.41
e. 0.46

Questions **Q6-21** through **Q6-24** refer to the Heat Input formula below:

$$\text{Heat Input (J/in)} = \frac{\text{Amperage} \times \text{Voltage} \times 60}{\text{Travel Speed (in/min.)}}$$

Q6-21 The FCAW process is being utilized to weld a 1 inch thick structural steel member to a building column. The welding is being done with a 3/32 inch diameter self-shielded electrode with a 150° minimum preheat and interpass temperature. The welding parameters are adjusted to 30 volts, 250 amperes and 12 in/min. What is the heat input?

a. 375 J/in
b. 37,500 J/in
c. 375 kJ/m
d. a and b above
e. b and c above

Q6-22 GMAW (short circuiting) welds are produced at 18 volts, 100 amperes and 22 in/min. What is the heat input?

a. 238 J/in
b. 7333 J/in
c. 4909 J/in
d. 30 J/in
e. none of the above

Q6-23 The GMAW process is mechanized for welding 1/8" thick stainless steel sheets against a copper backing bar. The process is operated at 300 amperes, 28 volts and 15 in/min. What is the resulting heat input?
 a. 650 kJ/in
 b. 650,000 J/in
 c. 165,000 J/in
 d. 16,500 J/in
 e. none of the above

Q6-24 The GTAW process is being used for welding 1/16" thick titanium using DCEN at 110 amperes, 15 volts and 6 in/min. What is the heat input?
 a. 21 000 J/in
 b. 21 kJ/in
 c. 16,500 J/in
 d. a and b above
 e. b and c above

CHAPTER 7:
DESTRUCTIVE TESTING

Introduction

In Chapter 6, there was a detailed discussion of many of the important properties of metals. Once it is recognized that these properties may he important to the suitability of a metal or a weld, it then becomes important to determine the actual values for those properties. That is, now the designer would like to put a number on each of these important properties so he can effectively design some structure using materials having the desired characteristics.

As mentioned in Chapter 6, there are numerous tests that are used to determine the various mechanical and chemical properties of metals. While some of these tests provide values for more than one property, most are designed to determine the value for a specific characteristic of the metal. Therefore, it may he necessary to perform several different tests to determine all the desired information.

It is important to understand each of these tests, what results they provide and how to determine if test results are in compliance with specification requirements. It may also be helpful for the welding inspector to understand some of the methods used in testing, even though one may not be directly involved with the actual testing.

As a group, these tests are referred to as destructive. They bear this name, because usually they render the test piece useless for service once the test has been performed. Generally, they destroy, or fall, the material to learn how it behaves in resisting failure. Some of these tests are referred to as destructive tests, even though they do not ruin the part to the extent it cannot he used (e.g., hardness testing). However, that

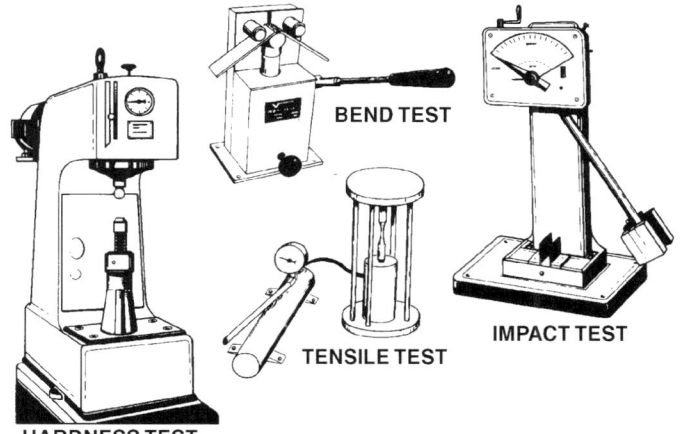

Inspectors must be able to interpret data from a variety of tests

depends on the shape, size and required surface condition of the part and its end-use.

Throughout this discussion there will be little mention made whether a specific destructive test is being used to determine a base metal or weld metal property. For the most part, this does not represent a significant change in the manner in which the test is performed. There will be occasion when testing is performed specifically to test the base or weld metal, but the mechanics of the testing operation will vary little, if any.

As these various methods are discussed, it is important to relate the actual test method with the appropriate material property(ies). In that way, it will be easier to understand and relate the values determined from the test with the properties that they describe.

The results of destructive testing are often indicated on Mill Test Reports. An inspector may be required to compare the Mill Test Report values with the applicable material specification to determine conformance. Consequently, the following discussion should enable the inspector to understand the information appearing on a typical Mill Test Report.

Tensile Testing

The first destructive test method to be discussed will be the tensile test. This one test provides us with a great amount of information about a metal. Some properties that can be determined as a result of tensile testing include:

- Ultimate Tensile Strength
- Yield Strength
- Ductility
- Percent Elongation
- Percent Reduction of Area

Some of these values can be determined through a direct reading of a gage, while others can be quantified only after analysis of the stress-strain diagram that is produced during the test. The values for ductility can be found by making comparative measurements of the tensile specimen before and after testing. The percentage of that difference indicates the amount of ductility present.

When we perform a tensile test, one of the most important aspects of that test involves the preparation of the tensile specimen. If this part of the test operation is conducted carelessly, the validity of the test results will be severely reduced. Small imperfections in the surface finish, for example, can result in significant reductions in the apparent strength and ductility of the tensile specimen.

Sometimes, the sole purpose of the tensile test will be to show if the weld zone will perform as well as the base metal. All that is necessary for such an evaluation is removal of a specimen (sometimes referred to as a strap) transverse to the longitudinal axis of the weld. The weld is roughly centered in the specimen. The two cut sides should be parallel using a saw or cutting flame, but no further surface treatment is essential, including the removal of any existing weld reinforcement.

This approach is used for procedure and welder qualification testing in accordance with API 1104. A successful tensile test is there described as a specimen that falls either:

1) In the base metal, or

2) In the weld metal (As long as it fails at some strength level above the specified base metal strength.)

For most cases in which the tensile test is required, however, there is a need to determine the actual values for strength and other properties exhibited by that metal, not just if the weld is as strong as the base metal. When the determination of these values is necessary, our specimen must he prepared in some configuration that provides a reduced section somewhere near the center of the length of the specimen.

This reduced section is intended to localize the failure. Otherwise the failure might tend to occur preferentially near the tensile test machine grips, making the measurements for percent elongation extremely difficult. Also, this reduced section results in the increased uniformity of the stresses throughout the cross section of the specimen.

This reduced section must exhibit the following three features so valid results can be obtained:

1) The entire length of the reduced section must be of a uniform cross section.

2) That cross section should be of some configuration that can be easily measured

so that a cross sectional area can be calculated.

3) The surfaces of the reduced section should be free of surface irregularities, especially if they lie perpendicular to the longitudinal axis of the specimen and the applied stress.

For these reasons, as well as considering the actual mechanics of preparing a specimen, the two most common cross sectional configurations for tensile specimens are circular and rectangular. Both are readily prepared and measured.

If you are required to perform a tensile test, you may have to calculate the actual cross sectional area of the reduced section of the tensile specimen. Examples 1 and 2 below show how these calculations are done for both common cross sections.

Example 1: Area of a Circular Cross Section
Diameter (d) = 0.505 in
Radius (r) = d/2 = 0.2525 in.
Area (A) = πr^2 or $\pi d^2 \div 4$
A = 3.1416 (0.2525)^2
A = 0.2
or
A = 3.1416(0.505$^2 \div 4$)
A = 0.2 in^2

Example 2: Area of Rectangular Cross Section
Width (w) = 1.5 in
Thickness (t) = 0.5 in
Area (A) = w x t
Area (A) = 1.5 x 0.5
Area (A) = 0.75 in^2

The determination of this cross-sectional area before testing is critical because that value will be used to determine the strength of the metal. This strength will be calculated by dividing the applied load by the original cross sectional area. Example 3 shows this calculation for a standard circular cross section specimen.

Example 3: Calculation of Tensile Strength
Load = 12,500 lb
Area = 0.2 in^2 (See Example 1)
Tensile Strength = Load/Area
Tensile Strength = 12,500/0.2
Tensile Strength = 62,500 psi (lb/in^2)

The example above shows a typical tensile strength calculation for a standard circular specimen. This is a standard specimen because it yields an area of 0.2 in. This is convenient since dividing some number by 0.2 is the same as multiplying that number by 5. Therefore, if this standard tensile specimen is used, the calculation for tensile strength can be performed as shown in Example 4.

Example 4: Alternate Tensile Strength Calculation
Load = 12,500 pounds
Area = 0.2 in^2
Tensile Strength = 12,500 x 5
Tensile Strength = 62,500 psi (lb/in^2)

As can be seen, the result of this calculation is identical to that of Example 3. The use of a standard size tensile specimen was very, popular years ago before the arrival of modern calculators. Then, it appeared easier to machine a tensile specimen to exact size, than it was to determine arithmetically the strength by dividing the load by some complicated number. However, today we can easily calculate the exact tensile strength no matter what the actual area happens to be.

Another operation that must be performed before testing is marking a gage length on the reduced section. This gage length is normally marked using a pair of center punch marks spaced at some prescribed distance apart. The most common gage lengths are 2 and 8 inches. After testing, the new distance between these marks is compared to the original distance to enable us to determine the amount of elongation, or stretch, exhibited by that specimen when stressed.

Once properly measured and marked, the specimen is then placed securely in the appropriate

COUPON TEST REPORT

Date _____ Test No. _____
Location _____
State _____ Weld Position: Roll ☐   Fixed ☐
Welder _____ Mark _____
Welding time _____ Time of day _____
Mean temperature _____ Wind break used _____
Weather conditions _____
Voltage _____ Amperage _____
Welding machine type _____ Welding machine size _____
Filler metal _____
Reinforcement size _____
Pipe type and grade _____
Wall thickness _____ Outside diameter _____

	1	2	3	4	5	6	7
Coupon stenciled							
Original specimen dimensions							
Original specimen area							
Maximum load							
Tensile strength per square inch of plate area							
Fracture location							

☐ Procedure   ☐ Qualifying test   ☐ Qualified
☐ Welder   ☐ Line test   ☐ Disqualified

Maximum tensile _____   Minimum tensile _____   Average tensile _____

Remarks on tensile-strength tests _____
1. _____
2. _____
3. _____
4. _____
Remarks on bend tests
1. _____
2. _____
3. _____
4. _____
Remarks on nick-break tests
1. _____
2. _____
3. _____
4. _____

Test made at _____   Date _____
Tested by _____   Supervised by _____

Notes: Use back for additional remarks. This form can be used to report either a procedure qualification test or a welder qualification test.

Sample Coupon Test Report Form (Source API 1104, 17th edition)

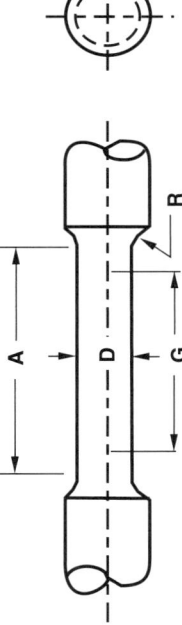

Dimensions

	Standard Specimen		Small-size specimens proportional to standard							
Nominal Diameter	in. 0.500	mm* 12.5	in. 0.350	mm* 8.90	in. 0.250	mm* 6.35	in. 0.160	mm* 4.05	in. 0.113	mm* 2.85
G - gage length	2.000 ± 0.005	51.0 ± 0.125	1.400 ± 0.005	35.5 ± 0.125	1.000 ± 0.005	25.5 ± 0.125	0.640 ± 0.005	16.0 ± 0.125	0.450 ± 0.005	11.5 ± 0.125
D - diameter (Note 1)	0.500 ± 0.010	12.5 ± 0.255	0.350 ± 0.007	8.90 ± 0.180	0.250 ± 0.005	6.35 ± 0.125	0.160 ± 0.003	4.05 ± 0.075	0.113 ± 0.002	2.85 ± 0.05
R - radius of fillet min.	3/8	9.5	1/4	6.5	3/16	5.0	5/32	4.0	3/32	2.5
A - length of reduced section min. (Note 2)	2-1/4 min.	57.0 min	1-3/4 min.	44.5 min	1-1/4	32.0	3/4	19.0	5/8	16.0

Standard 0.500 in. round tension test specimens with 2 in. gage length and examples of small size specimens with 2 in. gage length and examples of small size specimens proportional to the standard specimen.

*Rounded to the nearest 0.5 mm or 0.05 mm

Notes:
1. The reduced section may have a gradual taper from the ends toward the center with the ends not more than 1 percent larger in diameter that the center (controlling dimension).
2. If desired, the length of the reduced section may be increased to accommodate an extensometer of any convenient gage length. Reference marks for the measurement of of elongation should nevertheless be spaced at the indicated gage length.

Round Tensile Specimens - Source: ANSI/AWS B4.0

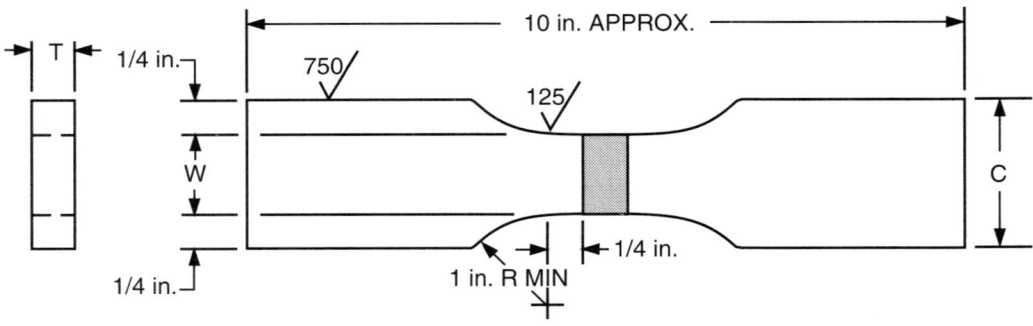

T	W
≤ 1 in	1.50 = 0.01 in.
≥ 1 in	1.00 = 0.01 in.

$\overset{125}{\vee}$ = 4 MICROMETERS

$\overset{750}{\vee}$ = 24 MICROMETERS

in.	mm
0.01	0.3
1/4	6.5
1	25.5
1-1/2	38.0
2	51.0
10	254.0

Notes:
1. Thin sheet metal being tested tends to tear and break near the shoulder. In such cases, dimension C shall be no greater than 1-1/3 times W.
2. Weld reinforcement and backing strip, if any, shall be removed flush with the surface of the specimen.
3. When the thickness, t, of the test weldment is such that it would not provide a specimen within the capacity limitations of the available test equipment, the specimen shall be parted through its thickness into as many specimens as required.
4. The length of reduced sections shall be equal to the width of the widest portion of weld, plus 1/4 in. (6.5 mm) minimum on each side.

SOURCE: ANSI/AWS B4.0-92

TRANSVERSE RECTANGULAR TENSION TEST SPECIMEN (PLATE)

grips of the stationary and moving heads of the tensile machine.

Once in place, the tensile load is then applied at some steady rate. Differences in this rate of loading could result in testing inconsistencies. Before this load application, a device known as an extensometer is connected to the specimen at the gage length marks. During the application of the load, the extensometer will measure the amount of elongation that results from a certain load. Load and elongation data are fed into a strip chart recorder resulting in a plot of the variation in the elongation as a function of the applied load. This is referred to as a load versus deflection curve. However, we normally see tensile test results expressed in terms of stress and strain.

Stress is equivalent to strength, since it is the applied load at any time divided by the cross sectional area. The strain is simply the amount of stretch apparent in a given length. Stress is expressed in terms of psi (lb/in^2) while strain is expressed as in/in.

When these values are plotted from the test data, the result is referred to as a stress-strain diagram. Looking at this stress-strain diagram, there are several important features that should be discussed. The test begins with the stress and strain both equal to zero. As the load is applied, the amount of strain increases linearly with stress. This shows what we refer to as elastic behavior, where the stress and strain are proportional. For any given material, the slope of this line is some known value. This slope is referred to as the modulus of elasticity.

For steel, the modulus of elasticity (or Young's modulus) is approximately equal to

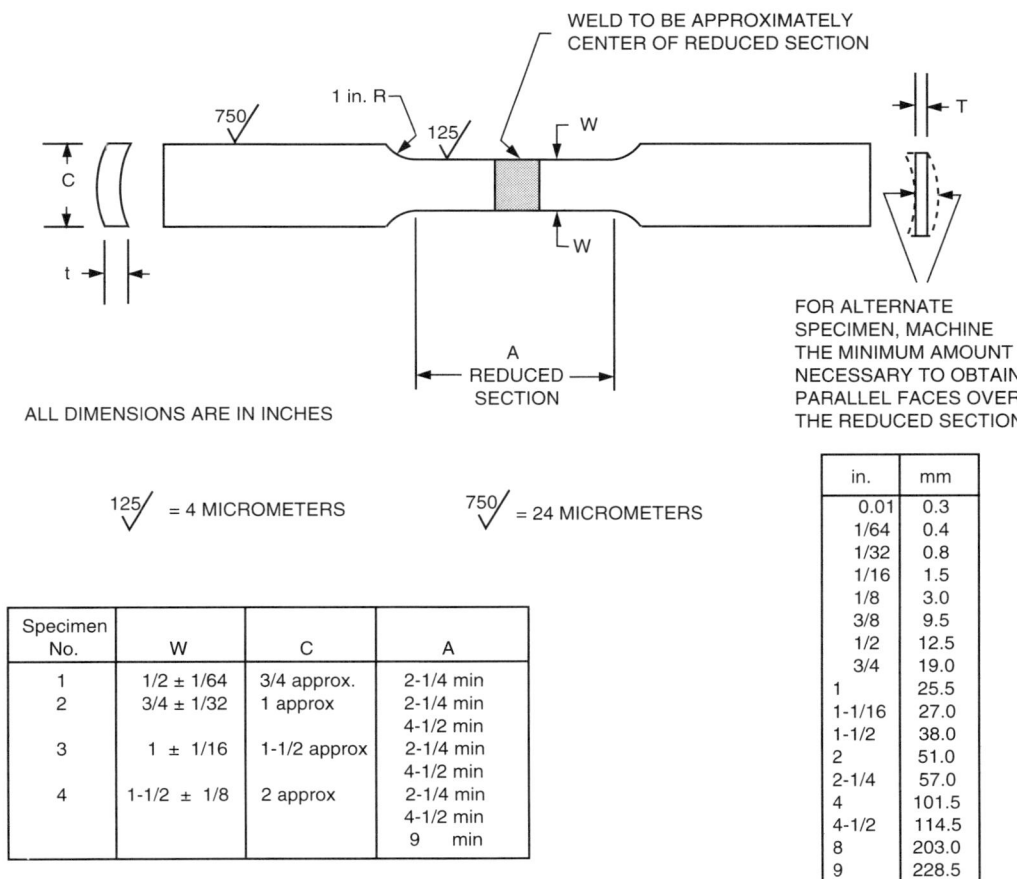

SOURCE: ANSI/AWS B4.0-92

TENSION SPECIMEN FOR PIPE SIZE GREATER THAN 2 INCHES NOMINAL

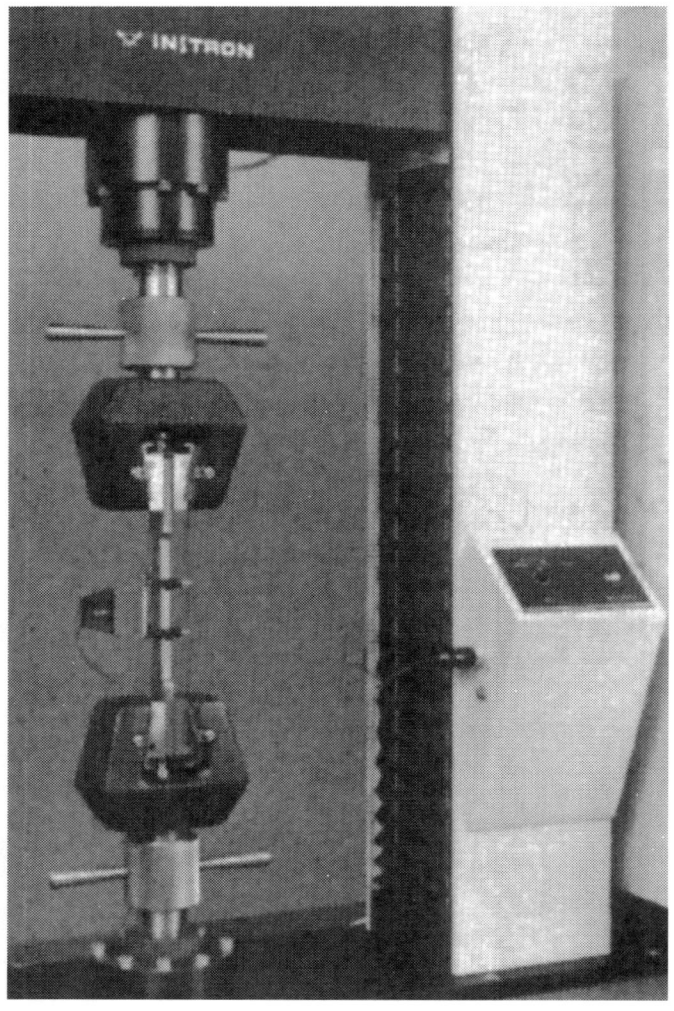

TENSILE MACHINE

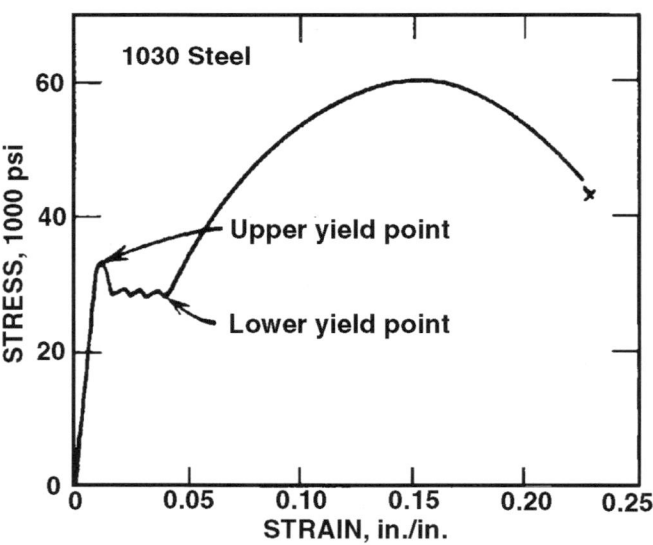

Abrupt Yielding or Yield Point in a mild steel

30,000,000 psi as compared to 10,500,000 psi for aluminum. What this number indicates is the stiffness of the metal; that is, the higher the modulus of elasticity, the stiffer the metal.

Eventually, the strain will begin to increase faster than the stress, meaning that the metal is stretching more for a given amount of applied stress. This change marks the end of elastic behavior and the onset of plastic or permanent deformation. The point on the curve showing the extent of the linear behavior is referred to as the elastic, or proportional, limit. If the load were removed at any time up to this point, the specimen would return to its original length.

Many metals tend to exhibit a drastic departure from the initial elastic behavior. With these metals, not only are stress and strain no longer proportional, but the stress may drop or remain steady while the strain increases. This phenomenon is characteristic of yielding in ductile steel. The stress increases to some maximum limit and then drops to some lower limit.

These limits are referred to as the upper and lower yield points. The upper yield point is that stress at which there is a noticeable increase in strain, or plastic flow, without an increase in stress. The stress then drops and remains relatively constant at the lower yield point while the strain continues to increase during what is known as the yield point elongation.

In a metal that exhibits this behavior, we describe the yield strength as the stress corresponding to the upper yield point or some point midway between the upper and lower yield points. During the tensile test, the yield point can be seen as a drop in the gage or recording device. We therefore can determine the yield strength simply by observing and noting this load reduction. When this method is utilized, we refer to it as the "drop-of-beam" technique.

During this yielding phenomenon, the plastic flow of the metal is increasing at such a rate that stresses are being relieved faster than they are formed. When this plastic flow occurs at room temperature, we refer to it as cold working. This action causes the metal to become stronger and harder and it is said to be work hardened. The yielding will therefore continue until the metal becomes work hardened to the extent that it now requires additional stress to produce any further elongation. Corresponding to this, the curve begins to climb in some nonlinear fashion.

The stress and strain continue to increase at varying rates until some maximum stress is reached. We refer to this point as the maximum stress, or ultimate tensile strength. It can be noted that this maximum stress is reached, followed by an apparent decrease in stress even though the strain continues to increase. This phenomenon is due to the fact that the specimen is "necking down" so that the actual cross section resisting the applied stress is less than the original area. Since the stress is calculated based on the original area this gives the appearance that the load is dropping when actually it continues to increase.

For less ductile metals, there may not be a pronounced change in behavior between elastic and plastic deformation. Therefore, we cannot use the drop-of-beam method to determine the yield strength. An alternative method is referred to as the offset technique.

When the offset method is used, a line is drawn parallel to the modulus of elasticity at some prescribed amount of strain. The amount of strain is usually described in terms of some percentage. The most common offset is 0.02%, of the strain; however, other amounts may also be specified. The stress corresponding to the intersection of this offset line with the stress-strain curve is therefore the yield stress. It should be reported as a 0.02% offset yield stress so that others know how it was determined.

Following the actual testing, it is now necessary to make a determination of the metal's ductility. This is done in one of two ways, both of which involve making measurements both before and after testing. We remember that the two ways

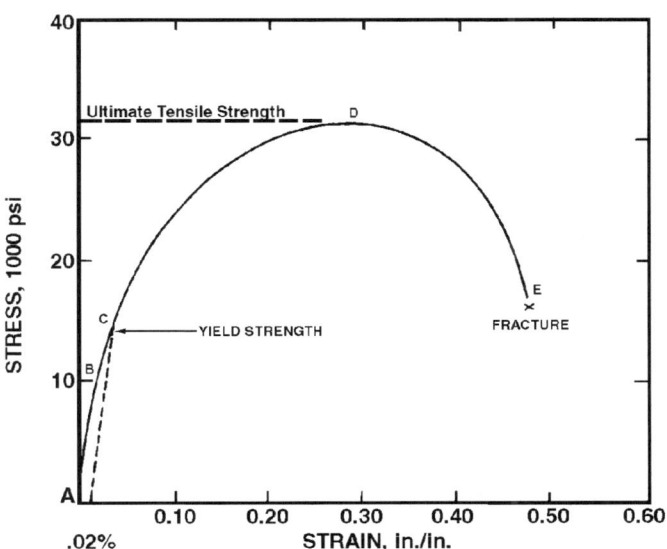

Engineering Stress-Strain Diagram for polycrystalline copper

in which we express ductility are in terms of percent elongation and percent reduction of area.

To determine the percent elongation, it is necessary to have placed gage marks on the specimen before pulling. After the specimen has failed, the two pieces are placed together and the new distance between the gage marks is measured.

Knowing this information, plus the original gage length, it is possible to calculate the percent elongation as shown in Example 5.

Example 5: Determination of Percent Elongation
Original length = 2.0 in
Final length = 2.6 in
%Elongation = (Final length - Original length)/Original length x 100
%Elongation = (2.6 - 2.0)/2.0 x 100
%Elongation = 0.6/2.0 x 100
%Elongation = 0.3 x 100
%Elongation = 30%

The metal's ductility can also be expressed in terms of how much it necks down during the tensile test. This is referred to as percent reduction of area, where the original and final areas of the tensile specimen are measured and compared. It can be seen in Example 6 how this calculation is performed.

Example 6: Determination of Percent Reduction of Area:
Original Area = 0.2 in^2
Final Area = 0.1 in^2
%Reduction of Area = (Original Area - Final Area)/Original Area x 100
%Reduction of Area = (0.2 - 0.1)/0.2 x 100
%Reduction of Area = 0.1/0.2 x 100
%Reduction of Area = 0.5 x 100
%Reduction of Area = 50%

While both percent elongation and percent reduction of area represent expressions for the amount of ductility exhibited by a tensile specimen, their values will seldom, if ever, be equal. Typically, the value for percent reduction of area may be as much as twice the value for percent elongation.

COMMERCIALLY USED HARDNESS TESTS

Test	Indenter	Shape of Indentation
Brinell	10mm sphere of steel or tungsten carbide	
Vickers	Diamond pyramid	
Knoop microhardness	Diamond pyramid	
Rockwell A, C, D	Diamond cone	
Rockwell B, F, G	1/16 in. diameter steel sphere	
Rockwell E	1/8 in. diameter steel sphere	

Shapes and types of indenters used with various hardness tests.

Hardness Testing

Strength is described as the ability of a material to transmit some load. It can be determined directly by performing a tensile test, or it can be approximated by conducting a hardness test on the material. This is permissible because the hardness and strength of many metals are directly related. Hardness testing is most often employed to provide us with a measurement of the metal's hardness, which is described as the ability of a metal to resist penetration.

Consequently, hardness tests are performed, for the most part, using some type of penetrator that is forced against the surface of the test object. Depending upon the type of hardness test being used, either the diameter or depth of the resulting indentation is measured. The following discussion will point out some of the important features and methods involved in the various hardness tests.

Hardness is one of the most measurable features of a metal. This is primarily due to the vast variety of methods that can be used to determine a metal's hardness. There will be a discussion of three of the more common types of hardness tests: Brinell, Rockwell and microhardness. In general, the three groups differ from one another in the size of indentation which is produced, with the Brinell being largest and microhardness the smallest.

The Brinell test is commonly used for determining the hardness of metal stock. It is well suited for this purpose because the indentation covers a relatively large area, eliminating problems associated with localized hard or soft spots in the metal. The characteristically higher loads used for Brinell tests also assist in reducing errors produced by surface irregularities.

Before Brinell testing, it is necessary to prepare the surface so that test error will be minimized. This includes grinding or sanding the surface to remove scale, rust, paint, etc., and to achieve a relatively flat test area. The surface should also be of sufficient smoothness so that

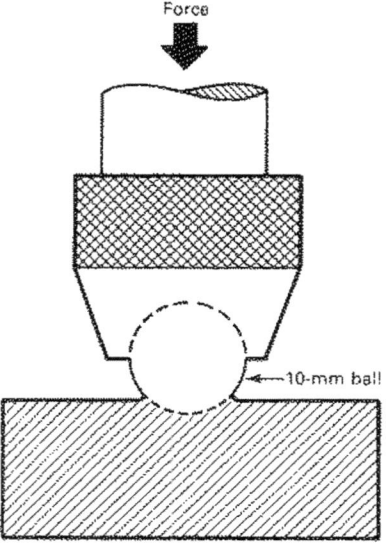

(a) Schematic of the principle of the Brinell indentation process.

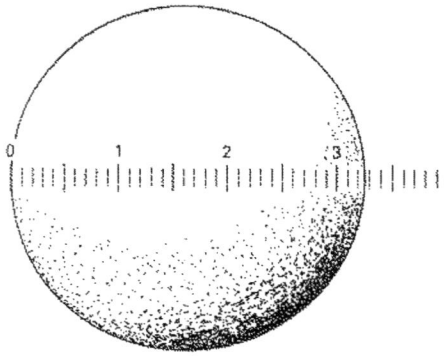

(b) Brinell indentation with measuring scale in millimeters.

Indenter Measurements

the size of the indentation can be accurately measured. To perform a Brinell test, a penetrator is forced into the surface of the test object at some prescribed load. Once this load is removed, the diameter of the indentation is then measured using some graduated magnifier. On the basis of the size and type of the indenter, the applied load, and the resulting diameter of the impression, a Brinell Hardness Number (BHN) can be determined. Since this is a mathematical relationship, a BHN can be determined for a variety of indenter types and loads. As mentioned above, strength and hardness are directly related. For steels, the approximate tensile strength can be determined by multiplying the BHN times 500.

The most commonly used Brinell test uses a 10 mm hardened steel ball and a 3000 kg load. However, some test conditions, such as specimen hardness and thickness, will require variations in the type of ball and the amount of the applied load. Other types of balls that can be used include the 5 mm hardened steel ball and the 10 mm tungsten carbide ball. For soft metals, loads as low as 500 kg may be used. Other loads between 500 and 3000 kg can also be used with equivalent results.

Normally the BHN can be determined by simply measuring the diameter of the impression and reading the value from some table that has results already calculated.

For additional information regarding Brinell testing, please refer to ASTM E10, *Standard Test Method for Brinell Hardness of Metallic Materials*.

Quite often, there is a need for the testing of objects too large to be placed in a stationary Brinell test machine. In such cases, portable test machines can be employed. These come in a variety of types and configurations, but the basic test principles are similar to field testing procedures.

The next hardness test to be discussed will be the Rockwell hardness test. This is a group of tests encompassing numerous different variations of the same basic principle. Like the Brinell test, the basic test can be modified by using different indenters as well as different test loads. As mentioned, the Rockwell tests result in smaller indentations than would be expected for Brinell testing. This allows for the localized testing of relatively small areas of a metal.

Like the Brinell test, Rockwell testing uses different indenters for different hardness ranges. The indenters used are the diamond Brale, 1/16", 1/8", 1/4", and 1/2" diameter hardened steel balls.

BRINELL HARDNESS NUMBERS
Ball diameter 10 mm

Ball impression, diam, mm	Brinell hardness number — Load, kgf						Ball impression, diam, mm	Brinell hardness number — Load, kgf					
	500	1000	1500	2000	2500	3000		500	1000	1500	2000	2500	3000
2.00	158	316	473	632	788	945	4.25	33.6	67.2	101	134	167	201
2.05	150	300	450	600	750	899	4.30	32.8	65.6	98.3	131	164	197
2.10	143	286	428	572	714	856	4.35	32.0	64.0	95.9	128	160	192
2.15	136	272	408	544	681	817	4.40	31.2	62.4	93.6	125	156	187
2.20	130	260	390	520	650	780	4.45	30.5	61.0	91.4	122	153	183
2.25	124	248	372	496	621	745	4.50	29.8	59.6	89.3	119	149	179
2.30	119	238	356	476	593	712	4.55	29.1	58.2	87.2	116	145	174
2.35	114	228	341	456	568	682	4.60	28.4	56.8	85.2	114	142	170
2.40	109	218	327	436	545	653	4.65	27.8	55.6	83.3	111	139	167
2.45	104	208	313	416	522	627	4.70	27.1	54.2	81.4	108	136	163
2.50	100	200	301	400	500	601	4.75	26.5	53.0	79.6	106	133	159
2.55	96.3	193	289	385	482	578	4.80	25.9	51.8	77.8	104	130	156
2.60	92.6	185	278	370	462	555	4.85	25.4	50.8	76.1	102	127	152
2.65	89.0	178	267	356	445	534	4.90	24.8	49.6	74.4	99.2	124	149
2.70	85.7	171	257	343	429	514	4.95	24.3	48.6	72.8	97.2	122	146
2.75	82.6	165	248	330	413	495	5.00	23.8	47.6	71.3	95.2	119	143
2.80	79.6	159	239	318	398	477	5.05	23.3	46.6	69.8	93.2	117	140
2.85	76.8	154	230	307	384	461	5.10	22.8	45.6	68.3	91.2	114	137
2.90	74.1	148	222	296	371	444	5.15	22.3	44.6	66.9	89.2	112	134
2.95	71.5	143	215	286	358	429	5.20	21.8	43.6	65.5	87.2	109	131
3.00	69.1	138	207	276	346	415	5.25	21.4	42.8	64.1	85.6	107	128
3.05	66.8	134	200	267	334	401	5.30	20.9	41.8	62.8	83.6	105	126
3.10	64.6	129	194	258	324	388	5.35	20.5	41.0	61.5	82.0	103	123
3.15	62.5	125	188	250	313	375	5.40	20.1	40.2	60.3	80.4	101	121
3.20	60.5	121	182	242	303	363	5.45	19.7	39.4	59.1	78.8	98.5	118
3.25	58.6	117	176	234	293	352	5.50	19.3	38.6	57.9	77.2	96.5	116
3.30	56.8	114	170	227	284	341	5.55	18.9	37.8	56.8	75.6	95.0	114
3.35	55.1	110	165	220	276	331	5.60	18.6	37.2	55.7	74.4	92.5	111
3.40	53.4	107	160	214	267	321	5.65	18.2	36.4	54.6	72.8	90.8	109
3.45	51.8	104	156	207	259	311	5.70	17.8	35.6	53.5	71.2	89.2	107
3.50	50.3	101	151	201	252	302	5.75	17.5	35.0	52.5	70.0	87.5	105
3.55	48.9	97.8	147	196	244	293	5.80	17.2	34.4	51.5	68.8	85.8	103
3.60	47.5	95.0	142	190	238	285	5.85	16.8	33.6	50.5	67.2	84.2	101
3.65	46.1	92.2	138	184	231	277	5.90	16.5	33.0	49.6	66.0	82.5	99.2
3.70	44.9	89.8	135	180	225	269	5.95	16.2	32.4	48.7	64.8	81.2	97.3
3.75	43.6	87.2	131	174	218	262	6.00	15.9	31.8	47.7	63.6	79.5	95.5
3.80	42.4	84.8	127	170	212	255	6.05	15.6	31.2	46.8	62.4	78.0	93.7
3.85	41.3	82.6	124	165	207	248	6.10	15.3	30.6	46.0	61.2	76.7	92.0
3.90	40.2	80.4	121	161	201	241	6.15	15.1	30.2	45.2	60.4	75.3	90.3
3.95	39.1	78.2	117	156	196	235	6.20	14.8	29.6	44.3	59.2	73.8	88.7
4.00	38.1	76.2	114	152	191	229	6.25	14.5	29.0	43.5	58.0	72.6	87.1
4.05	37.1	74.2	111	148	186	223	6.30	14.2	28.4	42.7	56.8	71.3	85.5
4.10	36.2	72.4	109	145	181	217	6.35	14.0	28.0	42.0	56.0	70.0	84.0
4.15	35.3	70.6	106	141	177	212	6.40	13.7	27.4	41.2	54.8	68.8	82.5
4.20	34.4	68.8	103	138	172	207	6.45	13.5	27.0	40.5	54.0	67.5	81.0

Using one or the other of these indenters, various loads can also be used. These loads are much lower than those used for Brinell testing, ranging from 60 to 150 kg.

Just as with the Brinell test, the test surface must be properly prepared before applying a Rockwell test. Once prepared, the proper scale should be selected based on the approximate range of hardness expected. The "B" and "C" scales are by far the most commonly used scales for steel, with the "B" scale chosen for softer alloys and the "C" scale used for the harder types. When in doubt as to which scale might be chosen for some unknown alloy, the "A" scale could be employed because it includes a range of hardness covering both the "B" and "C" scales. Once the proper scale has been selected, the test object is placed on the anvil. The anvil can be of various shapes depending upon the configuration of the test piece. The object must be adequately and fully supported. Otherwise, test errors will result. The Rockwell method relies on the accurate depth of penetration measurement of the indenter. So, if the test object is not properly supported, this measurement could be inaccurate. A variation in this depth measurement of only 0.00008 inch will result in a change of one Rockwell number. After the specimen has been prepared and placed in the Rockwell machine, the load is applied and the results read directly from the dial on the machine. For further information regarding the Rockwell tests, refer to ASTM E18, *Standard Test Methods for Rockwell Hardness and Rockwell Superficial Hardness of Metallic Materials.*

Like Brinell testing, there are also portable devices that can be used to determine the Rockwell hardness of a metal. Although their operation may vary slightly from that of the stationary models, the results will be equivalent.

The next group of hardness tests to be discussed are those which are referred to as microhardness tests. They bear this name because their impressions are so small that high magnification is required to facilitate their measurement. Microhardness testing is very beneficial in the investigation of metal microstructures, because these hardness tests can

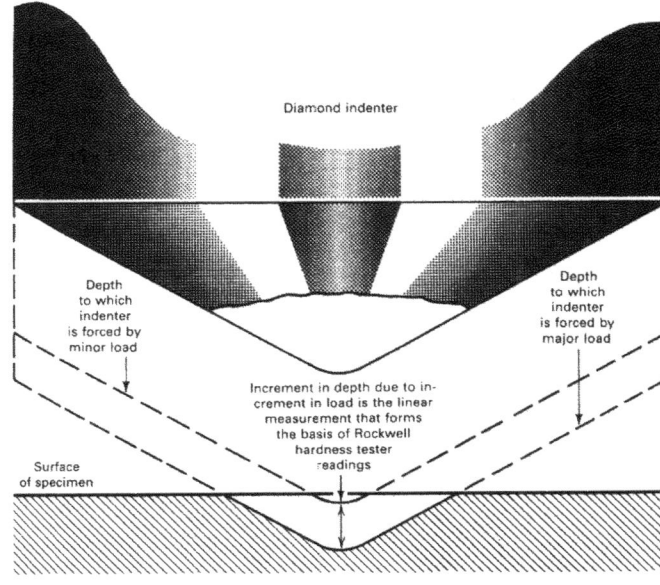

PRINCIPLE OF THE ROCKWELL TEST
Although a diamond indenter is illustrated, the same principle applies for steel ball indenters and other loads.

be performed on single grains of a metal to determine the hardness in that microscopic region. Therefore, the metallurgist is primarily interested in this type of hardness testing.

There are two major types of microhardness tests: Vickers and Knoop. Both use diamond indenters, but their configurations are slightly different. The square-based Vickers indenter provides an indentation in which the two diagonals are approximately equal. The Knoop indenter, however, makes an indentation having a long and a short dimension.

As with the other test methods, the tester has a selection of test loads as well as indenter types. The term microhardness implies that the applied loads will range from 1 to 1000 g. However, the majority of microhardness tests use loads in the range of 100 to 500 g.

ROCKWELL STANDARD HARDNESS

Scale symbol	Indenter	Major load, kgf	Typical applications
A	Diamond (two scales—carbide and steel)	60	Cemented carbides, thin steel, and shallow case-hardened steel
B	1/16-in. (1.588-mm) ball	100	Copper alloys, soft steels, aluminum alloys, malleable iron
C	Diamond	150	Steel, hard cast irons, pearlitic malleable iron, titanium, deep case-hardened steel, and other materials harder than HRB 100
D	Diamond	100	Thin steel and medium case-hardened steel and pearlitic malleable iron
E	1/8-in. (3.175-mm) ball	100	Cast iron, aluminum and magnesium alloys, bearing metals
F	1/16-in. (1.588-mm) ball	60	Annealed copper alloys, thin soft sheet metals
G	1/16-in. (1.588-mm) ball	150	Phosphor bronze, beryllium copper, malleable irons. Upper limit HRG 92 to avoid possible flattening of ball
H	1/8-in. (3.175-mm) ball	60	Aluminum, zinc, lead
K	1/8-in. (3.175-mm) ball	150	Bearing metals and other very soft or thin materials. Use smallest ball and heaviest load that do not produce anvil effect.
L	1/4-in. (6.350-mm) ball	60	Bearing metals and other very soft or thin materials. Use smallest ball and heaviest load that do not produce anvil effect.
M	1/4-in. (6.350-mm) ball	100	Bearing metals and other very soft or thin materials. Use smallest ball and heaviest load that do not produce anvil effect.
P	1/4-in. (6.350-mm) ball	150	Bearing metals and other very soft or thin materials. Use smallest ball and heaviest load that do not produce anvil effect.
R	1/2-in. (12.70-mm) ball	60	Bearing metals and other very soft or thin materials. Use smallest ball and heaviest load that do not produce anvil effect.
S	1/2-in. (12.70-mm) ball	100	Bearing metals and other very soft or thin materials. Use smallest ball and heaviest load that do not produce anvil effect.
V	1/2-in. (12.70-mm) ball	150	Bearing metals and other very soft or thin materials. Use smallest ball and heaviest load that do not produce anvil effect.

Fig. 1 Indentations made by Knoop and Vickers indenters in the same work metal at the same load

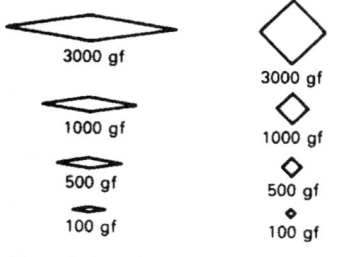

Fig. 2 Pyramidal Knoop indenter and resulting indentation in the workpiece

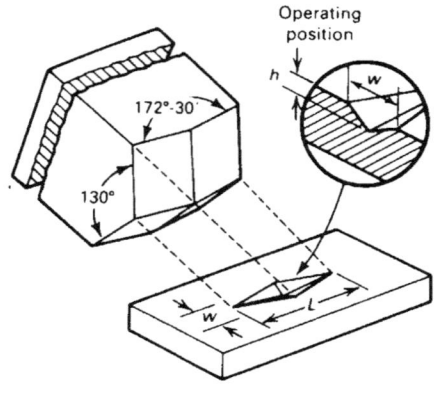

Fig. 3 Diamond pyramid indenter used for the Vickers test and resulting indentation in the workpiece

D is the mean diagonal of the indentation in millimeters.

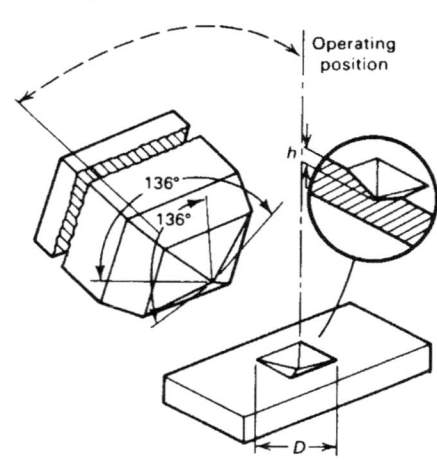

To perform either Vickers or Knoop microhardness testing, the preparation of the surface is of utmost importance. Since the area of interest is quite minute, even the smallest surface irregularity could cause inaccuracies. Normally, preparation of the surface for microhardness testing is identical to that for other metallographic investigations. The importance of this surface finish increases as the applied test load is reduced.

Once prepared, the specimen is securely clamped in some type of test fixture or holder so that the indentations can be accurately placed. Many microhardness machines employ some type of moving stage that facilitates accurate movement of the specimen without the need for its removal and adjustment. Such a device is handy when taking a number of readings across some region of the metal. An example of this type of application would be the determination of the hardness variation across the weld heat affected zone. The result would be referred to as a "microhardness traverse."

The use of hardness testing will provide us with a great deal of useful information about a metal. However, as can be seen from the preceding discussion, we must specify which hardness method will be used for a given application. If not, the desired results may not be obtained.

Toughness Testing

Another metal property of interest is toughness. Toughness is the property that describes the material's ability to absorb energy. Regarding this property, the most important information relates to how the material absorbs energy when a load is applied very rapidly, as in an impact, especially when there is a notch present on the surface. Both the rapid loading and the presence of a notch will cause the metal's behavior to change drastically compared to how it might behave under loading conditions similar to those present in a tensile test.

Therefore, when the load is applied rapidly, the concern is with notch toughness, or impact strength. The discussion that follows will concern itself with tests that can be used to determine this particular metal property. Therefore, the various tests used to determine the notch toughness of a metal will usually use a specimen containing some type of machined notch and the load will be applied in a very rapid manner. Another important factor in impact testing is test temperature. The temperature of the specimen has a drastic effect on

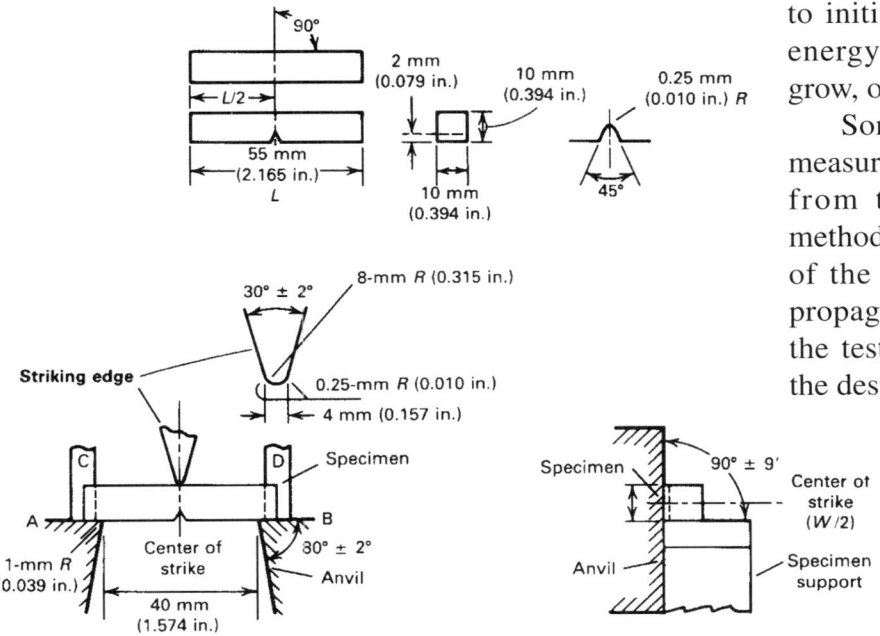

ASTM standard dimensions for the Type A Charpy V-Notch specimen and the striker-anvil arrangement.

to initiate a crack. Then, some additional energy is required to cause that crack to grow, or propagate.

Some of the notch toughness tests can measure the energy of propagation separate from the energy of initiation. Other methods simply provide us with a measure of the combined energy of initiation and propagation. The engineer should specify the test method that will provide him with the desired information.

Although numerous types of notch toughness tests exist, probably the most commonly used in this country is the Charpy V-Notch test. The standard specimen used for this test is a bar 55 mm long and 10 mm by 10 mm square. One of the long sides of the specimen has a carefully machined V shaped notch 2 mm deep. The base of this notch is radiused to precisely 0.25 mm. The machining of this radius is extremely critical, since tiny inconsistencies will result in drastic variations in test results.

Once the specimen has been carefully machined, it is then cooled to the prescribed test temperature, if it is some temperature below room temperature. This can be accomplished using either a liquid or gaseous medium. After the specimen is stabilized at the required temperature, it is then removed from the low temperature bath and quickly placed in the anvil of the testing machine.

The Charpy testing machine consists essentially of a pendulum with a striker head, an anvil, a release level, and a pointer and scale. Since the goal is to measure the amount of energy that is absorbed during the fracturing of a specimen, a given amount of energy is supplied by raising the pendulum to a specified height. Upon release, the pendulum will fall and continue through its stroke until it reaches a maximum height on the opposite side of its travel.

the test results. Consequently, impact testing must be performed at some prescribed temperature.

Since the advent of interest in the notch toughness of metals, numerous different tests have been developed to measure this important property. When talking about the energy absorption capabilities of a metal, it must be understood that the metal absorbs energy in steps. First, there is a certain amount of energy required

A. Pendulum
B. Release lever and brake lever
C. Pointer and scale to indicate energy absorbed
D. Drive arm which pushes pointer around the scale
E. Anvil on which Charpy specimen rests
F. Striker head

Schematic of Typical Charpy Testing Machine

If it meets no resistance, it will rise to some height that is designated as zero energy absorption. When it contacts the Charpy specimen, there is a certain amount of energy required to initiate and propagate a fracture. This causes the pendulum to rise to some level below that for zero energy absorption. The maximum height of this swing is indicated by the pointer on the scale. Since this scale is calibrated, the amount of energy required to break the specimen can be read directly from the scale.

This value, referred to as breaking energy, is the primary piece of information gained from the Charpy impact test. This energy is expressed in terms of foot-pounds of energy. While most Charpy results are expressed in terms of footpounds of energy absorption, there are other means of describing the notch toughness of a metal. They are determined by measuring various features of the failed Charpy specimen. These values are: lateral expansion and percent shear. Lateral expansion is a measure of the amount of lateral deformation produced during the fracturing of the specimen. It is measured in terms of mils, or thousandths of an inch. Percent shear is an expression for the amount of the fracture surface that failed in a ductile, or shearing fashion. Values change with temperature. No matter which of these methods of measurement are employed, the usual concern is with the results from a whole series of tests. Once a number of specimens have been tested at various temperatures, it is possible to determine how the values change with metal temperature. If these values are plotted versus temperature, the result will be a curve having upper and lower horizontal shelves. For each category, there is some temperature at which the values drop rather abruptly. These are referred to as transition temperatures, meaning, the behavior of the metal changes from ductile to brittle at that temperature. The designer then knows that the metal should behave satisfactorily above that temperature.

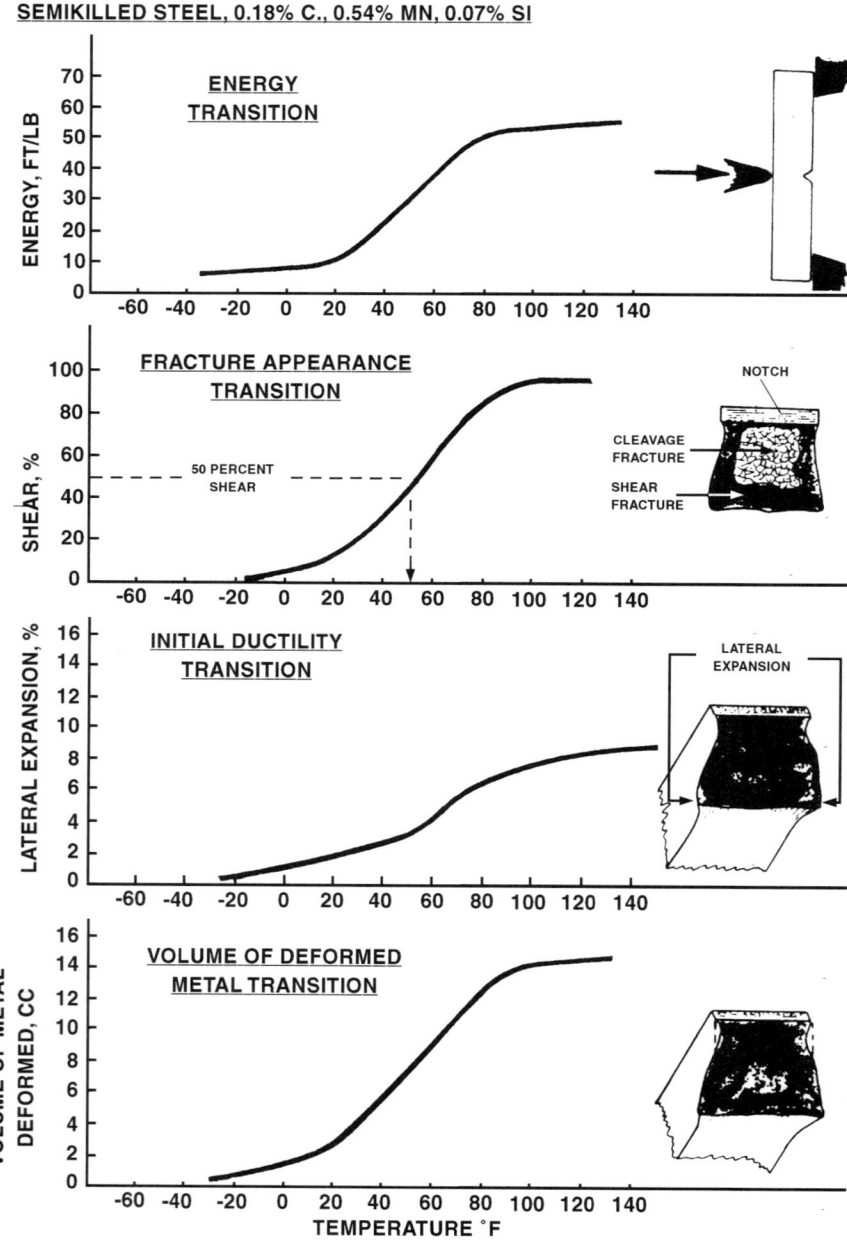

Relation between energy transition and fracture appearance transition in Charpy V-Notch impact specimens with changes in temperature

Besides the Charpy test there are others that can be employed for various applications. Other tests that may be used to measure a metal's notch toughness include: drop-weight nil-ductility, explosion bulge, dynamic tear, and crack tip opening displacement (CTOD). These tests employ different methods for applying the load to the specimen.

Soundness Testing

These groups of tests are designed to aid in the determination of the metal's soundness, or its freedom from imperfections. Soundness tests are routinely used for the qualification of welding procedures and welders. After a test plate has been welded, specimens are removed and then subjected to some soundness test to determine if the weld contained any imperfections or defects.

There are three general types of soundness tests: bend, nick-break and fillet break. Bend testing can be performed in a number of different ways. This is probably the most common test used to judge the adequacy of a welder's qualification test coupon.

There are several different types of bend tests, depending on the orientation of the weld with respect to the bending action. There are three types of transverse weld bend specimens: face, root and side bends. With these three types, the weld lies across the longitudinal axis of the specimen and its name refers to the side of the weld that is placed in tension during the test. That is, the face of the weld is stretched in a face bend, the root of the weld is stretched in a root bend, and the side of a cross section of the weld is stretched in a side bend.

Bend tests are normally performed using some type of bend jig. There are three basic types: guided-bend, roller-equipped guided-bend, and wraparound guided-bend. The standard guided bend test jig consists of a plunger (also referred to as mandrel or ram) and a matching die that forms the previously straight bend specimen into a U shape.

To perform a bend test, the specimen is placed across the shoulders of the die with the side to be placed in tension facing down toward the inside of the die. The plunger is then situated over the area of interest and forced toward the die causing the specimen to be bent 180° and become U-shaped. The specimen is then removed and evaluated.

The second type of guided-bend test jig is similar to the standard guided-bend jig, except it is equipped with rollers instead of hardened shoulders on the die portion. This reduces the friction against the specimen allowing for lower loads to achieve the bending.

The last common type of guided-bend jig is referred to as the wraparound jig. It bears this name because the specimen is bent by being wrapped around a stationary pin. This type is useful for bending specimens having different strengths of base and weld metal. If there is a great imbalance, there will be a tendency to bend preferentially in the weaker metal, resulting in a kink or a bend away from the area of intersection.

Most qualification tests for mild steel require that the specimen be bent around a mandrel having a diameter four times the thickness of the specimen. Therefore, a 3/8 inch specimen would be bent around a 1-1/2 inch diameter mandrel. This results in a 20% elongation of the outer surface of the bend specimen. If a smaller bend mandrel is used, the necessary amount of elongation would increase.

With any of these bend tests, the specimens must be carefully prepared to prevent test inaccuracies. Any grinding or sanding marks on the tension surface should be running parallel to the direction of bending so they don't provide stress raisers which could cause the specimen to fail prematurely. The corners of the specimen should be radiused to relieve that stress concentration as well. For specimens removed from pipe coupons, the side of the bend specimen against the ram may be required to be ground flat

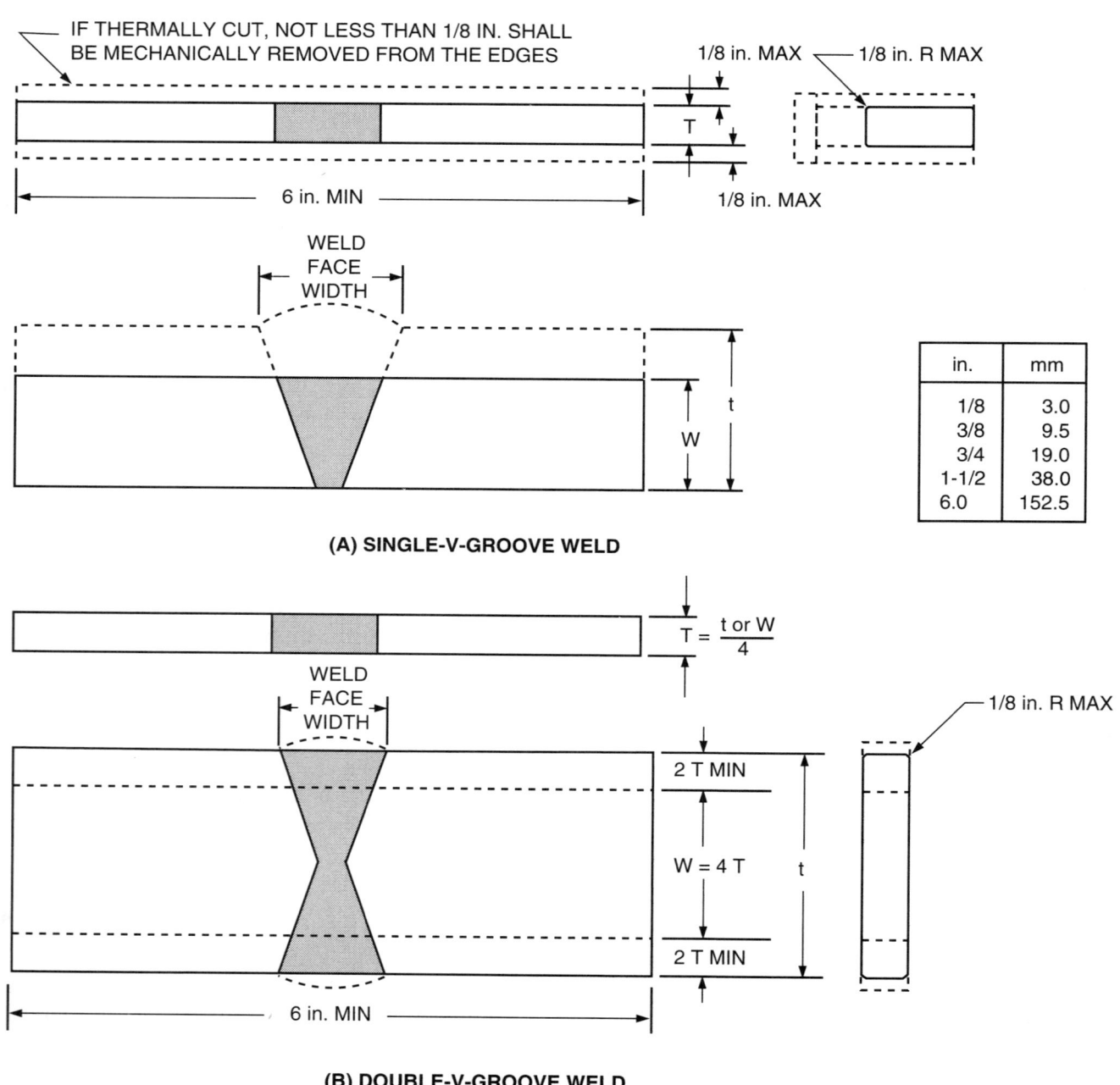

TRANSVERSE SIDE-BEND SPECIMENS (PLATE)

SOURCE: ANSI/AWS B4.0-92

7-19

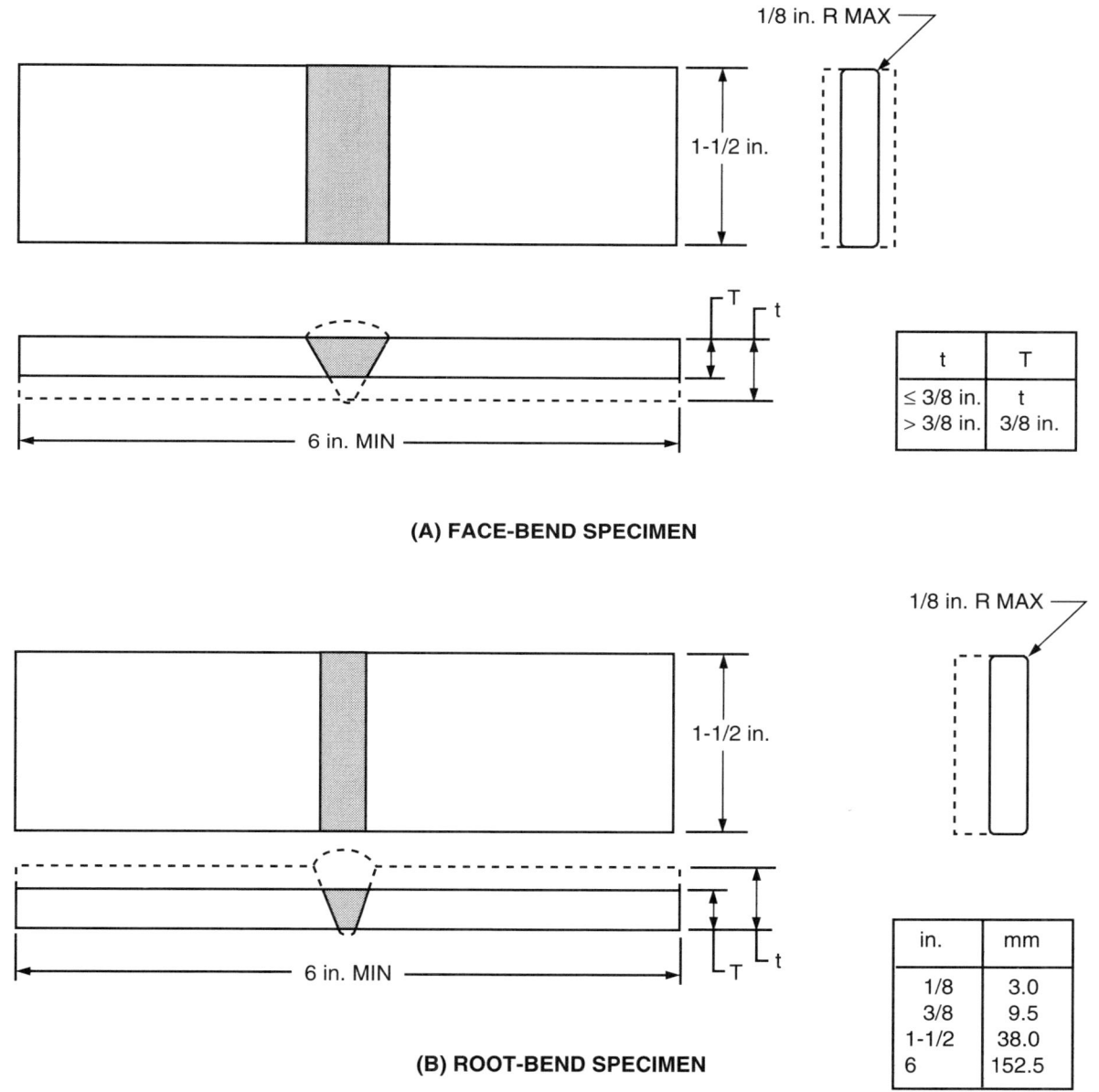

TRANSVERSE FACE-BEND AND ROOT-BEND SPECIMENS (PLATE)

SOURCE: ANSI/AWS B4.0-92

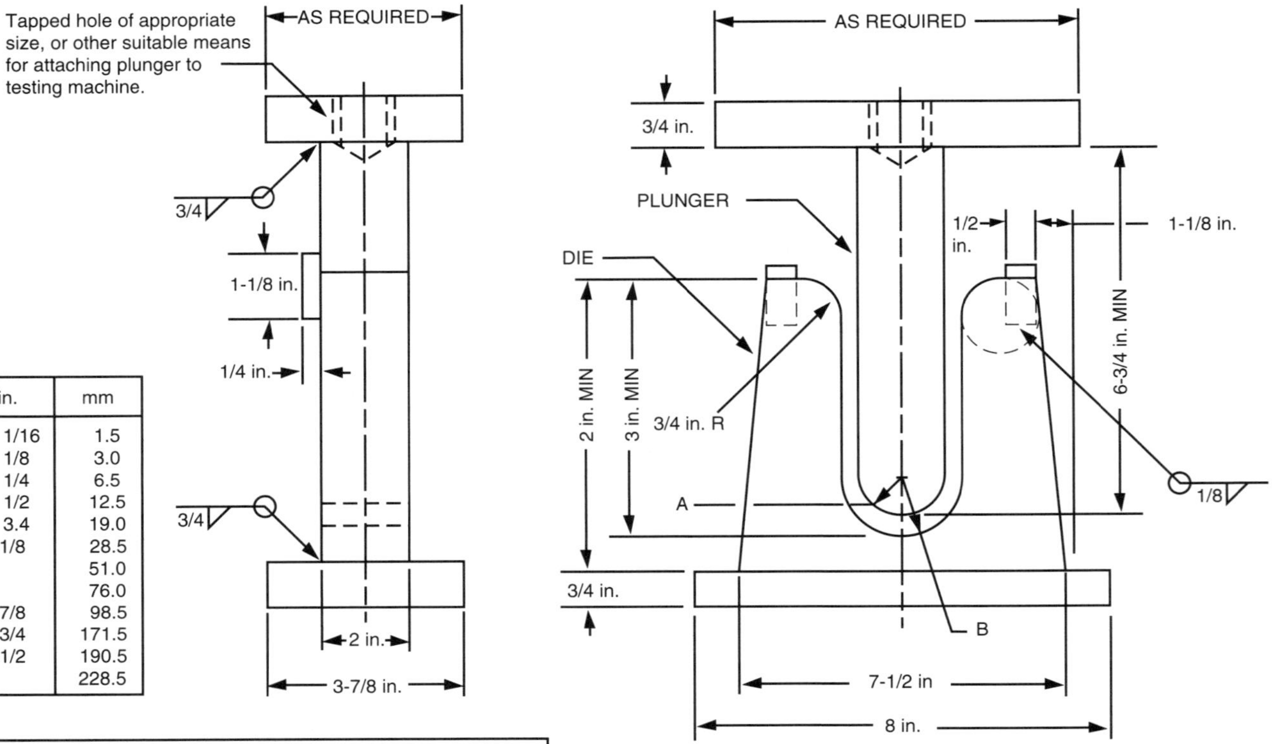

in.	mm
1/16	1.5
1/8	3.0
1/4	6.5
1/2	12.5
3.4	19.0
1-1/8	28.5
2	51.0
3	76.0
3-7/8	98.5
6-3/4	171.5
7-1/2	190.5
9	228.5

Jig Dimensions for 20% Elongation of Weld					
Specimen Thickness, T		Plunger Radius, A		Die Radius, B	
in.	mm	in.	mm	in.	mm
3/8 T	(9.5)	3/4 2T	(19.0)	1-3/16 A+T+1/16	(30.0) (A+T+1.5)

SOURCE: ANSI/AWS B4.0-92

TYPICAL BOTTOM GUIDED-BEND TEST FIXTURE

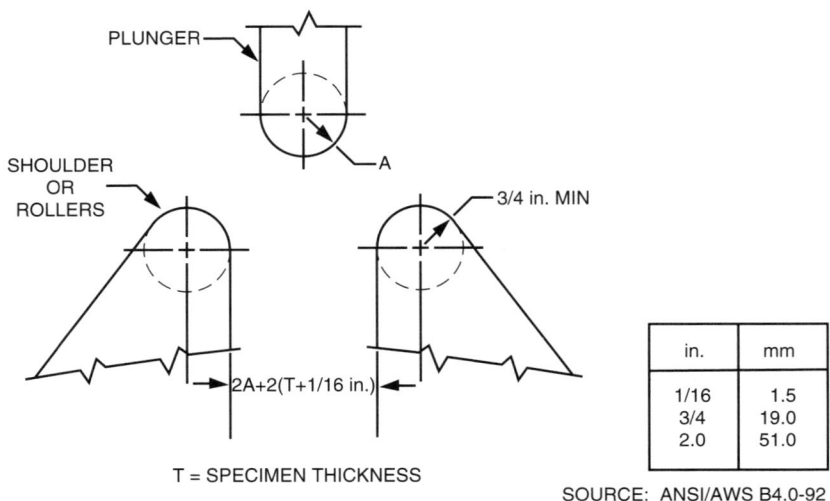

T = SPECIMEN THICKNESS

in.	mm
1/16	1.5
3/4	19.0
2.0	51.0

SOURCE: ANSI/AWS B4.0-92

TYPICAL BOTTOM EJECTING GUIDED-BEND TEST BEND FIXTURE

7-21

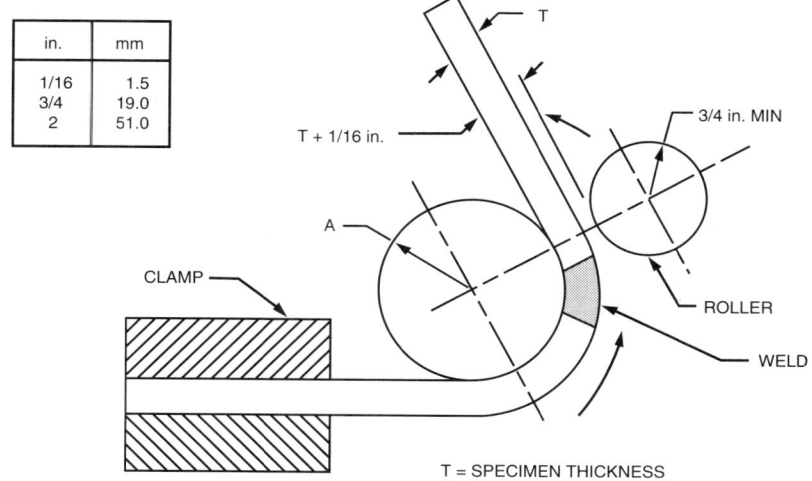

ALTERNATIVE WRAP-AROUND GUIDED BEND TEST BEND FIXTURE

SOURCE: ANSI/AWS B4.0-92

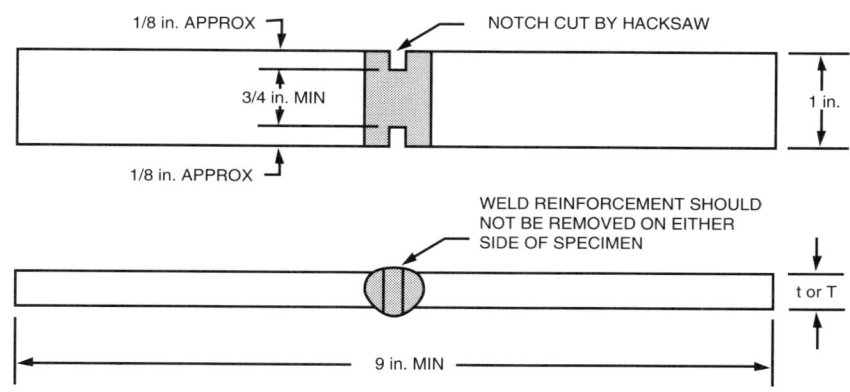

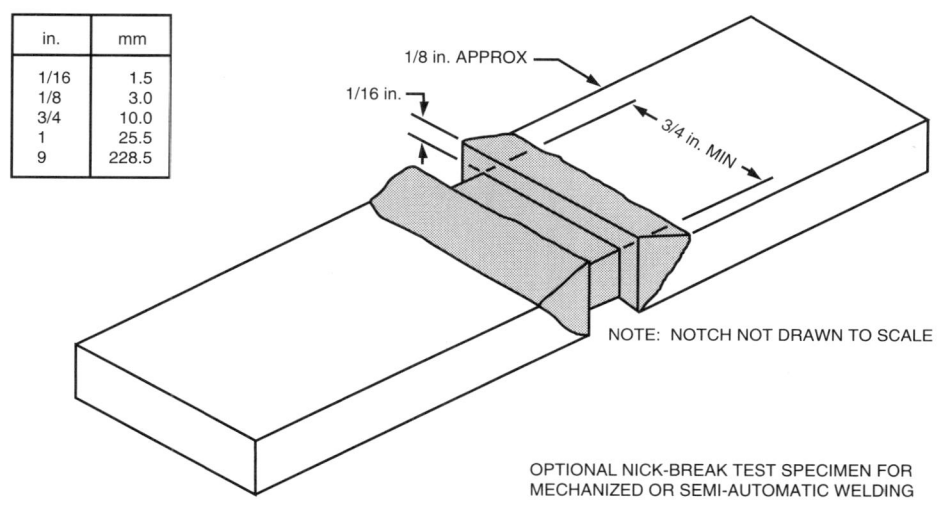

OPTIONAL NICK-BREAK TEST SPECIMEN FOR MECHANIZED OR SEMI-AUTOMATIC WELDING

SOURCE: ANSI/AWS B4.0-92

NICK BREAK TEST SPECIMEN

to eliminate the bending in the direction transverse to the bending direction.

The acceptability of bend test specimens is normally judged based on the size and/or number of defects that appear on the tension surface. The governing code or specification will dictate the exact acceptance/rejection criteria.

The next type of soundness test to be discussed is the nick-break test. This test is used almost exclusively by the pipeline industry as described in API 1104. This method judges the soundness of the weld by fracturing the specimen through the weld so that the fracture surface can be analyzed for the presence of discontinuities. The fracture is localized in the weld zone with saw cuts along two or three surfaces.

Once the specimen has been saw cut, it is then broken by pulling in a tensile machine, hitting the center with a hammer while supporting the ends or hitting one end with a hammer while the other end is held in a vise. The method of breaking is not significant because you are not interested in how much effort is required to fail the specimen. The sole purpose is to fail the specimen through the weld zone so it can be determined if any imperfections are present. The fracture surface is then examined for evidence of any areas of slag inclusions, porosity or incomplete fusion. If present, they are measured and accepted or rejected based on the code limitations.

The last soundness test to be mentioned here is the fillet weld break test. Like the other two types, this soundness test is used primarily in the qualification of welders. This is the only test required for the qualification of tackers in accordance with AWS D1.1, *Structural Welding Code- Steel*.

To perform this test, a welder places a fillet weld on one side of a T-joint. Once complete, the specimen is placed in a press and bent to produce a fracture at or near the weld. Again, the interest is not how much load is required for failure, but rather the condition of the fracture surface, with respect to the presence of discontinuities.

With this test, the inspector is looking for a weld having a satisfactory surface appearance. The fractured surface is examined to assure that the weld has evidence of fusion to the root and that there are no areas of incomplete fusion to the base metal or porosity larger than 3/32 inches in their greatest dimension.

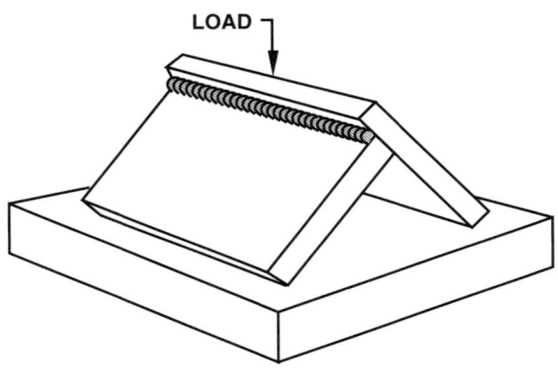

Method of testing fillet weld break specimen

These soundness tests are used routinely in many different industries. Their application and evaluation appear to be quite straightforward. However, the welding inspector should be aware that the evaluation of these tests may not be as simple as the various codes and' specifications might imply. For this reason, it is important that the inspector actually spend some time performing these tests to become familiar with their performance and interpretation.

Destructive Tests for Chemical Properties

The tests that have been previously discussed are used to determine the mechanical properties of a metal. However, there is also interest in the various chemical properties of a metal. In fact, the chemical makeup of a metal determines to a great degree the mechanical properties of that metal. Therefore, it is often necessary to determine the chemical composition of a metal. The three most common methods are: *spectrographic*, *combustion*, and *wet chemical analysis*.

The welding inspector will rarely be required to perform chemical analysis. However, one may have to review the results of an analysis to determine if a metal complies with a particular specification. For more information regarding chemical analysis of metals, refer to the ASTM specifications that cover this subject. The particular methods used for steel are listed in ASTM A751.

Another group of tests that can generally be classified as chemical tests are corrosion tests. These are specific tests designed to determine the corrosion resistance of a metal or combination of metals. Losses from corrosion of metals cost industry tens of billions of dollars annually. Therefore, designers are very concerned about how a metal will behave in a particular corrosive environment. The tests used to determine the degrees of corrosion resistance are designed to simulate as closely as possible the actual conditions that the metal will encounter during its service. Some of the considerations that must be addressed when setting up a corrosion test are: chemical composition, corrosive environment, temperature, presence of moisture, presence of oxygen, presence of other metals and amount of stress. If any of these features are ignored, the corrosion test may yield invalid results.

Metallographic Testing

Another way in which the characteristics of a metal can be determined is through the use of various metallographic tests. These tests basically consist of removing a section of a metal or a weld and polishing it to some degree. Once prepared, the specimen can then be evaluated with the unaided eye or with the use of magnification.

Metallographic testing is generally classified as either macroscopic or microscopic. They differ in the amount of magnification that is utilized. Macro tests are generally performed using magnifications of 10X or lower. Micro tests, on the other hand, use magnifications greater than 10X.

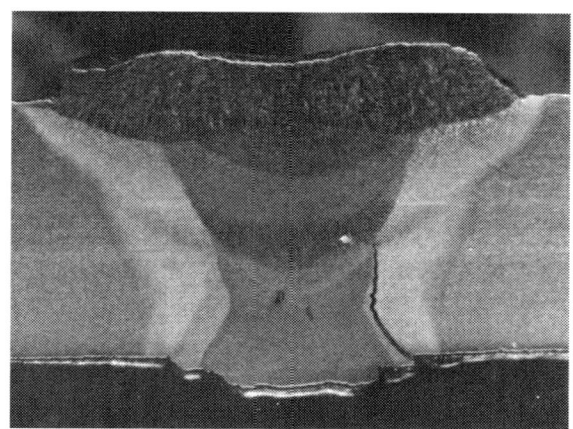

Typical Weld Photomacrograph (Crack adjacent to the weld)

A number of different features can be observed on a typical macro specimen. Often the cross section of a weld can be examined to determine such things as: depth of fusion, depth of penetration, effective throat, weld soundness, degree of fusion, presence of weld discontinuities, weld configuration, number of weld passes, etc. A picture of a macro specimen is referred to as a photomacrograph.

Micro tests can be used to determine various features as well. Included are: microstructural constituents, presence of inclusions, presence of microscopic defects, nature of cracking, etc. Similarly, pictures of micro specimens are called photomicrographs.

These various metallographic tests can be very helpful in such matters as failure analysis, weld procedure and welder qualification, and process control testing.

The two types of specimens also differ in the amount of preparation required. Some macro specimens need only be rough ground, whereas micro specimens require fine grinding and even polishing to produce a mirror finish.

There is so much information which can be gained about the properties of a metal by making simple macro and micro evaluations that metallographic testing is an important tool for both the welding inspector and the engineer.

Selection of Destructive Testing Samples

Another important factor related to these various destructive tests, as well as the nondestructive test methods to be mentioned later, is the number of tests that will be required for a particular situation. That is, will all welds be subjected to a certain test or will only a few be selected for evaluation? Sampling is the selection of some of the welds, weldments or materials from a production run. Conclusions about the quality of welding in the entire production run are drawn from the results of these inspections.

The welding inspector is often the key person responsible for this activity. The purpose of sampling is presumably acceptance inspection. This is an examination performed for the buyer of weldments to satisfy him that what he is offered meets the purchasing specifications. To reach that decision, the complete lot (100 percent sampling) or a portion of the lot (partial sampling) must be examined. The welding inspector will also sample the shop output if acting as a fabricator's inspector, as part of his quality assurance program.

It is hoped that the purchaser and the fabricator have noted at the outset the impossibility of obtaining 100 percent assurance from partial sampling. The reason for this is twofold.

Primarily, the tests will be nondestructive examinations because destructive testing becomes nonproductive when it approaches 10 percent. Destructive tests of the mechanical reliability of a few out of many welds permits only a statistical judgment as to how many of the remaining welds not tested may meet specifications. The assurance is less than 100 percent reliable because of the inherent variability of welding.

Secondly, complete testing by nondestructive examination methods provides only indirect evidence of weld soundness. One hundred percent nondestructive examination will never provide assurance that all the welds are perfect. In fact, statisticians recognize that 100 percent inspection of any item, even by simple go/no-go tests, is seldom 100 percent efficient. It should be realized how easily the inspector tires and how often his attention wanders. The monotony of testing dulls the mind. The purchaser must understand that there will always be an uncertainty in the resulting conclusions.

The engineers must consider whatever final testing and examinations are required in the specifications for the job. The welding inspector does only as directed, of course. Actual specification of the percentage, order, and frequency of sample selection is well outside the welding inspector's normal inspection function.

Summary of Sampling Plans

Sampling welds for evaluation tests or for acceptance testing is one of the welding inspector's important functions. Acceptance inspection is the usual purpose. The number of welds or weldments to be sampled and the tests to be applied should be designated in the purchase specifications or the applicable code.

The welding inspector will use complete sampling where weldments of the highest quality are required for critical services. One or more methods of nondestructive or destructive testing, above and beyond visual inspection, will probably be specified for the critical joints (such as complete penetration welds in butt joint).

The less important welds may require less critical inspection. For the average job, the inspection will generally involve a combination of complete visual inspection plus random partial sampling. The random partial sampling will usually be inspected by one or more of the various methods of nondestructive examination or destructive testing.

Complete 100 percent sampling and testing is usually restricted to critical joints in critical weldments. It must be recognized, however, that the detection of defects by 100 percent inspection is inherently less than 100 percent reliable.

Units Pertaining to Welding		
Property	Unit	Symbol
area dimensions	square millimeter	mm^2
current density	ampere per square millimeter	A/mm^2
deposition rate	kilogram per hour	kg/h
electrical resistivity	ohm meter	$\Omega \cdot$ m
electrode force	newton	N
flow rate (gas and liquid)	liter per minute	L/min
fracture toughness	meganewton meter$^{-3/2}$	MN$\cdot$ m$^{-3/2}$
impact strength	joule	J = N $\cdot$ m
linear dimensions	millimeter	mm
power density	watt per square meter	W/m2
pressure (gas and liquid)	kilopascal	kPa = 1000 N/m^2
pressure (vacuum)	pascal	Pa = N/m2
tensile strength	megapascal	MPa = 1 000 000 N/m^2
thermal conductivity	watt per meter kelvin	W / (m $\cdot$ K)
travel speed	millimeter per second	mm/s
volume dimensions	cubic millimeter	mm^3
wire feed speed	millimeter per second	mm/s

Metric Conversions

In recent years, there has been an effort to convert our U. S. customary system of measurement to an international system referred to as the *International System of Units* (In French, *"Le Systeme Internationale d'Unites"*) or *SI*. The welding inspector may have occasion to inspect parts and interpret specifications containing SI units. In addition, the inspector may be asked to convert the SI units into U. S. customary units.

This is part of the job math which the welding inspector may be required to perform. As a minimum, one will be asked to perform some of these conversions on the AWS CWI examination. The following discussion is meant to provide a basis for the welding inspector to understand how these conversions are performed.

Before actually looking at the conversion process, it is important to understand some of the notations which will be used to express different numeric values. One of the techniques used to express numbers which may be very large or very small, is called scientific notation. What this method does is reduce some number to an expression which includes a number multiplied by some power of ten. For example, the number 234,567 would be expressed as 2.34567×10^5 in scientific notation.

It should be obvious that the same digits are used but the decimal place has been moved. The decimal place is always moved to where there is only one number appearing to its left. The number of spaces which the decimal place was moved becomes the power of ten in the scientific notation expression. If the decimal point was moved in the opposite direction, as would be the case for a number less than one, then the power of ten becomes a negative number. The examples below show how scientific notation is used.

Scientific Notation Examples
$234 = 2.34 \times 10^2$
$0.0234 = 2.34 \times 10^{-2}$
$5.678 \times 10^3 = 5,678$
$5.67 \times 10^{-4} = 0.000567$

From this exercise, it is evident that movements of the decimal point one space is equivalent to multiplying or dividing by ten, depending on the direction in which it is moved.

Another type of notation with which the welding inspector should become familiar are the various prefixes which are used to indicate powers of ten. These are simply abbreviations to reduce the number of digits required. As an example, "kilo" means 1000, so a kilometer is 1000 meters. Similarly, "milli" means one-thousandth, so a millimeter is one-thousandth of a meter or there are 1000 millimeters in one meter. Some examples of the use of these prefixes appear below.

Examples of Prefixes for Powers of Ten
456,000,000 Pa = 456 MPa
56 km = 56,000 m
234,000 mm = 234 m
456 g = 0.456 kg

With this background, it is now possible to begin a discussion of how to actually perform numeric conversions from US to SI and SI to US.

SI Prefixes (Source: ANSI/AWS A1.1)			
Exponential Expression	Multiplication Factor	Prefix	Symbol
10^{18}	1 000 000 000 000 000 000	exa	E
10^{15}	1 000 000 000 000 000	peta	P
10^{12}	1 000 000 000 000	tera	T
10^{9}	1 000 000 000	giga	G
10^{6}	1 000 000	mega	M
10^{3}	1 000	kilo	k
10^{2}	100	hecto*	h
10	10	deka*	da
10^{-1}	0.1	deci*	d
10^{-2}	0.01	centi*	c
10^{-3}	0.001	milli	m
10^{-6}	0.000 001	micro	μ
10^{-9}	0.000 000 001	nano	n
10^{-12}	0.000 000 000 001	Pico	p
10^{-15}	0.000 000 000 000 001	femto	f
10^{-18}	0.000 000 000 000 000 001	atto	a

*Nonpreferred. Prefixes should be selected in steps of 10^3 so that the resultant number before the prefix is between 0.1 and 1000. These prefixes should not be used for units of linear measurement, but may be used for higher order units. For example, the linear measurement, decimeter, is nonpreferred, but square decimeter is acceptable.

The initial point to understand is that the welding inspector is not intended to memorize all of the conversion factors. Every factor which will be needed for the AWS CWI examination will be provided. The CWI candidate must be capable of manipulating the numbers to arrive at some solution.

A listing of some of the more common conversion factors is provided here to illustrate how these conversion factors will be noted on the CWI examination. You see that the table on the facing page is arranged in four columns, entitled: *"Property," "To convert From," "To,"* and *"Multiply by."* These columns will be used in the same order as they are listed.

For any conversion exercise, the first step is to decide what particular property is described by the units which are to be converted. Once the proper category has been chosen from the "Property" column, look at the second column ("To convert from") and locate the line which contains the unit that is given in the exercise. This is the unit which will be converted. Moving straight across to the right, look for the unit that matches the unit to which the conversion will be made. When the line which contains both the known and desired units is located, the value found in the last column ("Multiply by") is the appropriate conversion factor. At this point, simply multiply the number of the known units by this conversion factor. The result is the number of desired units. Several examples appear below to show how to use this table to perform typical conversions.

Conversion Example 1: An oxygen gage shows a pressure of 550 kPa. That is how many psi?
1) Property = pressure (gas and liquid)
2) Known unit= 550 kPa
3) Desired unit= psi
4) Conversion factor = 1.450×10^{-1}
Solution: 550 kPa x .1450 = ?psi
550 kPa = 79.75 psi

Conversion Example 2: A tensile specimen was pulled and displayed an ultimate tensile strength of 655 MPa. This corresponds to how many psi?
1) Property = tensile strength
2) Known unit = 655 MPa
3) Desired unit = psi
4) Conversion factor = 1.450×10^{2}
Solution: 655 MPa x 145.0 = ? psi
655 MPa = 94,975 psi

Conversion Example 3: What is the diameter in millimeters of a 5/32 inch electrode?
1) Property = linear measurements
2) Known unit = 5/32 inch (0.156 in)
3) Desired unit = mm
4) Conversion factor = 2.540 x 10
Solution: 0.156 in x 25.4 = ?mm
0.156 in = 3.96 mm

Conversion Example 4: Welding parameters were adjusted to produce a weld metal deposition rate of 7.3 kg/h. What is that deposition rate in terms of lb/h?
1) Property = deposition rate
2) Known unit = 7.3 kg/h
3) Desired unit = lb/h
4) Conversion factor = 2.2
Solution: 7.3 kg/h x 2.2 = ? lb/h
7.3 kg/h= 16.06 lb/h

Conversions for Common Welding Terms* (Source ANSI/AWS A1.1-)

Property	To Convert From	To	Multiply by
area dimensions (mm^2)	in^2 mm^2	mm^2 in^2	$6.451\ 600 \times 10^2$ $1.550\ 003 \times 10^{-3}$
current density (A/mm^2)	A/in^2 A/mm^2	A/mm^2 A/in^2	$1.550\ 003 \times 10^{-3}$ $6.451\ 600 \times 10^2$
deposition rate** (kg/h)	lb/h kg/h	kg/h lb/h	0.45** 2.2**
electrical resistivity $(\Omega \cdot m)$	$\Omega \cdot cm$ $\Omega \cdot m$	$\Omega \cdot m$ $\Omega \cdot cm$	$1.000\ 000 \times 10^{-2}$ $1.000\ 000 \times 10^2$
electrode force (N)	pound-force kilogram-force N	N N lbf	4.448 222 9.806 650 $2.248\ 089 \times 10^{-1}$
flow rate (L/min)	ft^3/h gallon per hour gallon per minute	L/min L/min L/min	$4.719\ 475 \times 10^{-1}$ $6.309\ 020 \times 10^{-2}$ 3.785 412
fracture toughness $(MN \cdot m^{-3/2})$	$ksi \cdot in^{1/2}$ $MN \cdot m^{-3/2}$	$MN \cdot m^{-3/2}$ $ksi \cdot in^{1/2}$	1.098 855 0.910 038
heat input (J/m)	J/in J/m	J/m J/in	$3.937\ 008 \times 10$ $2.540\ 000 \times 10^{-2}$
impact energy	foot pound force	J	1.355 818
linear measurements (mm)	in. ft mm mm	mm mm in. ft	$2.540\ 000 \times 10$ $3.048\ 000 \times 10^2$ $3.937\ 008 \times 10^{-2}$ $3.280\ 840 \times 10^{-3}$
mass (grams)	lb kg	kg lb	0.454** 2.205**
power density (W/m^2)	W/in^2 W/m^2	W/mm^2 W/in^2	$1.550\ 003 \times 10^3$ $6.451\ 600 \times 10^{-4}$
pressure (gas and liquid) (kPa)	psi lb/ft^2 N/mm^2	kPa kPa kPa	6.894 757 $4.788\ 026 \times 10^{-2}$ $1.000\ 000 \times 10^3$
pressure (gas and liquid) (kPa)	kPa kPa kPa	psi lb/ft^2 N/mm^2	$1.450\ 377 \times 10^{-1}$ $2.088\ 543 \times 10$ $1.000\ 000 \times 10^{-3}$
pressure (vacuum) (Pa)	torr (mm Hg at 0°C) micron (μ m Hg at 0°C) Pa Pa bar	Pa Pa torr micron psi	$1.333\ 220 \times 10^2$ $1.333\ 220 \times 10^{-1}$ $7.500\ 640 \times 10^{-3}$ 7.500 640 $1.450\ 377 \times 10^1$
tensile strength (MPa)	psi lb/ft^2 N/mm^2 MPa MPa MPa	MPa MPa MPa psi lb/ft^2 N/mm^2	$6.894\ 757 \times 10^{-3}$ $4.788\ 026 \times 10^{-5}$ 1.000 000 $1.450\ 377 \times 10^2$ $2.088\ 543 \times 10^4$ 1.000 000
thermal conductivity $(W/[m \cdot K])$	$cal/(cm \cdot s° \cdot C)$	$W/(m \cdot K)$	$4.184\ 000 \times 10^2$
travel speed,	in./min	mm/s	$4.233\ 333 \times 10^{-1}$
wire feed speed (mm/s)	mm/s	in/min	2.362 205

* Preferred units are given in parenthesis
** Approximate conversion.

REVIEW - CHAPTER 7 DESTRUCTIVE TESTING

Q7-1 Which property cannot be determined from a tensile test?
a. ultimate tensile strength
b. percent elongation
c. percent reduction of area
d. impact strength
e. yield strength

Q7-2 The property which describes the ability of a metal to deform or resist some applied load is:
a. strength
b. toughness
c. hardness
d. ductility
e. none of the above

Q7-3 The point at which a metal's behavior changes from elastic to plastic (onset of permanent deformation) is referred to as:
a. yield strength
b. ultimate tensile strength
c. modulus of elasticity
d. Young's modulus
e. none of the above

Q7-4 Which of the following is an expression for a metal's ductility?
a. percent elongation
b. percent reduction of area
c. proportional limit
d. a and b above
e. b and c above

Q7-5 What is the percent elongation of a specimen whose original gage length was 2 inches and final gage length was 2.5 inches?
  a. 30%
  b. 25%
  c. 50%
  d. 40%
  e. none of the above

Q7-6 The property of metals which describes their resistance to indentation is called:
  a. strength
  b. toughness
  c. hardness
  d. ductility
  e. none of the above

Q7-7 The type of testing which is used routinely for the qualification of welding procedures and welders is:
  a. tensile strength
  b. hardness
  c. soundness
  d. impact strength
  e. all of the above

Q7-8 Of the following, which properties can be determined as a result of tensile testing?
  a. ultimate tensile strength
  b. ductility
  c. percent elongation
  d. yield strength
  e. all of the above

Q7-9 The family of hardness tests which uses both a minor and major load is called:
  a. Brinell
  b. Vickers
  c. Rockwell
  d. Knoop
  e. none of the above

Q7-10 Which of the following tests are referred to as microhardness tests?
a. Rockwell
b. Vickers
c. Knoop
d. a and b above
e. b and c above

Q7-11 What type of test uses a weighted pendulum which strikes a notched test specimen?
a. Brinell test
b. fatigue test
c. tensile test
d. crack opening displacement (COD)
e. Charpy impact test

Q7-12 Of the following, which is one of the most measurable features of a metal?
a. fatigue
b. hardness
c. soundness
d. tension
e. none of the above

Q7-13 Which of the following is not considered a soundness test?
a. tensile
b. face bend
c. fillet break
d. root bend
e. nick-break

Q7-14 The type of testing used to evaluate the type of microstructure present in a metal is called:
a. tensile
b. hardness
c. toughness
d. metallographic
e. none of the above

Q7-15 A 50 lb can of welding electrodes weighs approximately how many kg?
   a. 227 kg
   b. 2.3 kg
   c. 22.7 kg
   d. 23,000 kg
   e. none of the above

Q7-16 Which two metal properties are directly related for many steels?
   a. impact strength and fatigue strength
   b. tensile strength and ductility
   c. tensile strength and hardness
   d. toughness and fatigue strength
   e. none of the above

Q7-17 What is the wire feed speed which is measured to be 175 in./min?
   a. .125 m/s
   b. 74 mm/s
   c. 7.4 mm/s
   d. a and b above
   e. b and c above

Q7-18 The property of metals which describes their ability to carry some type of load is:
   a. strength
   b. toughness
   c. hardness
   d. ductility
   e. none of the above

Q7-19 For less ductile metals, which method is used to determine the yield strength?

   a. drop of beam
   b. offset technique
   c. stress-strain curve
   d. abrupt yielding
   e. none of the above

Q7-20 The ability of a metal to absorb energy is called:
a. strength
b. ductility
c. hardness
d. toughness
e. none of the above

Q7-21 A weld joint is measured and found to be 345 mm long. How long is that joint in terms of inches?
a. 135.8 in
b. 13.58 in
c. 8760 in
d. 876 in
e. none of the above

Q7-22 Which of the following tests are used to verify the soundness of a weld?
a. nick break
b. fillet break
c. bend test
d. radiographic test
e. all of the above

Q7-23 With the SAW process we achieve a deposition rate of 19.7 kg/h. How many lb/h is this?
a. 434 lb/h
b. 43.4 lb/h
c. 87.5 lb/h
d. 8.9 lb/h
e. none of the above

Q7-24 Ultimate tensile strength can be determined using which of the following tests?
a. tensile
b. bend
c. Charpy
d. nick break
e. nil-ductility drop-weight

Q7-25 Calculation of percent elongation is determined after measuring the change in:
   a. percent reduction of area
   b. depth of indentation
   c. diameter of indentation
   d. cross sectional area
   e. length between gage marks

Q7-26 With the GMAW process we use a wire feed speed of 170 mm/s. How many in./min is this?
   a. 40.16 in/min
   b. 53.7 in/min
   c. 401.6 in/min
   d. 537 in/min
   e. none of the above

Q7-27 With the GTAW process, flow rates are measured at 22 L/min. How many ft^3/h is this?
   a. 10.4 ft^3/h
   b. 1.39 ft^3/h
   c. 46.6 ft^3/h
   d. 83.2 ft^3/h
   e. none of the above

Q7-28 Calculation of tensile strength is accomplished by dividing the tensile load by:
   a. cross sectional area
   b. percent elongation
   c. percent reduction of area
   d. gage length
   e. none of the above

Q7-29 The metal property describing its freedom from imperfections is:
   a. tensile strength
   b. soundness
   c. impact strength
   d. toughness
   e. ductility

Q7-30 If a metal exhibits a great deal of elongation prior to falling when a tensile load is applied is said to have high:
a. tensile strength
b. hardness
c. impact strength
d. toughness
e. ductility

Q7-31 A specimen approximately 2" long with a V-notch machined in the center of one of its sides is used for which of the following tests?
a. tensile
b. nil-ductility drop-weight
c. Charpy
d. bend
e. tuck break

Q7-32 A heat input of 1,500,000 J/m is how many J/in?
a. 381 J/in
b. 3,810 J/in
c. 38,100 J/in
d. 381,000 J/in
e. none of the above

Q7-33 Which of the following properties can be determined from a tensile test?
a. ultimate tensile strength, yield strength, ductility
b. yield strength, ductility, toughness
c. ductility only
d. toughness only
e. all of the above

Q7-34 The force required to bend a steel bar is measured to be 890 N. That is how many lbf?
a. 200.3 lbf
b. 2003 lbf
c. 20 lbf
d. 550 lbf
e. none of the above

Q7-35 A coating was removed from a SMAW electrode. After weighing, it was found to have a mass of 2.4 oz. That is how many grams?
  a. 68 g
  b. 6.8 g
  c. 0.008 g
  d. 0.8 g
  e. none of the above

CHAPTER 8: WELDING PROCEDURE AND WELDER QUALIFICATION

Introduction to Welding Procedure Qualification

A welding procedure details the steps by which the welding of a specific joint or weldment is to be accomplished. It gives the prescribed values or ranges of values for all the controllable variables in the process and specifies all materials to be used. A welding procedure determines the mechanical properties of a welded joint.

It must be realized from the outset that the generation of a welding procedure is the specific responsibility of the welding engineer and not the welding inspector. According to AWS QC1, the CWI "Verifies that the welding procedures are as specified and qualified and that the welding is performed in accordance with the applicable procedure." So, the information which follows is included to address the various contents of welding procedures and is not intended to imply that the CWI is responsible for the generation of that information.

Almost every welding job needs a welding procedure. Commonly, the contract or the specifications require that a written procedure be prepared. The requirements often are those of a general specification or standard, which the fabricator may be following as a quality control standard.

When the governing specifications or standards comprise a code that has been adopted by a governing agency such as a city, state, or province, then the welding procedure becomes a legal entity under that code.

The purpose of this section is fourfold:

- To define the welding procedure specification, including its types, content, documentation, and application.
- To explain the qualification of the welding procedure specification.
- To define circumstances when certain parts of the specification may require requalification.
- To define the welding inspector's responsibilities during the qualification process and later during the fabrication inspection process.

Welding Procedure Specification

The purpose of a welding procedure specification is to define and document the details that are to be carried out in welding specific materials or parts.

Many companies prepare a quality assurance manual that establishes the responsibilities for preparation, review, and approval of the welding procedures to be used, although the procedures themselves will not ordinarily be included in the quality assurance manual.

Contents

The written welding procedure specification should be arranged in accordance with the contract or purchase requirements. The specifications should be sufficiently detailed to ensure that the welding will meet all requirements of the applicable code, standard or specification.

The topics that follow in this section are the most common and generally the most essential in welding procedure specifications. Every item will not apply to every process or application, and some items that are familiar to you may not be listed at all. The list is given for illustration, to provide guidance to the welding inspector in reviewing welding procedure specifications or in determining whether or not production welding is being performed in accordance with welding procedure specifications. It must be emphasized that the welding inspector is not responsible for producing the welding procedure specification. The following can serve as an outline for a checklist for the welding inspector when reviewing a welding procedure specification produced by a welding engineer or other responsible individual.

Scope: Have the types of welding, the materials, and the governing specifications been clearly stated?

Base Metals and Applicable Specifications: Are suitable base metals specified? They should be identified by their chemical composition and applicable specifications. The procedure should indicate what condition the base metal should be in before welding (that is, normalized, annealed, quenched and tempered, solution treated, etc.). There may also be a requirement that the fabricator know the identity of all material. Full plates or sections can be identified by the mill numbers; small portions cut from full plates or sections should be marked with the same numbers. The rolling direction of the plate should also be identified.

Welding Process: What welding process is to be used?

Type, Classification and Composition of Filler Metals: The welding process should be clearly named, and the composition, identifying type, or classification designation of the filler metal should always be spelled out to ensure proper use. In addition, the sizes of filler metals or electrodes that can be used when welding different thicknesses of material in different positions must be designated. Some types of filler metals are even identified on each individual pass or layer.

AWS ELECTRODE CLASSIFICATION SYSTEM		
DIGIT	SIGNIFICANCE	EXAMPLE
1st two or 1st three	Min. tensile strength (stress relieved)	E-60XX = 60,000 p.s.i. (min) E-110XX = 110,000 p.s.i. (min) E-xx1x = all positions E-xx2x = horizontal and flat E-xx3x = flat
2nd last	Welding position	
Last	Power supply, type of slag, type of arc, amount of penetration, prescence of iron powder in coating	

Note: Prefix "E" (to left of a 4 or 5 - digit number) significes arc welding electrode

LAST DIGIT INTERPRETATION AWS ELECTRODE CLASSIFICATION SYSTEM		
DESIGNATION	CURRENT	COVERING TYPE
E6010	DCEP only	Organic
E6011	DCEP or AC	Organic
E6012	DCEN or AC	Rutile
E6013	DCEN, DCEP, or AC	Rutile
E7014	DCEN, DCEP, or AC	Rutile, Iron-powder (approx 30%)
E7015	DCEP only	Low hydrogen
E7016	DCEP or AC	Low hydrogen
E7018	DCEP or AC	Low hydrogen, iron powder (approx 25%)
E6020	DCEN, DCEP or AC	High iron oxide
E7024	DCEN, DCEP OR AC	Rutile, Iron powder (approx 50%)
E6027	DCEN, DCEP or AC	Mineral, Iron powder (approx 50%)
E7028	DCEP or AC	Low hydrogen, Iron powder (approx 50%)

Identification of filler metals may be lost when original containers are discarded. Electrode marking, moreover, does not guarantee that the electrode is in satisfactory condition. For example, low hydrogen electrodes that have been exposed to the atmosphere must be baked in an oven to restore their low moisture content. Such baking requirements should be included in the welding procedure, following the manufacturer's specification.

Type of Current and Current Ranges: What type of current is to be used? Some electrodes work well on either ac or dc. If dc is needed, the proper polarity should be specified (DCEP or DCEN). In addition, current ranges for different electrode sizes, different procedure positions, and various thicknesses of materials should be listed.

Welder Qualification Requirements: The procedure specification may designate the requirements for welder or welding operator qualification. Applicable welder qualification specifications or paragraphs of the governing specifications may be referenced in the welding procedure specification. The latter portion of this Chapter is devoted to the topic of welder qualification.

Joint Designs and Tolerances: Permissible joint design details as well as the designated welding sequence should be identified. Use of cross-sectional sketches that show the thickness of material and details of the joint or references to standard drawings or specifications are suitable ways of expressing this information. Tolerances for all dimensions must be listed.

Joint Preparation and Cleaning of Surfaces to be Welded: What methods may be used to prepare the joints? How are the surfaces to be cleaned? Joint preparation methods such as oxyfuel cutting, air carbon arc cutting, and plasma arc cutting (with or without after-cleaning) should be called out. Any required machining or grinding, and any special cleaning such as vapor, ultrasonic, dip, or lint-free cloth cleaning must be specified. There may also be mention of whether or not weld anti-spatter compounds may be used. Be sure that methods or practices specified for production work are specified for qualifying the welding procedure.

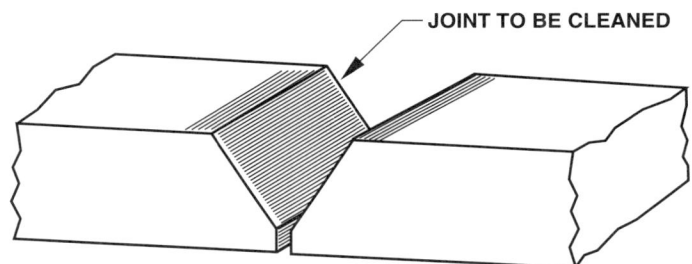

Joint Preparation and Cleaning of Surfaces for Welding.

Tack Welding: What tack welding practices are to be followed? The applicable code should be referenced to determine whether tack welders must be qualified.

Joint Welding Details: Details of electrode sizes for the different portions of each welding position, the arrangement of weld passes for filling the joints, the thickness of each pass, pass width or electrode weave limitations, amperage ranges and whatever other details are important for each particular joint must be specified.

Positions of Welding: In which positions may welding be done? (See illustrations page 8-4, 8-5)

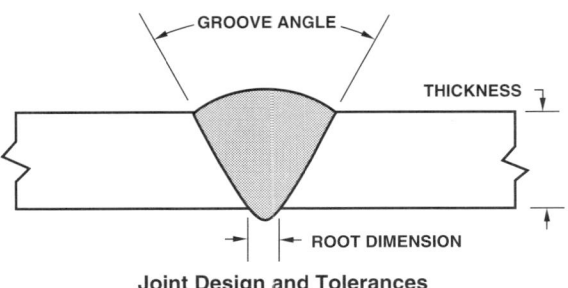

Joint Design and Tolerances

Tabulation of positions of groove welds			
Position	Diagram reference	Inclination of axis	Rotation of face
Flat	A	0° to 15°	150° to 210°
Horizontal	B	0° to 15°	80° to 150° 210° to 280°
Overhead	C	0° to 80°	0° to 80° 280° to 360°
Vertical	D E	15° to 80° 80° to 90°	80° to 280° 0° to 360°

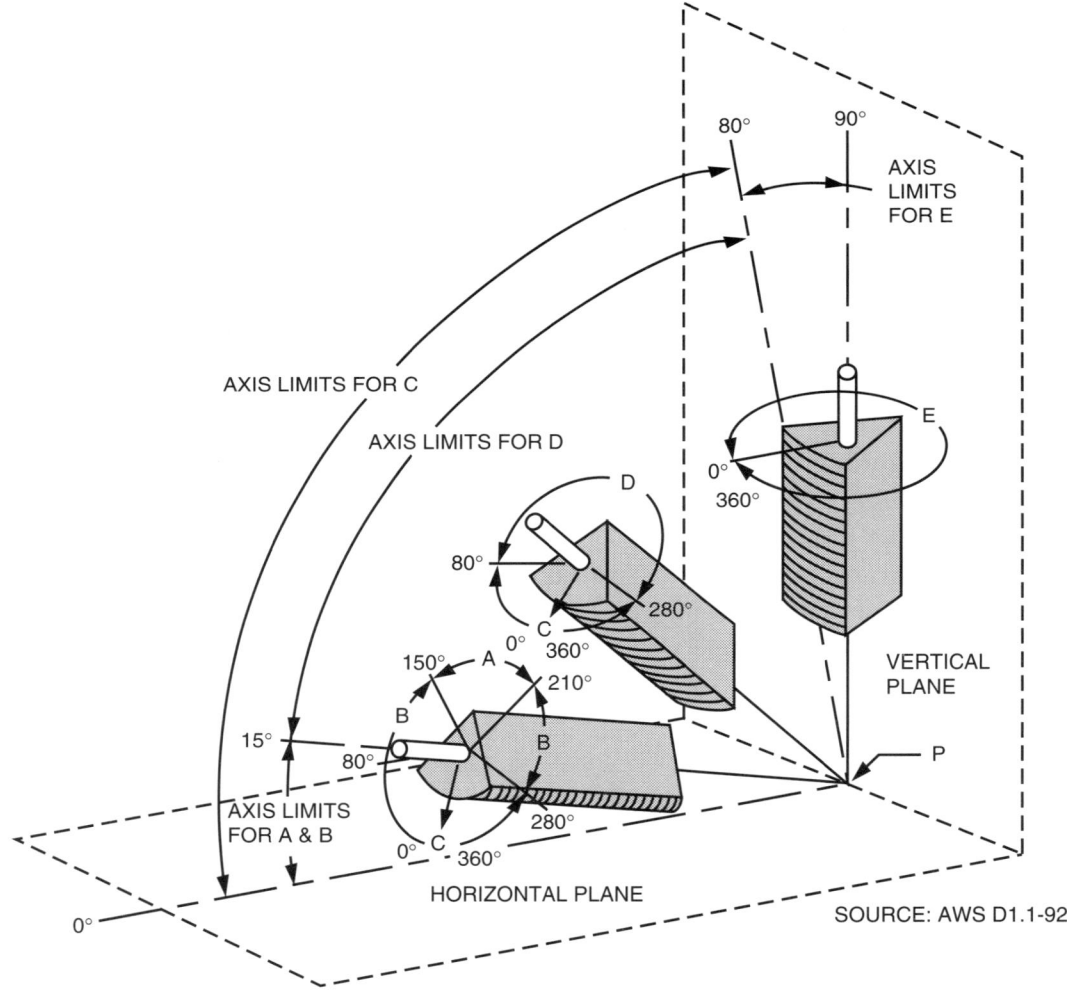

Notes:
1. The horizontal reference plane is always taken to lie below the weld under construction.
2. The inclination of axis is measured from the horizontal reference plane toward the vertical reference plane.
3. The angle of rotation of the face is determined by a line perpendicular to the theoretical face of the weld which passes through the axis of the weld. The reference position (0°) of rotation of the face invariably points in the direction opposite to that in which the axis angle increases. When looking at point P, the angle of rotation of the face of the weld is measured in a clockwise direction from the reference position (0°).

POSITIONS OF GROOVE WELDS

8-5

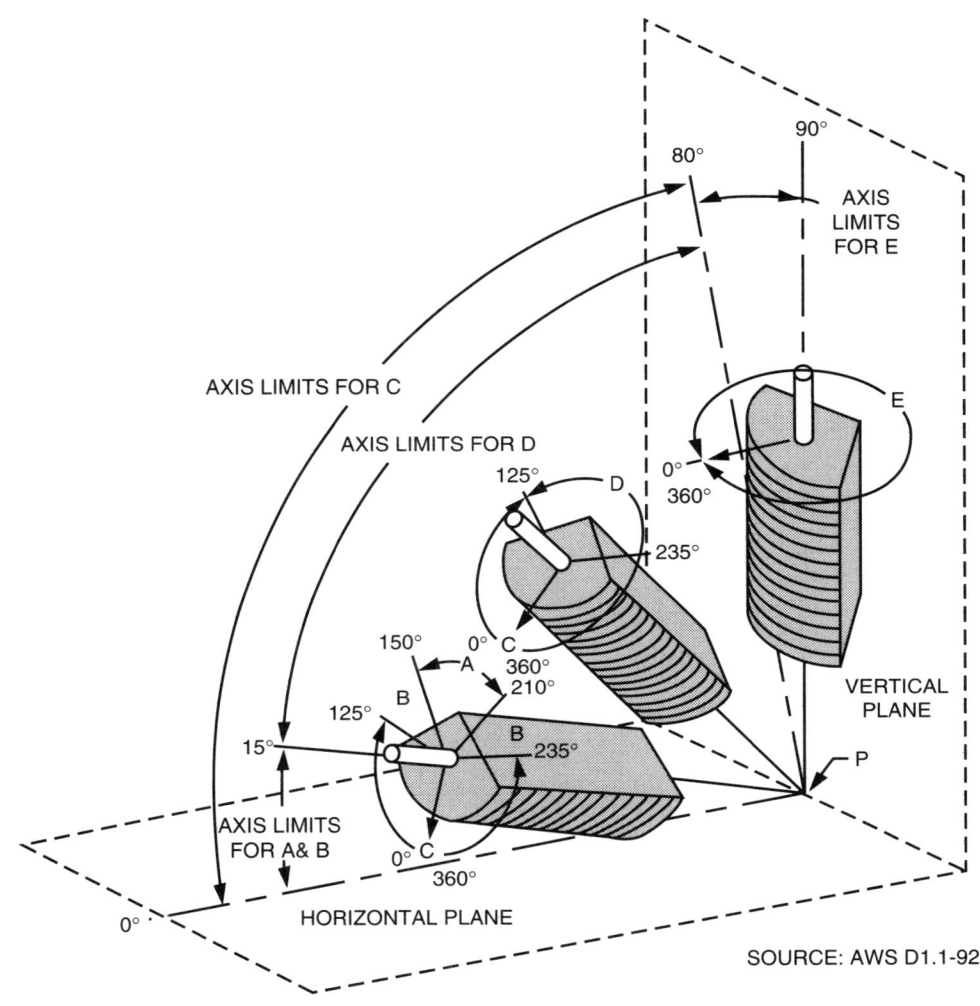

Tabulation of positions of fillet welds			
Position	Diagram reference	Inclination of axis	Rotation of face
Flat	A	0° to 15°	150° to 210°
Horizontal	B	0° to 15°	125° to 150° 210° to 235°
Overhead	C	0° to 15°	0° to 125° 235° to 360°
Vertical	D	15° to 80°	125° to 235°
	E	80° to 90°	0° to 360°

POSITIONS OF FILLET WELDS

Preheat and Interpass Temperatures: What are the temperature limits for any preheat or interpass temperature?

Peening: Peening is a mechanical treatment utilized to reduce the effects of welding heat cycles which could produce excessive residual stresses, distortion and even cracking. Indiscriminate use of peening should not be permitted; however, it is sometimes applied to highly restrained or thick welds to avoid warpage or cracking of the weld or base metal. If peening is to be used, details of its application must be specified in the welding procedure specification.

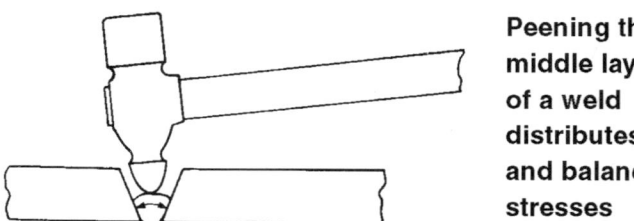

Peening the middle layer of a weld distributes and balances stresses

Heat Input: With heat-treated alloy steels and austenitic stainless steel alloys, the energy input during welding must not heat the work adjacent to the weld above certain temperatures. To control this, there must be specification of the preheat and interpass temperatures, arc voltage, current, and travel speed within well defined ranges, as determined during the procedure qualifications. This will allow for maintenance of the desirable properties in the heat-affected zone of the base metal.

Root Preparation Prior to Welding Second Side: In joints welded from both sides, there should be a description of how the root of the first weld is or may be prepared for back welding. The procedure specification should state whether chipping, grinding, air carbon arc cutting, oxyacetylene gouging, etc. is to be used.

Removal of Weld Sections for Repair: What methods are to be used for removing welds or sections of welds for repair? The methods may well be the same ones used for preparing the root of the first pass for welding from the second side.

Repair Welding: Details of any repair welding methods and procedures which may differ from the standard methods to be used to create a welded joint should be identified.

Examination: What type and extent of examination is each weld joint to receive? The examination may include radiography, magnetic particle, ultrasonic, penetrant, or other types of testing. Visual examination of every weld is routinely required, but that should be spelled out in the procedure.

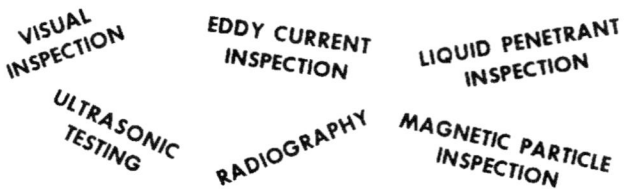

Postheat Treatment: What heat treatment or stress relief will be required after welding? It should be the same treatment that is applied to all procedure qualification test welds. A full description of the heat treatment should be specified or a suitable heat treating note, drawing or document should be referenced. Check to see that postheat treatment of heat treated alloy steels will not exceed the final tempering temperature that was given the base metal. If a full reheat treatment is intended, it may be desirable to do the welding with the metal in the annealed condition.

Marking: Some codes require that welder identification marks be made on or near each weld, or a record of welds made by the welders is maintained. The identification marking should be made with low-stress steel die stamps rather than sharp-edged stamps.

Marking a completed weldment.

Records: There should be details of what welding records will be required, and the specific requirements for these records.

Welding Procedure Specification Sample: Samples of a general welding procedure and several joint welding procedure specifications are illustrated in Appendix B. These should be considered as illustrating only the general form in which welding procedure specification details may be treated, not the specific details to follow for any particular application. The data given for electrode size, welding current and voltage, preheat and interpass temperature, location and sequence of weld passes, etc., have been filled in solely for illustration, not for actual use.

Welding Procedure Qualification

The purpose of procedure qualification testing is to demonstrate that the materials and methods prescribed in a procedure specification will in fact produce weld joint mechanical properties that meet the application and specification requirements.

There are four steps in the qualification of a welding procedure:

- Preparation and welding of suitable samples.
- Testing of representative specimens.
- Evaluation of overall preparation, welding, testing, and end results.
- Approval (if the results are favorable).

Preparation of Procedure Qualification Sample Joints: Test assemblies usually have a representative joint in their middle. The size, type and thickness are related to the type and thickness of material to be welded in production and the number, type and size of specimens to be removed for testing. The materials to be used as well as the welding details are governed by the particular welding procedure specifications that are to be qualified.

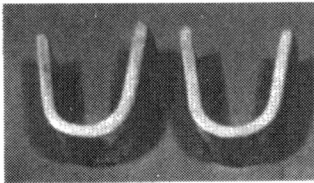
FACE-BEND

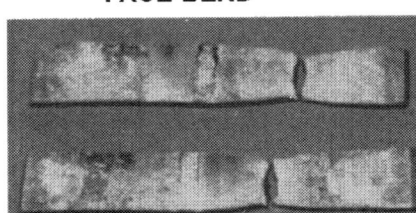
TENSILE

ROOT-BEND

NICK-BREAK

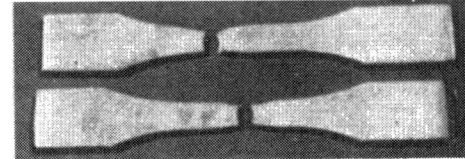
REDUCED TENSILE

Testing of Procedure Qualification Welds: Specified tests and examinations are made on the sample joints. The type and number of specimens to be removed for destructive tests will depend upon the requirements of the particular application or specification. Such tests may include: tensile, bend, nick-break, charpy, fillet-break, etc. Often nondestructive examinations will also be applied. The welding inspector should be certain that the records show how the procedure qualification welds were made and tested.

Evaluation of Test Results: The test results for a procedure qualification sample weld, with the records of joint preparation, welding and testing, should be made available for review. These results will be analyzed by the responsible parties to determine whether the test details and results meet the requirements of the applicable specification.

Approval of Qualification Tests and Procedure Specifications: As a rule, the inspection agency or customer must approve the procedure qualification tests, the test results, and the procedure specifications before any production welding is done.

Qualification is accomplished when the required tests have been completed and approval has been obtained. However, authentic documentary evidence must be available to show that the joints were indeed satisfactory. The welding inspector should witness the welding and testing of all specimens, if possible.

At any time during the use of some qualified procedure, the welding inspector may request requalification of that procedure if production use shows that it is not producing consistently reliable results.

Code Qualification Requirements

AWS D1.1 Structural Welding Code - Steel: AWS D1.1 covers the welding of various structures, including buildings, bridges and tubular structures. It features a unique welding procedure concept referred to as prequalified weld joints. As long as the welding is performed in accordance with the design, workmanship and technique requirements set forth in the Code, no actual procedure qualification testing is required. Section 5.1 outlines the limitations of these prequalified procedures.

Use of procedures operating outside these prescribed limitations will require actual qualification testing. Section 5.5 places limitations on the variables for these qualified procedures. Section 5.10 describes the type and number of specimens that must be tested to qualify a welding procedure. Section 5.11 talks about the methods to be use for testing the various types of specimens, and the results of these tests must meet the requirements of Section 5.12.

Sample forms approved by the AWS Structural Welding Committee for recording

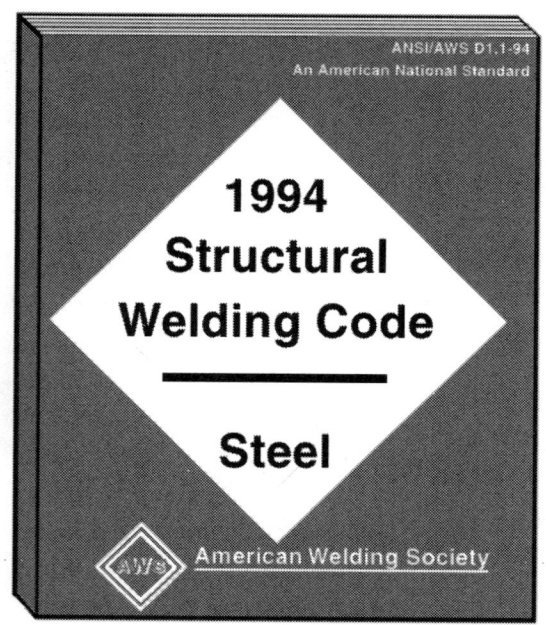

procedure qualifications are reproduced in Appendix B.

ASME Boiler and Pressure Vessel Code - Section IX: The ASME Boiler and Pressure Vessel Code covers the welding of pressure vessels. Any welding performed in accordance with this Code must be done using welding procedures which have been qualified using the guidelines set forth in Section IX of that Code. It is entitled *"Welding and Brazing Qualifications."* This Code section also applies to qualification for welding pressure piping in accordance with ANSI B31.1.

Unlike the AWS Structural Welding Code, Section IX always requires procedure qualification testing. It covers both the welding and brazing of virtually all types of construction alloys using a wide variety of processes.

Documentation for a welding procedure qualified in accordance with ASME Section IX consists of both a Welding Procedure Specification (QW-482) as well as a Procedure Qualification Record (QW-483). Copies of these two forms appear in Appendix B.

API Standard 1104: The welding of cross-country pipelines and other types of petroleum equipment is controlled by the requirements set forth in API 1104, *Standard for Welding Pipelines and Related Facilities*. This Standard describes how each welding procedure will be qualified. It is somewhat unique in that it also describes the inspection and quality requirements for the production welding. Like ASME Section IX, each procedure requires actual qualification testing. Appendix B contains a sample welding procedure qualification form approved by API 1104.

Changes in a Qualified Welding Procedure

If a fabricator who has qualified a welding procedure desires at some later date to make a change in that procedure, it may be necessary to conduct additional qualifying tests. Requalification is necessary when any one of the essential variables listed in the governing standard or code is changed beyond the limits which have been established.

For example, one such variable is the heat treatment that follows welding. Heat treatment has a profound effect on most welds. Its omission (when called for) or its addition to a welding procedure (when not called for) would be a change in an important essential variable, requiring requalification of the procedure.

In ASME Section IX, there is reference to both essential and nonessential variables. Changes in essential variables requires requalification of the procedure. However, a change in a nonessential variable only requires modification of the documentation to reflect the desired change, so no retest is necessary.

Welding Inspector Responsibilities

Welding Procedure Qualification: Before any production welding is started under a contract, it is the welding inspector's duty to verify that welding procedures have been established and that they are capable of producing welded joints of the type and quality required. To be certain of this, the welding inspector should witness the welding and testing of the qualification weld specimens. There must be authentic documentary evidence available that the joints were satisfactory. As mentioned earlier, the welding inspector should try to witness the tests to increase his familiarity with the details of the procedure.

Welding Inspection: It should be kept in mind that the passing of a qualification test does not ensure proper application of the procedure. Adequate inspection is necessary to verify how the qualified procedure is being applied. To aid in the performance of an inspection, the welding inspector will find it advantageous to prepare a checklist for each procedure specification.

Once the welding procedures have been reviewed and found to be acceptable, the task now becomes one of assuring that the production welding is performed in accordance with those procedures. Qualified welding procedures are useless if not followed. The welding inspector should also note any changes in any welding procedure in excess of the limits prescribed for the essential variables. In that case, see that the procedure specification is changed accordingly and determine whether requalification of the procedure is required. If requalification is required, see that the modified procedure is not applied to production work until requalification testing has been completed and the change has been proven satisfactory.

Make certain that weldments needing repair are corrected using qualified welding procedures. The welding and quality requirements for the repair are the same as those for the original welding.

Welder and Brazer Qualification

Performance qualification tests determine the ability of welders, welding operators and brazers to produce acceptably sound welds and brazes with the processes, materials and techniques defined in the qualified welding or brazing procedures. Qualification of these personnel is the legal responsibility of the employer of that individual.

The manufacturer, contractor, fabricator, erector, and owner are responsible for the quality of their work. They should establish quality assurance programs that utilize in-house and outside experience and qualifications to assure that they meet or exceed the minimum specified requirements.

Qualification requirements for these personnel are usually defined in the governing standard or the contract specification. The welding inspector's responsibility is to verify that every welder, welding operator or brazer who works under the standard or specification has been properly qualified for the work to be done.

Qualification Testing

Performance qualification tests improve the probability of obtaining satisfactory welds in finished products. However, test welds are made with special attention and effort. They show whether the welder, welding operator or brazer can produce acceptably sound joints, but they do not tell whether he or she normally will do so under every production condition. For that reason,

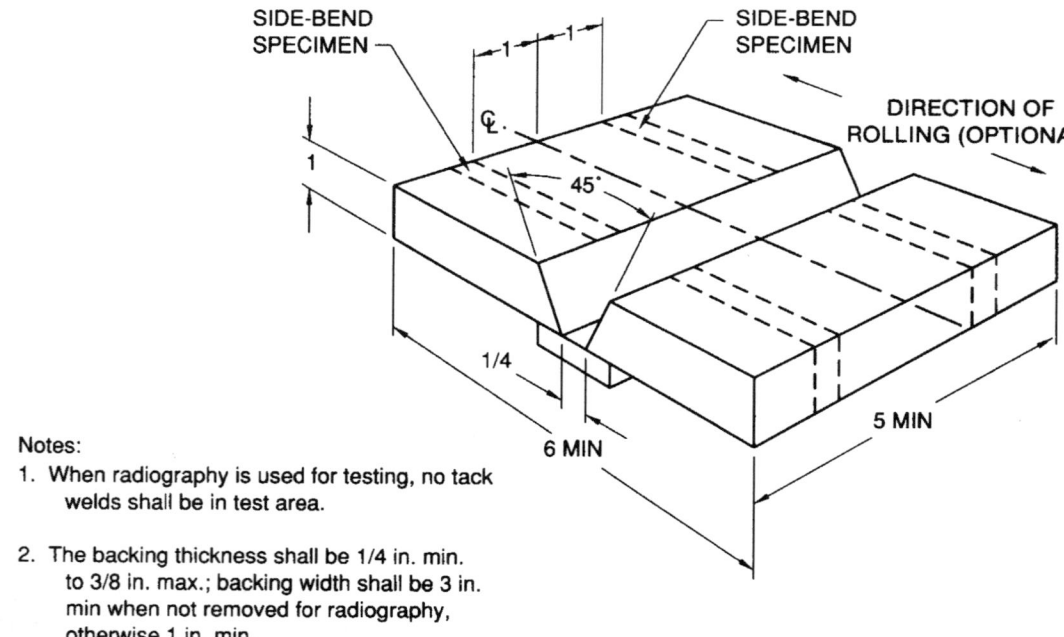

Notes:
1. When radiography is used for testing, no tack welds shall be in test area.
2. The backing thickness shall be 1/4 in. min. to 3/8 in. max.; backing width shall be 3 in. min when not removed for radiography, otherwise 1 in. min.

Unlimited Thickness Welder Qualification Test Plate (Source: AWS D1.1-92)

in.	mm
1/4	6
3/8	10
1	25
3	75
5	125
6	150

complete reliance should not be placed on qualification testing of welders, welding operators and brazers. Production welds and brazes should be inspected during, and following, the actual welding.

Tests prescribed by most codes, specifications and governing rules are similar for the most part. The types of tests that are most common will be described for the following applications:

- Plate and structural member welding
- Pipe welding
- Sheet metal welding
- Brazing

In addition, there will also be mention of the various test methods used to examine the qualification test welds or braze joints. Another factor which has a great impact on the ability of a welder or brazer to produce a satisfactory weld or braze is the position in which the test is performed. This will depend on the position, or positions, in which production welding or brazing will be done.

Plate and Structural Member Welding: Qualification requirements for welders of plate and structural parts usually have the welder or welding operator make one or more test welds on plate or pipe in accordance with the requirements of the qualified welding procedure. Each qualification weld is tested in a specific manner, often both destructively and nondestructively. The requirements prescribe the thicknesses of material and the test positions that qualify for production work.

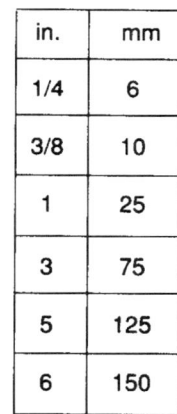

in.	mm
1/4	6
3/8	10
1	25
3	75
5	125
6	150

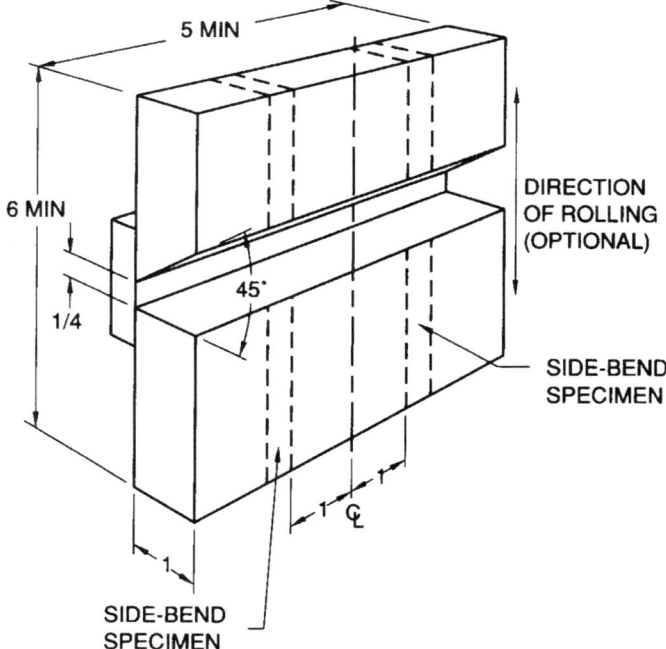

Notes:
1. When radiography is used for testing, no tack welds shall be in test area.
2. The backing thickness shall be 1/4 in. min to 3/8 in. max; backing width shall be 3 in. min when not removed for radiography, otherwise 1 in. min.

Optional Test Plate for Unlimited Thickness - Horizontal Position - Welder Qualification (Source: AWS D1.1-92)

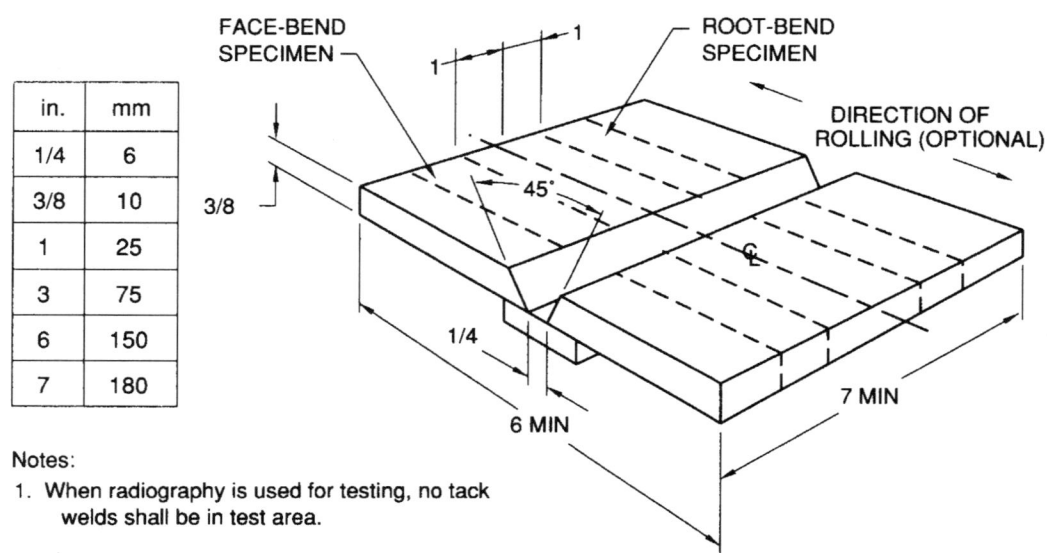

in.	mm
1/4	6
3/8	10
1	25
3	75
6	150
7	180

Notes:
1. When radiography is used for testing, no tack welds shall be in test area.
2. The backing thickness shall be 1/4 in. min to 3/8 in. max; backing width shall be 3 in. min when not removed for radiography, otherwise 1 in. min.

Test Plate for Limited Thickness - All Positions - Welder Qualification (Source: AWS D1.1-92)

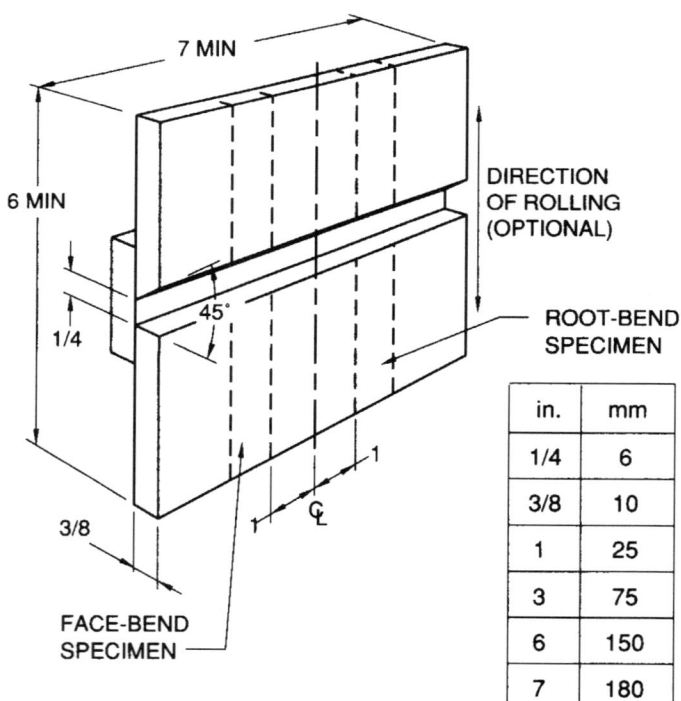

in.	mm
1/4	6
3/8	10
1	25
3	75
6	150
7	180

Notes:
1. When radiography is used for testing, no tack welds shall be in test area.
2. The backing thickness shall be 1/4 in. min to 3/8 in. max; backing width shall be 3 in. min when not removed for radiography, otherwise 1 in. min.

Optional Test Plate for Limited Thickness - Horizontal Position - Welder Qualification (Source: AWS D1.1-92)

8-13

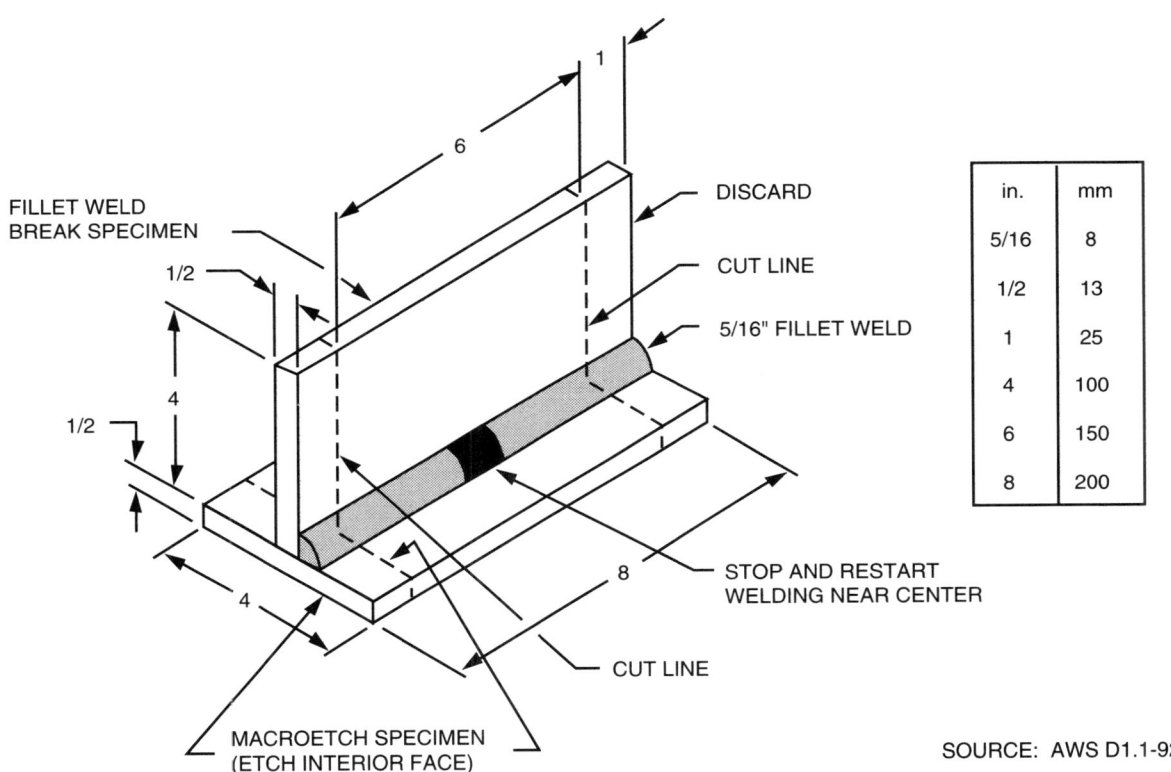

**FILLET WELD BREAK AND MACROETCH TEST PLATE
WELDER QUALIFICATION - OPTION 1**

8-14

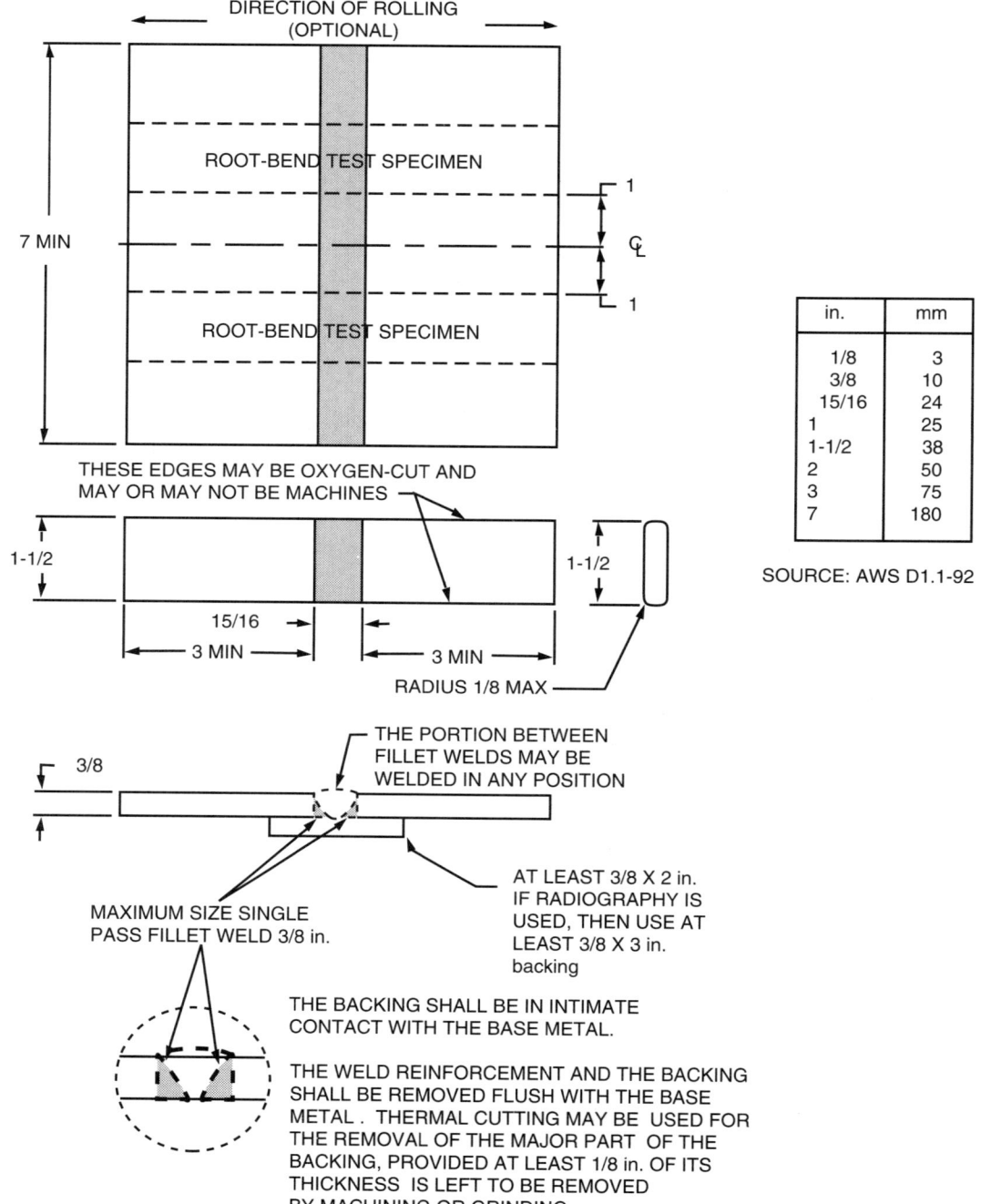

**FILLET WELD ROOT-BEND TEST PLATE
WELDER QUALIFICATION - OPTION 2**

8-15

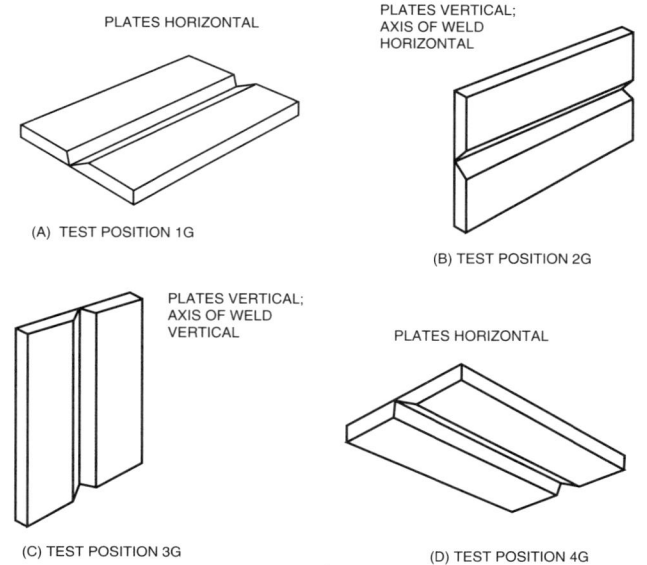

POSITIONS OF TEST PLATES FOR GROOVE WELDS

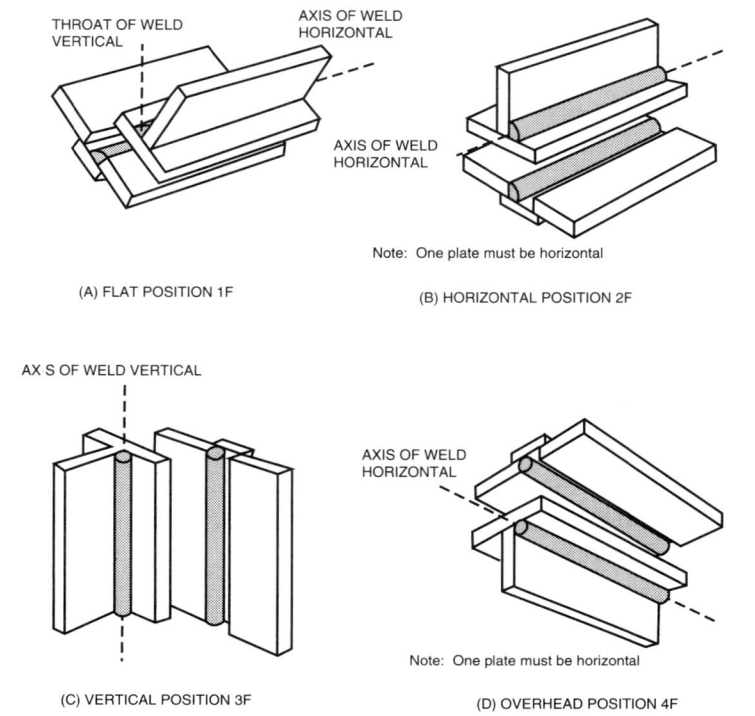

POSITIONS OF TEST PLATE FOR FILLET WELDS

Other details cover groove welds with or without backing and the direction of welding when

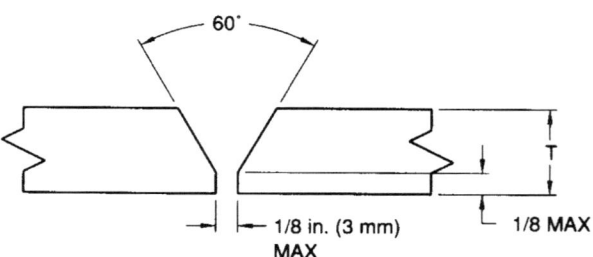

Typical Tubular Butt Joint Welder Qualification (without backing)

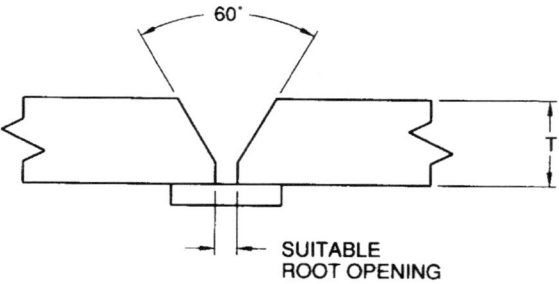

Typical Tubular Butt Joint Welder Qualification (with backing)

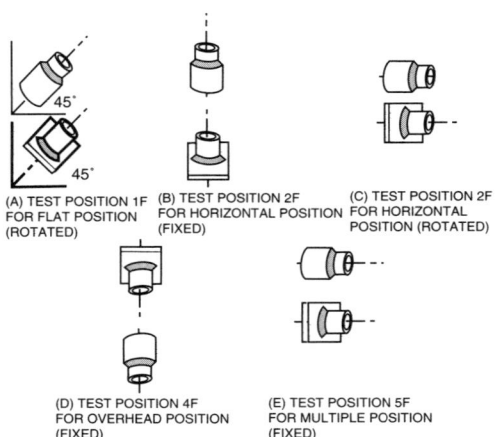

POSITIONS OF TEST PIPES FOR FILLET WELDS (SOURCE: AWS D1.1-92)

Pipe Welding: Qualification requirements for welding pipe differ from those for welding plate and structural members chiefly in the type of test assemblies and test positions. Another major difference with pipe welding is the fact that often there is no practical access to the root surface, requiring the use of some backing ring or consumable insert or the production of a one-side weld with an open root (no backing). This requires a more skilled individual than the welding of plate or structural joints with backing.

To simulate the difficulties of production welding, the pipe weld qualification tests utilize pipe coupons which are welded in the position, or positions, for which the welder wishes to be qualified. There may also be space restrictions placed on the welder during the test which measures the individual's ability to produce a satisfactory weld in locations where joint access is limited. (Illustrations above, at right and on page 8-17.)

Sheet Metal Welding: The welding of sheet metal requires special skills simply because the thin members tend to melt rapidly, which could result in burning holes rather than joining parts together. Consequently, the qualification tests will examine the ability of the welder to produce sound welds in these thin sheet metal thicknesses. You will note that all codes place limits on the minimum thicknesses which a welder can weld in production. Quite often, the minimum thickness qualified is that thickness used during qualification. (See illustrations page 8-18)

Brazing: Like welders, brazers also require some type of qualification testing. The test usually consists of the production of some type of joint arrangement which is positioned in a manner similar to that expected in production. The evaluation of the test results is normally accomplished by cross sectioning the brazed joint and measuring the amount of bonding that has resulted.

Positions of Welding and Brazing: Welder and brazer qualification tests usually must be made in the most difficult positions that will be encountered in production (for example, vertical, horizontal and overhead) if the production work involves other than flat position welding and brazing. Qualification in a more difficult position usually qualifies for welding or brazing in less

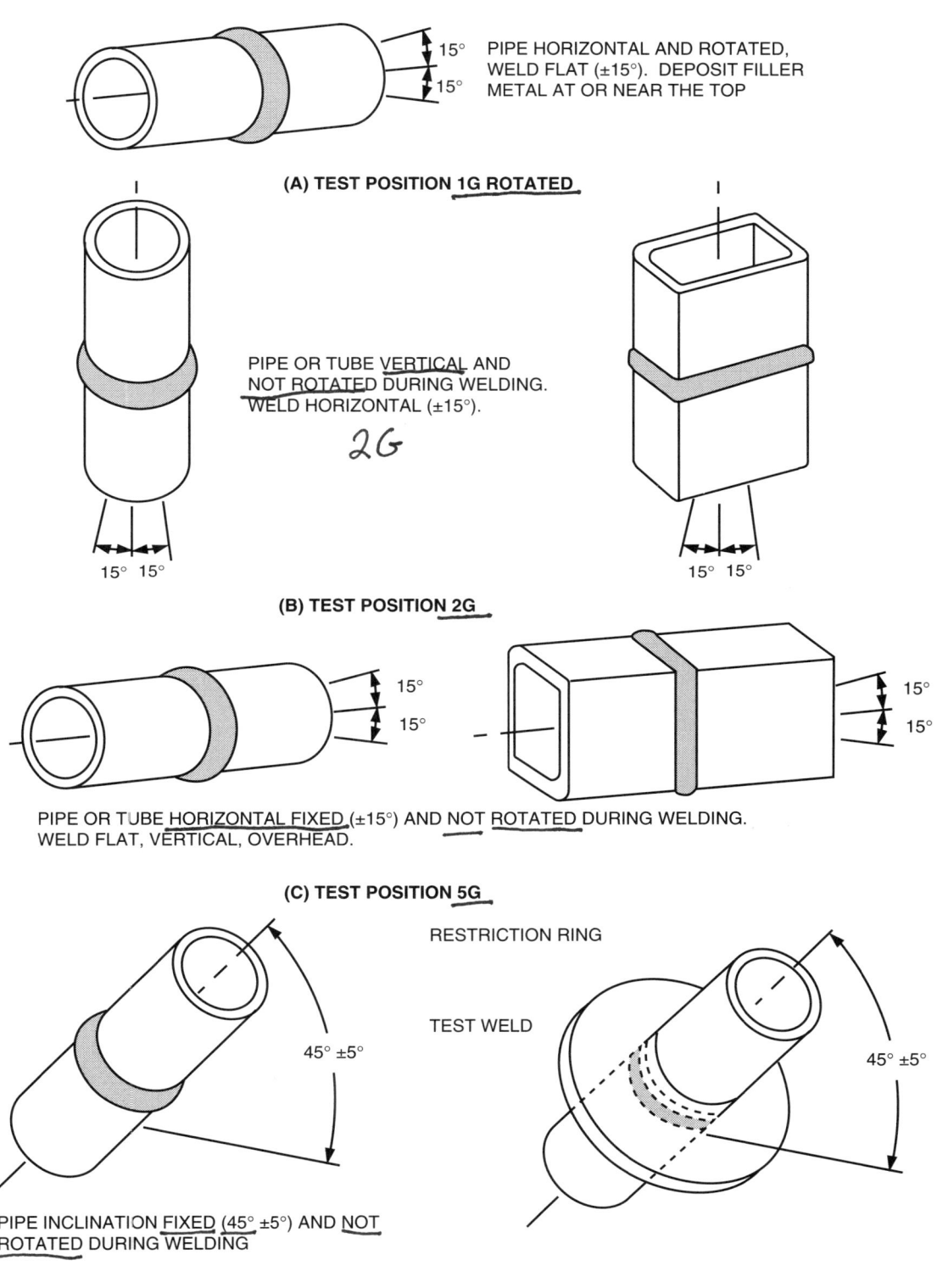

POSITIONS OF TEST PIPE OR TUBING FOR GROOVE WELDS

SOURCE: AWS D1.1-92

Table 6.2 - Welder Qualification Tests (See 6.8.2)							
Test Assemblies Shown in Figure:	Type of:		Qualifies for:			Number of Tests	Type of Test
	Welded Joint	Welding Position	Welding Position	Type of Welded Joint	Thickness		
6.1	Square groove butt joint sheet to sheet	F H V OH	F FH FHV FH OH	Square groove butt joint, sheet to sheet	Thickness tested (except as modified by 6.8.2.2(2)	2	Bend
6.2	Arc spot weld, sheet to supporting member	F	F	Arc spot weld and arc seam weld, sheet to supporting member	Thickness tested	2	Twist
6.3A	Arc seam weld, sheet to supporting member	F	F	Arc seam weld, sheet to supporting member	Thickness tested	2	Bend
6.3B	Arc seam weld, sheet to sheet	H	H	Arc seam weld, sheet to sheet	Thickness tested	2	Bend
6.4	Fillet welded lap joint, sheet to sheet	F H V OH	F F H F H V F H OH	Fillet welded lap joint, sheet to sheet, or sheet to supporting member	Thickness tested and thicker	2	Bend
6.4	Fillet welded lap joint, sheet to supporting member	F H V OH	F F H F H V F H OH	Fillet welded lap joint, sheet to, to supporting member	Thickness tested and thicker	2	Bend
6.4	Fillet welded T-joint, sheet to sheet	F H V OH	F F H F H V F H OH	Fillet welded T- or lap joint, sheet to sheet, or sheet to supporting member	Thickness tested and thicker	2	Bend
6.4	Fillet welded T-joint, sheet to supporting member	F H OH	F F H F H OH	Fillet welded T- or lap joint, sheet to supporting member	Thickness tested and thicker	2	Bend
6.4A	Flare bevel, sheet to sheet	F H V OH	F F H F H V F H OH	Flare-bevel-groove weld, sheet to sheet, or sheet to supporting member; or flare-V-groove weld, sheet to sheet	Thickness tested and thicker	2	Bend
6.4B	Flare-bevel, sheet to supporting member	F H V	F F H F H V	Flare-bevel-groove weld, sheet to supporting member.	Thickness tested and thicker	2	Bend
6.4C	Flare-V, sheet to sheet	F H V OH	F F H F H V F H OH	Flare-V-groove weld, sheet to sheet; or flare-bevel-groove weld, sheet to sheet, or sheet to supporting member.	Thickness tested and	2	Bend

SOURCE: ANS/AWS D1.3-89

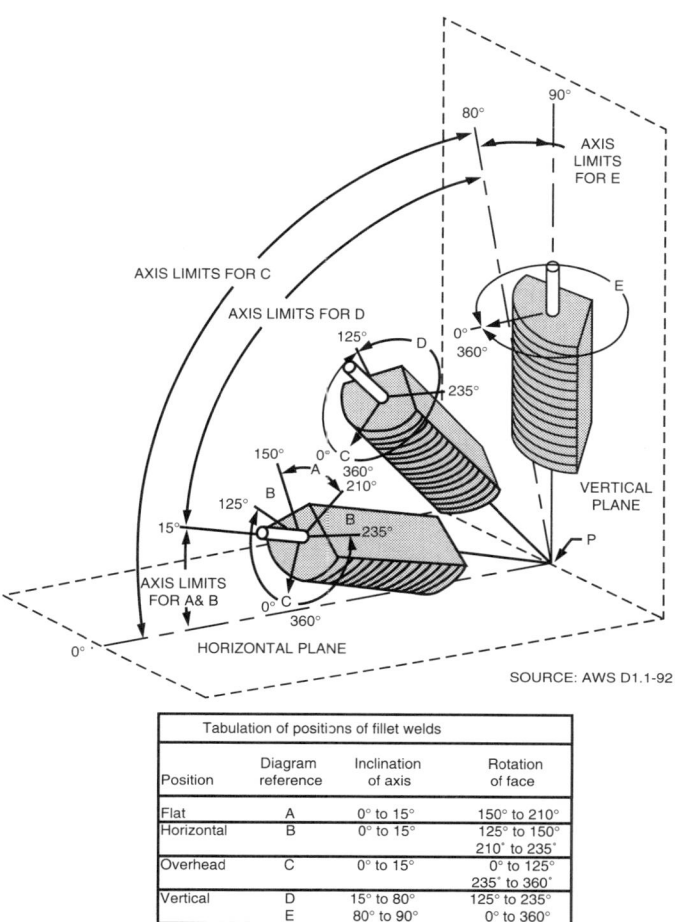

POSITIONS OF FILLET WELDS

difficult positions; e.g., qualification in the vertical, horizontal or overhead positions usually qualifies for welding or brazing in the flat position. Also, qualification tests on groove welds will normally qualify that welder for the production of fillet welds in the same position. The code in force will dictate the exact limits on production welding and brazing positions, depending on the qualification test position(s).

Testing of Qualification Welds and Brazes: All codes and specifications have definite rules for testing qualification welds and brazes. Most frequently, for welds, mechanical bend tests are made on specimens cut from specific locations in the welds. Some codes will permit welder qualification testing using radiography instead of mechanical bend testing. This radiographic examination may be permitted, either alone or in conjunction with mechanical or other tests. Other types of nondestructive test methods, such as penetrant testing may also be applied to measure the apparent soundness of the qualification weldment. All codes will require that the test welds be sound and thoroughly fused to the base metal.

Examination of braze qualification coupons is normally accomplished by cutting through the joint and measuring the degree of bonding between the members being joined.

Other properties required of procedure qualification welds, such as tensile strength and fracture toughness of the weld metal, are not usually specified in welder qualification.

Welders or brazers whose test welds or brazes meet the prescribed requirements are qualified to apply the process and to weld or braze with filler metals and procedures similar to those used in testing. The prescribed limits of similarity of the substitute filler metals and procedures must be stated.

Retests

When do welders or welding operators require retesting? Some of the circumstances are as follows:

- Failure of the initial test welds.

- A significant change in the welding procedure.

- A welder has not been engaged in a particular welding process for an extended period (usually three or six months).

- There is reason to question a welder's or welding operator's ability.

Refer to the applicable code or specification for the specifics of such retests.

Standardization of Tests

The purpose of welder and welding operator qualification testing is to determine whether this person will be able to make satisfactory welds in production. The test is not perfect, nor will it guarantee that all production welds will be defect-free. The specimen welds used for qualification do not usually correspond in detail to all of those that will be encountered in production welding. They are standardized welds, which do therefore permit uniform evaluation of the welder's skill. However, additional tests add little information about a welder's or welding operator's ability.

Relation of Qualification Tests to Welder Training

Welders and welding operators should not be trained merely to the extent necessary to pass a prescribed qualification test. Because those tests are standardized, training should cover a broader and more extensive range of procedures and joint details to prepare for conditions that will be encountered in production. Training is the responsibility of the manufacturer or fabricator for whom the welder works.

Code and Standard Requirements

AWS D1.1, Structural Welding Code - Steel: The AWS Structural Welding Code (AWS D1.1) requires that all welders, welding operators and tackers to be employed under this code shall have been qualified by tests as specified in Section 5, Parts C, D and E. The tests are often administered by you, the welding inspector. If the welder, welding operator or tacker has previously demonstrated his qualification under other acceptable supervision, you may, at your option, make certain of that fact and then consider the applicant qualified for the present job.

ASME Boiler and Pressure Vessel Code: The ASME Boiler and Pressure Vessel Code specifies that only welders, welding operators and brazers who are qualified in accordance with Section IX shall be used. In addition, Section III, Division 1 (Nuclear Components) and Section VIII, Division 2 (Pressure Vessels, Alternate Rules) specify that tack welds must also be made by qualified welders.

The ASME qualification tests for welders and welding operators call for the same kinds of bend specimens as used for procedure qualification.

Any welder, welding operator or brazer who makes a procedure qualification test that passes satisfactorily is automatically qualified for that process, within the limitations prescribed in Section IX. Welders and welding operators making repair welds shall be qualified in accordance with Section IX. Qualification is limited in duration, to the extent that it is dependent upon continued satisfactory use of the process by the welder with no lapses exceeding 90 days, or it may be extended for an additional 90 days if the welder has welded with another process.

API Standard 1104: The API Standard 1104 relies on visual examination and destructive tests (optionally, on radiographic examination) to qualify welders. Since this standard deals with the requirements for welding pipelines and related equipment, all qualification tests are performed on pipe. No limitations are placed on the persons doing the testing, except the radiographer. The company, represented in practice by the inspector, is required to keep records of qualified welders and welding operators.

Destructive Test Qualification

Destructive tests and examinations are made in-house or outside by personnel who are qualified through training, education or the manufacturer's certification standards. Many fabricators do not have properly trained or qualified people nor the proper testing equipment, so they call on a contract laboratory to run the tests. On the other hand, fabricators with the necessary equipment may arbitrarily declare that certain of their technicians, inspectors or engineers are qualified, and the tests are then run in-house.

Destructive tests may be made without qualification of personnel. It is the welding inspector's responsibility to verify that the tests were conducted according to the governing code, specification or other reference documents. For example, the AWS Structural Welding Code requires the all-weld-metal tension test (paragraph 5.11.4) to be performed in accordance with ASTM A370, *"Mechanical Testing of Steel Products."*

Summary

The welding inspector will likely be involved to some degree in the qualification of welding procedures and welders. He may be responsible for administering the tests or simply reviewing the documentation to determine if the qualifications are applicable and accurate.

The most important aspect of the welding inspector's job relative to qualification is the evaluation of production welding to determine whether or not the procedures and welders are producing welds of acceptable quality.

REVIEW - CHAPTER 8 WELDING PROCEDURE AND WELDER QUALIFICATION

Q8-1 Who is normally responsible for the qualification of welding procedures and welders?
  a. welder
  b. architect
  c. welder's employer
  d. independent test lab
  e. Code body

Q8-2 Which of the following destructive testing methods may be used for procedure qualification testing?
  a. tensile
  b. nick-break
  c. charpy
  d. bend
  e. all of the above

Q8-3 What is the pipe welding position where the pipe remains fixed with its axis horizontal, so the welder must weld around the joint?
  a. 1G
  b. 2G
  c. 5G
  d. 6G
  e. 6GR

Q8-4 What is the pipe welding position where the axis of the pipe lies fixed at a 45 degree angle?
  a. 1G
  b. 2G
  c. 5G
  d. 6G
  e. none of the above

Q8-5 What is the necessary pipe position test for welders who are trying to qualify to weld T-, K-, and Y- connections?
   a. 1G
   b. 2G
   c. 5G
   d. 6G
   e. 6GR

Q8-6 With relation to procedure and welder qualification, what is the most important part for the welding inspector?
   a. watching the welding qualification test
   b. identifying samples
   c. cutting test specimens
   d. testing specimens
   e. monitoring production welding

Q8-7 For most codes, if a welder continues to use a particular procedure, how long does his qualification remain in effect?
   a. indefinitely
   b. 6 months
   c. 1 year
   d. 3 years
   e. until he produces a rejectable weld

Q8-8 What document describes the requirements of welder qualification in accordance with ASME?
   a. ASME Section III
   b. ASME Section II, Part A
   c. ASME Section IX
   d. ASME Section XI
   e. ASME Section V

Q8-9 Qualification to weld cross country pipelines is normally done in accordance with:
   a. ASME Section III
   b. AWS D1.1
   c. AWS D14.3
   d. API 1104
   e. API 650

CHAPTER 9: WELDING, BRAZING AND CUTTING PROCESSES

Introduction

This chapter covers the relevance of welding, brazing and cutting processes to inspection. The information in the previous chapters will be summarized for each of the welding, brazing and cutting processes named in ANSI/AWS QC-1, *Standard for AWS Certification of Welding Inspectors*. Other limiting factors associated with each process will also be covered. Each of the processes will be treated as follows:

- Weldability
- Welding metallurgy
- Welding chemistry
- Limiting factors
- Discontinuities
- Inspection processes

Limiting Factors

In addition to the considerations covered in previous chapters, each welding process has certain other limiting factors which make it a

MASTER CHART OF WELDING AND ALLIED PROCESSES

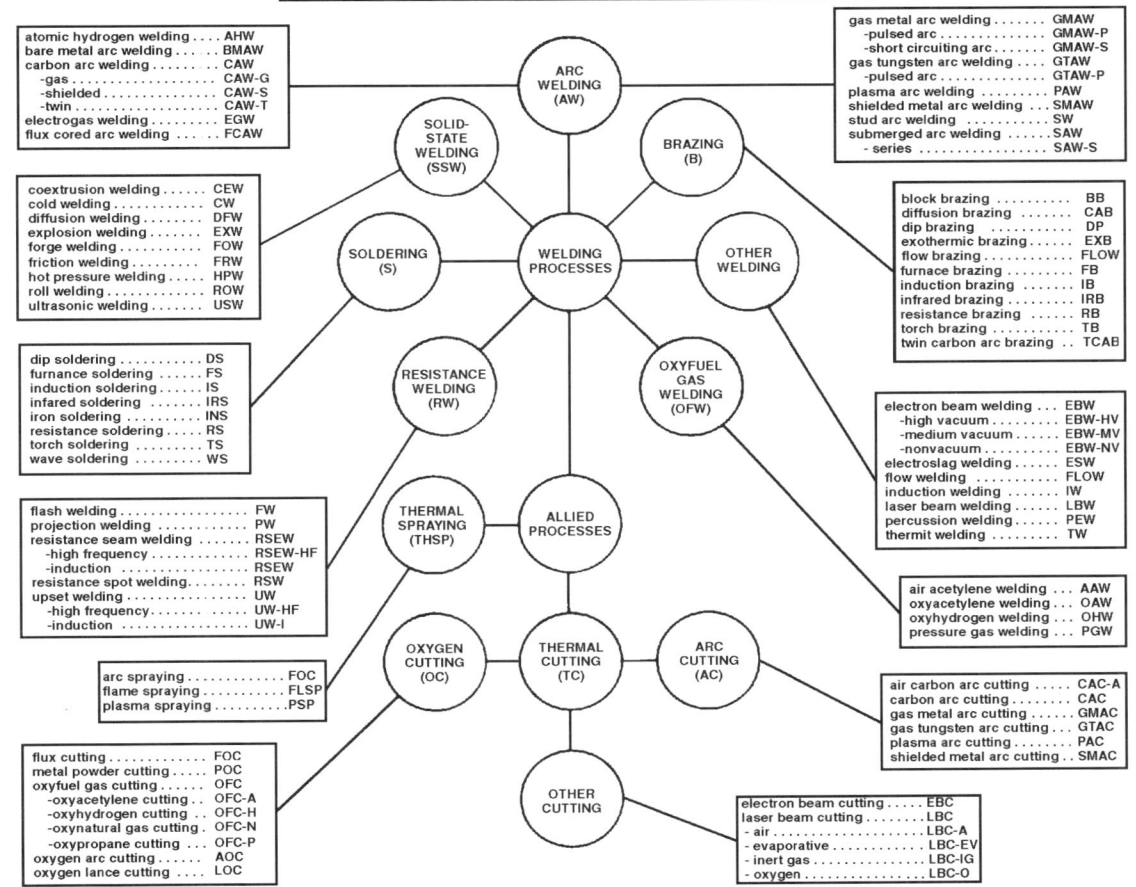

better choice for some applications than others. It may have been the obvious choice, but usually the selection is a compromise chosen because of cost and quality factors.

Whatever process is specified by the engineer, both the welding inspector and the welder must also be aware of the limiting factors for the specific job, so that both can anticipate and overcome the factors that could cause discontinuities or other more serious problems. In addition to weldability, these other factors influence the selection of a welding process:

- Dimensions of material being welded, especially thickness, and its shape and form
- The position in which welding must be done
- Requirements for the weld root
- Back side accessibility
- Joint preparation
- Availability of welding equipment, power sources, and fixtures
- Availability of welding consumables
- Availability of qualified welders and operators
- Quality level required
- Economics
- Safety

Dimensions of Material

The dimensions of a metal limit the use of some processes and encourage use of others. For heavy sections, the engineer may select flux cored arc, submerged arc or electroslag welding. However, sheet metal is brazed, soldered, oxy-acetylene welded, resistance welded, gas tungsten arc welded, or gas metal arc welded.

Welding Position

When the position of the joints cannot be located for flat position welding, limits on the application of arc and gas welding processes are imposed by the force of gravity. The small weld pool that must be used has little penetration, and the inspector may find slag inclusions and incomplete fusion as a result.

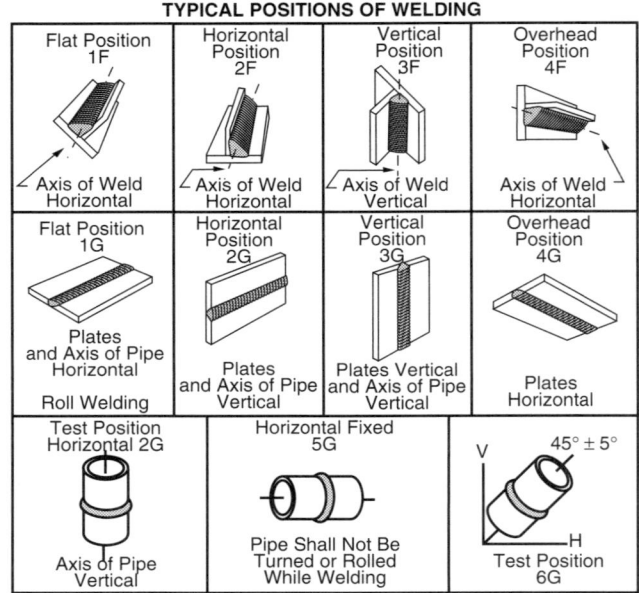

Root Requirements

The root of the weld joint must be fused by one technique or another, unless partial penetration welds have been specified by the design. Processes with deep penetration such as submerged arc, flux cored arc and spray transfer gas metal arc welding are chosen by the engineer in preference to shielded metal arc welding if that will simplify the joint preparation.

Root penetration in a tight butt joint welded by submerged arc welding in one pass from each side cannot be visually inspected. A joint which is backgouged to sound metal can be inspected with more confidence that the new root pass will fuse properly. The inspector relies on radiography and ultrasonic examination for volumetric inspection of such welds.

Back Side Accessibility

The accessibility of the back side of a joint strongly influences the choice of process. If inadequate aid is provided to the welder, he will need superb skill in making a root pass from one side of the joint. Such a weld made carelessly commonly displays incomplete fusion, incomplete joint penetration, excessive melt-thru, or other discontinuities.

Inaccessible roots are often welded using backing rings, consumable inserts, or automatic welding equipment. The inspector should expect to see good fit of the parts prior to welding from one side of the joint. A familiar example is the pipeline weld, traditionally deposited with E6010 type shielded metal arc electrodes. Radiographic inspection detects discontinuities in the root of many such welds.

UNION CARBIDE LINDE DIVISION

Joint Preparation

The suitability of available joint preparation also influences the process choice. The square-groove butt joint of an electroslag weld is unsuited for manual welding. Suppose the equipment fails midway through an electroslag weld. Does the fabricator have standby equipment? A double-V-groove preparation may be welded manually with no difficulty.

The groove angle of V-groove welds in heavy plate must be opened wider for gas metal arc welding than for covered electrodes, to accommodate the nozzle of the gun. Unsuitable joint preparations in any of these cases spell trouble for the welder and possibly will create imperfections which the inspector must catch.

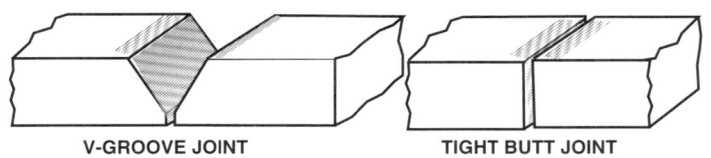

Availability of Welding Equipment

The availability of welding equipment obviously limits the choice of welding or cutting processes that can be considered for use, particularly in the field. Portability of the equipment would determine the availability in field situations. In remote areas, the process to be used is dictated by the availability of fixtures, power sources and consumables.

To name some specific examples, gas tungsten arc welding without high frequency starting requires another welder skill level, using copper starting tabs or striking the arc within the joint. Arc decay is more of a problem without a foot-controlled rheostat. Gas metal arc welding in the short circuiting mode requires a special power supply. Pulsed arc welding requires a pulse generator. Thin sheets subject to warping need hold-down fixtures and chill bars unless welding speeds are high.

Consumables

Consumable materials, such as bare electrodes or proper chemical composition and shielding gases needed for gas metal arc welding, may not be available. Covered electrodes are more readily supplied, with alloying elements added through the coating; they require no external shielding.

Availability of Welding Talent

The availability of the best welding equipment does not guarantee the success of a weldment.

Welders trained and qualified to use the equipment and apply the best techniques must also be available. If the desired talent is not available, a less sophisticated process will have to do; for example, manual shielded metal arc instead of gas tungsten arc. Maintenance work frequently is done by oxyacetylene welding. Root pass gas tungsten arc welds may require less skill than gas metal arc root pass welds.

Quality Level Required

The quality level required for a weldment may rule out certain processes. For this reason, it is not practical to weld reactive metals with the oxyacetylene welding process, because excessive oxidation would result in an unacceptable weldment. In conditions where the highest quality level is required gas tungsten arc welding would be used for root passes in lieu of shielded metal arc or gas metal arc welding.

Economics

Cost analysis is the "bottom line" in selecting a welding process. Economics favor the efficiency gained from automatic and semiautomatic equipment over manual processes.

Safety

Although safety is the last factor to be covered here, it should be the first to be considered. Safety is linked to economic considerations because of the cost of unsafe situations in loss of life and equipment, and because of insurance and legal considerations.

Some of the safety factors to be considered include the mandatory use of various types of protective equipment such as: safety glasses, hardhats, steel-toed safety shoes, welding and cutting filter lenses, leather welding garments, etc.

Welding Processes

Welding is a process which results in the joining of metals through their melting together by application of heat, with or without the application of pressure, and with or without the addition of filler metal. The welding processes discussed here are those with which the welding inspector must be familiar for the AWS CWI examination. These processes include:

- Shielded Metal Arc Welding (SMAW)
- Gas Metal Arc Welding (GMAW)
- Flux Cored Arc Welding (FCAW)
- Gas Tungsten Arc Welding (GTAW)
- Plasma Arc Welding (PAW)
- Submerged Arc Welding (SAW)
- Electroslag Welding (ESW)
- Oxyacetylene Welding (OAW)
- Stud Welding (SW)

LENS SHADE SELECTOR	
Operation	Shade number
Soldering	2
Torch brazing	3 or 4

Oxygen cutting
  Up to 25.4 mm (1 in.)
  25.4 mm to 152.4 mm (1 to 6 in.)
  152.4 mm (6 in.) and over
	3 or 4
	4 or 5
	5 or 6

Gas welding
  Up to 3.2 mm (1/8 in.)
  3.2 mm to 12.7 mm (1/8 in. to 1/2 in.)
  12.7 mm (1/2 in.) and over
	4 or 5
	5 or 6
	6 or 8

Shielded metal arc welding 1.6 mm (1/16 in.), 2.4 mm (3/32 in.), 3.2 mm (1/8 in.), 4.0 mm (5/32 in.) electrodes	10
Gas tungsten arc welding (nonferrous) Gas metal arc welding (nonferrous) 1.6 mm (1/16 in.), 2.4 mm (3/32 in.), 3.2 mm (1/8 in.), 4.0 mm (5/32 in.) electrodes	11
Gas tungsten arc welding (ferrous) Gas metal arc welding (ferrous) 1.6 mm (1/16 in.), 2.4 mm (3/32 in.), 3.2 mm (1/8 in.), 4.0 mm (5/32 in.) electrodes	12

Shielded metal arc welding
  4.7 mm (3/16 in.), 5.6 mm (7/32 in.), 6.4 mm (1/4 in.) electrodes
  7.9 mm (5/16 in.), 9.5 mm (3/8 in.) electrodes
	12
	14

Atomic hydrogen welding	10 to 14
Carbon arc welding	14

Shielded Metal Arc Welding (SMAW)

Shielded metal arc welding uses the heat of an electric arc between a covered metal electrode and the work. Shielding comes from the decomposition of the electrode flux coating. Filler metal is supplied by the electrode core wire and covering (iron powder and alloys).

This process is usually manually applied, but it has sometimes been automated with a machine or gravity stick feeder. The basic equipment comprises a power source, cables, an electrode holder, a work clamp, and the electrode. Electrodes for shielded metal arc welding operate variously on alternating current, direct current electrode positive (reverse polarity) or direct current electrode negative (straight polarity)

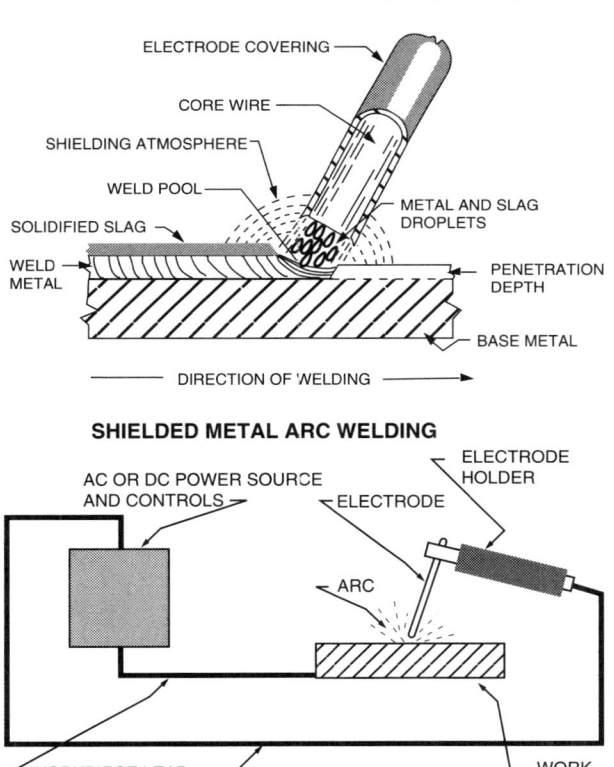

SHIELDED METAL ARC WELDING

TYPICAL WELDING CIRCUIT FOR SHIELDED METAL ARC WELD

Welding Metallurgy: A shielded metal arc weld is strengthened by adding alloying elements and by incorporating iron powder in the electrode covering. Unfortunately, some ingredients and the binder in the covering may attract and hold moisture, a source of hydrogen, which causes cracking in certain welds. A group of electrodes specifically formulated to result in weld deposits having very low levels of hydrogen are referred to as "low hydrogen." Those electrode classifications considered to be low hydrogen have identifications ending with the numbers 5, 6 or 8. Once removed from their sealed shipping containers, they require electrically-heated storage in a vented oven capable of holding the electrodes at temperatures between 250-350° F.

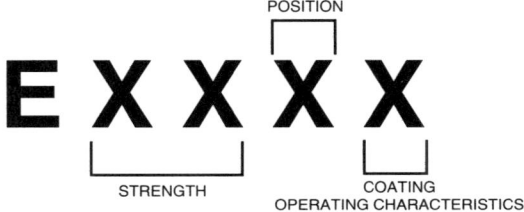

SMAW ELECTRODE IDENTIFICATION SYSTEM

Welding Chemistry: The inspector should understand the many functions of the electrode covering:

- Arc stabilization from ionizing elements (which dictate the usability of the electrode on AC, DCEP or DCEN)
- Gas shielding from decomposition of such ingredients as cellulose and limestone
- Fluxing of oxides from the work surface
- Deoxidation of deposited weld metal to reduce the tendency for porosity
- Slag protection over the cooling weld metal, resulting in slower cooling rates to improve weld properties
- Strengthening the weld metal by supplying alloying elements
- Increased weld metal deposition by incorporation of iron powder (up to 50% of covering)

Limiting Factors: All of the welding manipulations are controlled by the welder:

Classification	Current	Arc	Penetration	Covering & Slag	Iron Powder
F3 EXX10	DCEP	Digging	Deep	Cellulose-sodium	0-10%
F3 EXXX1	AC & DCEP	Digging	Deep	Cellulose-potasium	0
F2 EXXX2	AC & DCEN	Medium	Medium	Rutile-sodium	0-10%
F2 EXXX3	AC & DC	Light	Light	Rutile-potassium	0-10%
F2 EXXX4	AC & DC	Light	Light	Rutile-Iron powder	25%-40%
F4 EXXX5	DCEP	Medium	Medium	Low hydrogen-sodium	0
F4 EXXX6	AC or DEEP	Medium	Medium	Low hydrogen-potassium	0
F4 EXXX7	AC or DCEP	Medium	Medium	Low hydrogen-iron powder	25%-40%
F1 EXX20	AC or DC	Medium	Medium	Iron oxide-sodium	0
F1 EXX24	AC or DC	Light	Light	Rutile-iron powder	50%
F1 EXX27	AC or DC	Medium	Medium	Iron oxide-iron powder	50%
F1 EXX28	AC or DCEP	Medium	Medium	Low hydrogen-iron powder	50%

Note: Iron powder percentage based on weight of covering.

Significance of Last Digit of SMAW Identification

electrode inclination, arc length, travel speed, weaving, or backstepping. The welder must set the proper welding current at his power source and select the polarity (if direct current).

For safety, the inspector must not watch the electric arc without the protection of No. 10 or 12 filter lenses (No. 14 if the welder is using more than 400 amperes). Safety glasses with side shields are needed for protection from slag when being removed from the weld face. The inspector and the welder need forced ventilation when working in confined areas.

Discontinuities: Almost any discontinuity can be produced by the shielded metal arc welding process, if not applied properly. Porosity is most likely caused by faulty technique, or by contaminated base metal or electrode (moisture). Cluster porosity (sometimes referred to as starting porosity) often occurs at the initiation and termination of the arc. Back stepping at the start permits the experienced welder to remelt his cold start area and float the gas out of the weld metal. Piping porosity (also called wormhole or elongated porosity) is often found in both single pass fillet welds and groove welds produced at high welding speeds, especially if the joint is contaminated.

Slag inclusions are most often the result of improper welding technique, through insufficient interpass cleaning or faulty manipulation of the electrode. If the designer has provided insufficient access for welding within the joint, slag inclusions could result.

Incomplete fusion suggests inadequate welding current or difficult access to the joint, especially in multiple pass welds. It may also be a consequence of improper joint preparation, such as insufficient groove angle.

Incomplete joint penetration is normally the result of an excessive root face, insufficient welding current or insufficient root opening. Joint designs should not overestimate the penetration ability of the process.

Undercut, underfill and overlap are all due to welder errors. Traveling too fast will cause undercut, as will excessive welding currents. Traveling too slow will produce overlap.

Lamellar tearing is not the fault of the welder. This discontinuity normally results from low quality steel plates or an improper joint design for joining heavy sections. Use of massive amounts of weld metal in a joint design which results in high shrinkage stresses in the through-thickness direction of rolled plate or structural shapes will often cause lamellar tearing.

Shielded metal arc welds may fail by either hot or cold cracking. Hot cracks include crater, root and throat (or longitudinal) cracks. They result from the high stresses associated with weld shrinkage when cooling from welding temperatures. Materials susceptible to "hot-shortness" are particularly prone to these types of cracks. When they appear, changes in technique (perhaps a thicker bead) or changes in materials are needed.

Toe and underbead cracks are forms of cold cracks. Toe cracks often arise when a weldment is subjected to fatigue loads, especially when the presence of excessive weld reinforcement or convexity results in an acute reentrant angle at the toe. Underbead cracks (and toe cracks, at times) are the result of hydrogen being present in the weld zone. They are also referred to as delayed cracks because they may not appear until many hours after the weld has cooled to room temperature. Sources of the hydrogen that contribute to these cracks include damp electrodes, humid conditions and dirty surfaces.

Inspection Processes: All of the usual inspection processes are used to detect discontinuities in shielded metal arc welds.

Gas Metal Arc Welding (GMAW)

Gas metal arc welding (sometimes called metal inert gas or MIG) uses the heat of an electric arc between a continuous bare wire filler metal electrode and the work. Shielding is obtained entirely from an externally supplied inert gas (argon or helium) or reactive gases (CO_2 or O_2), or some combination thereof.

This can be a semiautomatic, machine, automatic, or automated process. In the semiautomatic mode, the welder controls the inclination and distance of his gun from the work, and the travel speed and manipulation of the arc. Arc length and electrode feed are controlled automatically by the power source and wire feeder controller.

The gas metal arc process deposits the weld metal in the joint by one of the following modes: spray transfer, globular, short circuiting transfer, and pulsed arc welding.

Spray transfer occurs with high current and voltage combinations, when there is at least 80% argon present in the shielding gas. For each electrode size, there is a threshold current above which the metal pinches off in fine droplets many times a second. The current propels the droplets axially down the center of the arc, away from the

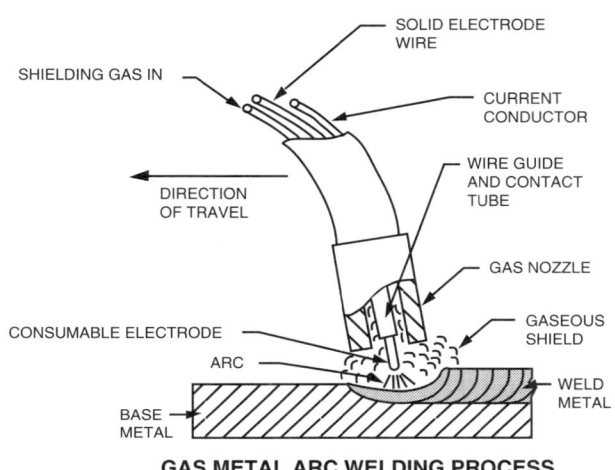

GAS METAL ARC WELDING PROCESS

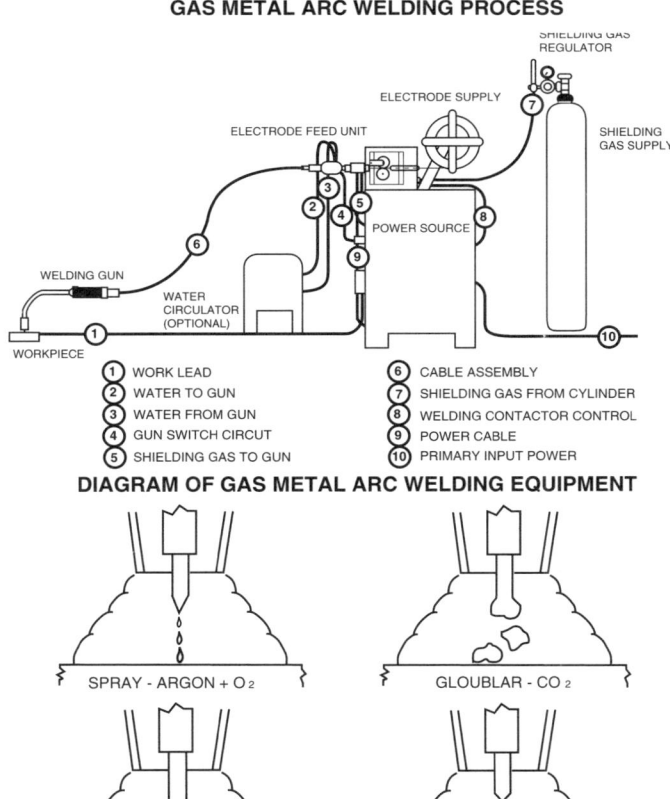

DIAGRAM OF GAS METAL ARC WELDING EQUIPMENT

THE FOUR TYPES OF METAL TRANSFER

electrode, straight into the pool. Spray transfer mode best defines the arc and the pool for the welder. It requires high current relative to the diameter of the electrode. Due to its high heat capacity, this mode of transfer is best suited for flat and horizontal position welding.

Globular transfer occurs at low currents compared to spray transfer—low, that is, in relation to the size of the electrode. Low current density at the electrode tip produces large, irregular drops of metal that transfer to the pool without much direction. The result is increased amounts of spatter compared to spray transfer.

Short circuiting transfer requires a power source having sufficient secondary inductance to prevent excessive spatter. As the electrode melts, each drop shorts out the arc during its transfer to the pool, but the electrical impedance built into the power source prevents the welding current from spattering the bridging drop. The short circuiting mode is a relatively cold process, and its misapplication may result in incomplete fusion. It readily bridges gaps. Sheet metal can be welded without excessive melt-through and welds may be made in all positions.

Pulsed Arc welding maintains a low current arc as the background condition and injects high current pulses upon that current. Transfer of filler metal is by spray of droplets during each pulse. Pulsing the power lowers the average heat input from the current. Out-of-position welding then becomes possible using larger wire sizes.

Welding Chemistry: Shielding gases protect gas metal arc welds from the atmosphere. Fluxes are not used in this process. All deoxidizers and alloying elements are incorporated into the electrode wire.

Limiting Factors: The semiautomatic features of gas metal arc welding simplify the training of the welder and should make his work more consistent. However, the short circuiting mode may lead to incomplete fusion. The welding inspector should recognize that joints welded by this process are not prequalified under AWS D1.1, Structural Welding Code- Steel. The contractor must demonstrate that the welding procedure to be used under short circuiting conditions will indeed produce satisfactory welds, with adequate fusion.

For safety, the inspector exposed to a nearby welding arc must have protective clothing against arc burn, and he must not watch the arc without eye protection from No. 12 filter lenses (No. 14 if welding current exceeds 400 amperes). The lens may be in a helmet or hand-held shield. When the welder is welding in a confined area, ventilation must be provided; however, that ventilation must not be such that it disrupts the gas shielding.

Discontinuities: Gas metal arc welding may result in any of the common weld discontinuities, with the exception of slag inclusions.

Porosity, which is caused by gas trapped in the weld, is often caused by improper gas shielding. The shielding gas must displace the surrounding atmosphere, which contains oxygen and nitrogen. Welding without adequate shielding permits atmospheric oxygen and nitrogen to dissolve in the molten metal. Use of excessive shielding gas flow rates will also lead to the production of porosity due to the vortex action produced which tends to draw atmospheric gases into the arc region.

Incomplete fusion is possible, especially in welds made with short circuiting transfer. It may occur with globular transfer or even with pulsed or spray transfer at low current levels if the welder fails to properly manipulate the welding arc.

Incomplete joint penetration is more likely to occur with short circuiting and globular transfer

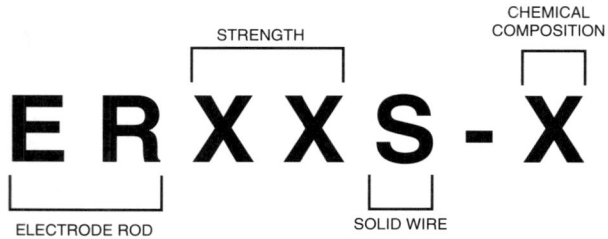

GMAW ELECTRODE IDENTIFICATION SYSTEM

than with spray conditions. The spray arc in argon results in good penetration when properly directed into the joint.

The presence of undercut and underfill reflect poor technique by the welder. Overlap is more prevalent in globular transfer and with the short circuiting arc.

Opening up of seams, laps, laminations, and lamellar tears is to be expected where welds are constrained. High stresses occur because of the rapid heating and cooling that accompany arc welding.

Cracks are to be expected in the weld metal if improper joint designs are used. Moreover, there is one type of crack peculiar to gas metal arc welding. It occurs beside the deeply penetrating finger of weld metal under a weld made with argon shielding, and it is strictly due to the unusual stress pattern created by the shape of the bead and the rapid heating and cooling that occur.

Inspection Processes: The inspection processes for evaluation of discontinuities in gas metal arc welds include nondestructive test methods for cracks in the heat affected zone, visual, magnetic particle and penetrant examination of cracks open to the surface, and radiographic, ultrasonic and magnetic particle inspection for sub-surface cracks. Radiography and ultrasonic testing are also utilized for the discovery of other types of subsurface discontinuities.

Flux Cored Arc Welding (FCAW)

Flux cored arc welding uses the heat of an arc between a continuous filler metal electrode and the work, as in gas metal arc welding. However, shielding is obtained, in whole or in part, from a flux contained within the tubular electrode. Self-shielded electrodes require no external gas protection, while other flux cored electrodes use additional external gas shielding, commonly carbon dioxide, supplied through the welding gun.

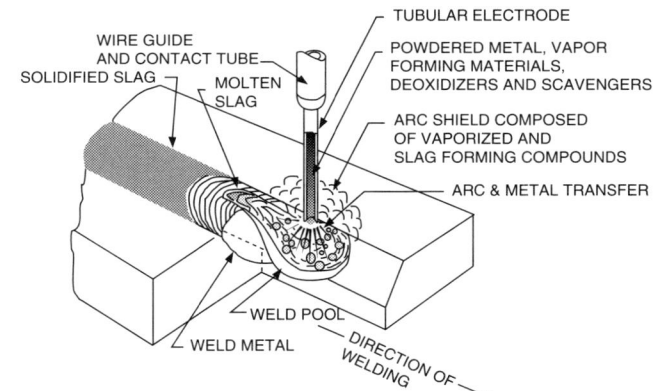

SELF-SHIELDED FLUX CORED ARC WELDING

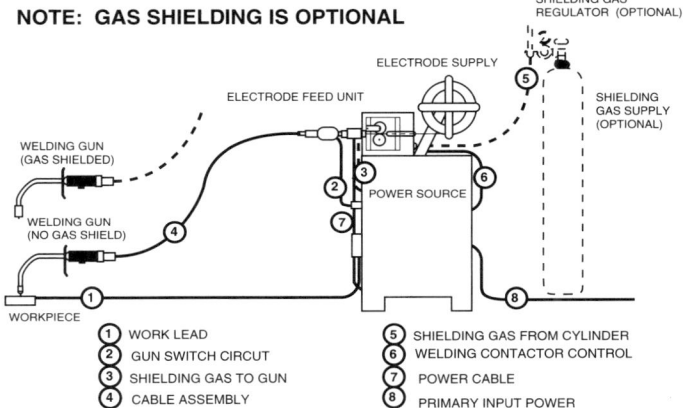

DIAGRAM OF FLUX CORED ARC WELDING EQUIPMENT

Welding Chemistry: The flux cored arc welding electrode contains flux, deoxidizers, alloying elements, and sometimes denitriders within the tubular wire. If external shielding is provided, the choice of shielding gas is usually carbon dioxide (CO_2) because of its low cost plus its ability to improve the penetration characteristics of the weld pass. Carbon dioxide is an active gas (rather than being inert like argon or helium); strong deoxidizers must be present in the electrode to produce a sound weld. Under self-shielding conditions even more protection must be

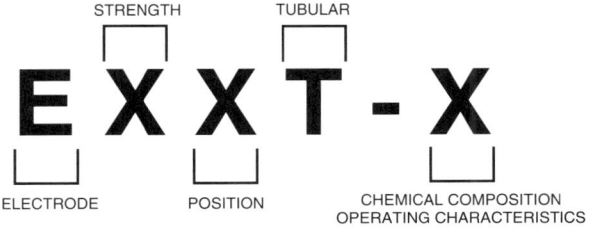

FCAW ELECTRODE IDENTIFICATION SYSTEM

built into the flux core. Gas mixtures are used to combine the separate advantages of two or more gases. Thus 75% argon-25% carbon dioxide may be used to improve the operating characteristics of the arc plus provide excellent mechanical properties of the finished weld.

Limiting Factors: Uniform quality is required in the flux cored electrodes. Any undesired irregularity in the structure of the flux cored electrode will introduce faults into the weld deposit. Since the wire is tubular and the absorption of moisture by the flux is possible, weld porosity may result if the electrode is improperly stored. Evidence of poor storage of the electrode coils in the shop should alert the inspector to this possibility.

Normal safety precautions, including eye protection (No. 12 filter lens) and sufficient ventilation, are necessary for the welding process. Safety goggles with side shields are also required when working in the vicinity where slag is being chipped from beads.

Discontinuities: Discontinuities mainly involve porosity and slag inclusions. Any unexpected porosity or bead irregularity that the welder cannot acknowledge as his fault may indicate a void in the flux core, deficiency of some ingredient in the flux, or improper storage conditions. The slag produced by the internal flux naturally increases the chance for slag inclusions, especially when the welder tends to travel too slowly for a given set of operating conditions.

Gas Tungsten Arc Welding (GTAW)

Gas tungsten arc welding (sometimes referred to as tungsten inert gas, or TIG welding) uses an electric arc between a nonconsumable electrode and the work. Shielding is obtained from an inert gas or inert gas mixture, except in the case where argon-hydrogen gas mixtures are used for some stainless steel applications. Filler metal may be added as needed. The torch is usually water-

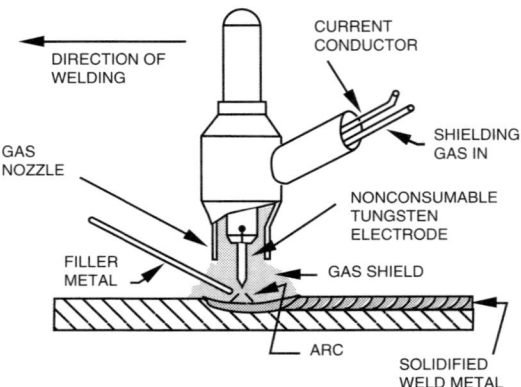

GAS TUNGSTEN ARC WELDING OPERATION

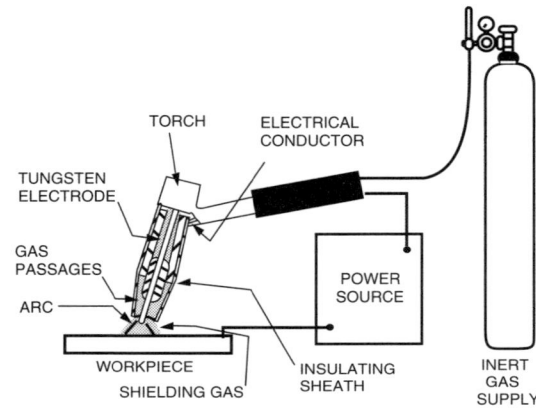

GAS TUNGSTEN ARC WELDING EQUIPMENT

cooled, but may be air-cooled for low current applications.

This type of welding may be accomplished by manual, mechanized or automatic methods. When filler metal is added, the process calls for a two-handed technique, as in oxyacetylene welding. Cold-wire and hot-wire feeds are automated versions of that technique.

Welding Metallurgy: The slow heating and lower temperatures combined with slower cooling rates characteristic of GTAW will result in improved weld metal and heat-affected zone mechanical properties. Welds may be made without filler metal when the edges of the joint provide sufficient material and deoxidizers.

Welding Chemistry: The tungsten electrode contributes neither deoxidation nor fluxing, so it is fortunate that the melting is essentially slow and

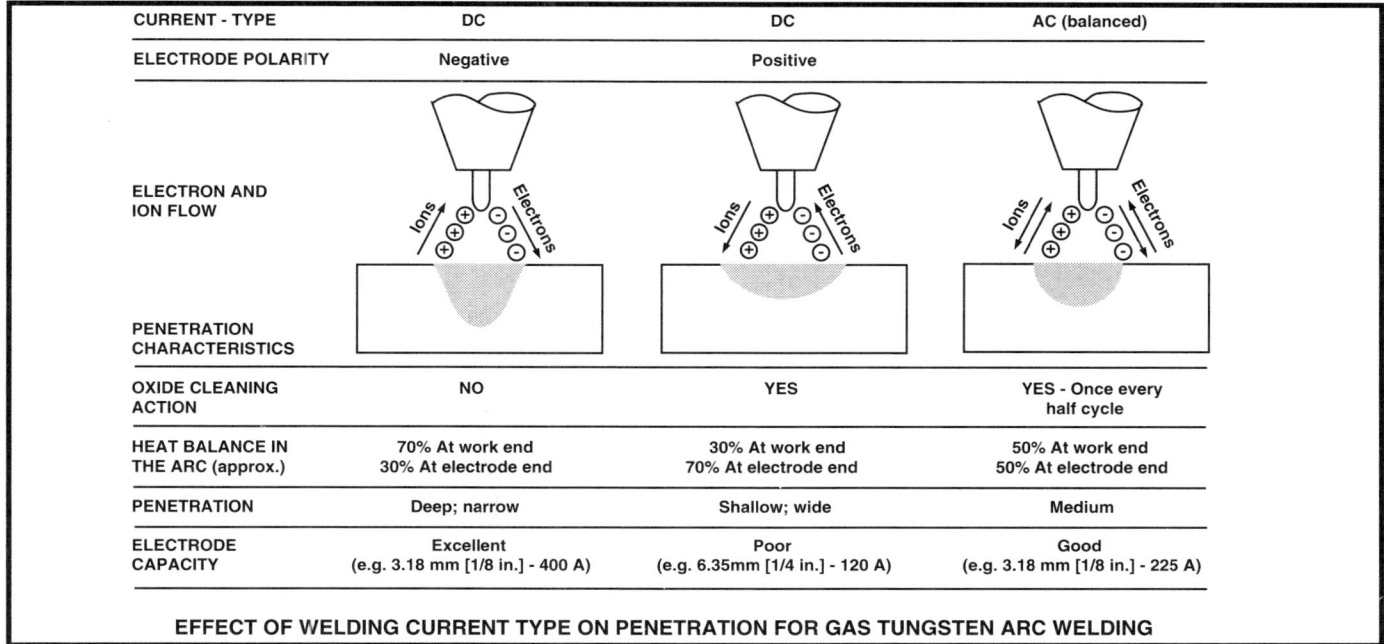

EFFECT OF WELDING CURRENT TYPE ON PENETRATION FOR GAS TUNGSTEN ARC WELDING

that most of the gases evolved can escape from the weld pool before it freezes. The corrosion resistant oxide film on aluminum and aluminum alloys, and on magnesium and its alloys, can be removed by the argon-shielded gas tungsten arc or gas metal arc using direct current, electrode positive. In alternating current operation, this occurs every half cycle, and arc re-ignition is normally accomplished by a superimposed high frequency current.

Filler rod containing the needed deoxidizers may be added manually or by the use of a wire feeder for mechanized applications.

When fully shielded by argon or helium, the tungsten electrode is not noticeably consumed, although some evaporation occurs while it is hot. The hot tip may become molten. Thoriated or zirconia-treated tungsten more readily emits electrons and has a smaller molten spot. This tends to improve the life expectancy of these types of electrodes compared to that of pure tungsten electrodes.

Limiting Factors: The outstanding feature of the gas tungsten arc welding process is the exceptional cleanliness that can be obtained in the weld deposits, producing crack-free welds in alloys that are difficult to weld by other processes. However, the process has its limiting factors. For example, the skill level necessary to produce high quality welds is only acquired by long experience in manipulating the electrode and feeding the filler wire (when used).

Without skillful control, incomplete fusion, incomplete joint penetration, overlap, and cracking are possible. Another drawback is the possibility of tungsten from the electrode entering the weld,

AWS TUNGSTEN ELECTRODE CLASSIFICATIONS

AWS CLASSIFICATION	ALLOY	COLOR
EWP	Pure Tungsten	Green
EWTh-1	0.8-1.2% Thoria	Yellow
EWTh-2	1.7-2.2% Thoria	Red
EWTh-3	0.35-0.55% Thoria	Blue
EWZr	0.15-0.40% Zirconia	Brown
EWCe-2	1.8%-2.2% Ceria	Orange

through inadvertent touching of the electrode to the molten puddle by the welder.

Specialized equipment and consumables are available for automated welding. Thoriated or zirconia-treated tungsten electrodes are used for improved arc stability. Argon and helium are the preferred shielding gases.

For safety from the brilliant visible arc of this process, the inspector must have protective clothing against arc burn and a shield with No. 12 filter lenses (No. 14 for over 400 amperes). Ventilation is needed in confined areas, but it should not disturb the shielding gas.

Discontinuities: All of the common types of discontinuities are possible with gas tungsten arc welding, with the exception of slag inclusions.

Porosity is a common discontinuity due to the process' low tolerance for contamination. Impurities on the surface or within the base metal readily lead to porosity if no deoxidizers are introduced into the weld pool. Loss of shielding gas will also result in the creation of porosity.

Tungsten inclusions may result from accidental touching of the tungsten electrode to the weld pool. The hot tip may have a molten spot to which the arc is attached. The molten drop is readily transferred to the pool, thus producing a tungsten inclusion in the weld. Use of excessive welding currents for a given electrode diameter can also contribute to the deterioration of the electrode and result in tungsten inclusions. The use of a "scratch-start" technique in the absence of high frequency current may also result in the production of this discontinuity. Whether such inclusions are rejectable depends on the governing specifications.

Incomplete fusion or incomplete joint penetration may be encountered if the welder's technique is poor. The penetration of the arc in gas tungsten arc welding is relatively low; arc voltage and penetration can be increased by helium additions to the shielding gas. For this reason, only those joints designed especially for gas tungsten arc welding should be specified.

Undercut, underfill and overlap as well as trouble with laminations, seams and laps can occur as with other welding processes.

Cracks are usually hot cracks in gas tungsten arc welds. Longitudinal cracking may occur in deposits made at high travel speeds. Crater cracks are prevalent due to improper control of the welding current at the termination of the weld. A hand- or foot-operated current control is usually used to provide proper welding and slope control.

Cold cracking due to hydrogen in the weld metal could be caused by the presence of moisture or contamination which produces hydrogen. This type of cracking could also result from the presence of high stresses in the weld zone.

Inspection Processes: All inspection processes are used, especially radiography when tungsten inclusions are to be detected.

Plasma Arc Welding (PAW)

Plasma arc welding is a process which utilizes a constricted arc between the electrode and the workpiece (transferred arc) or the electrode and the constricting nozzle (nontransferred arc). In many respects, this process is quite similar to gas

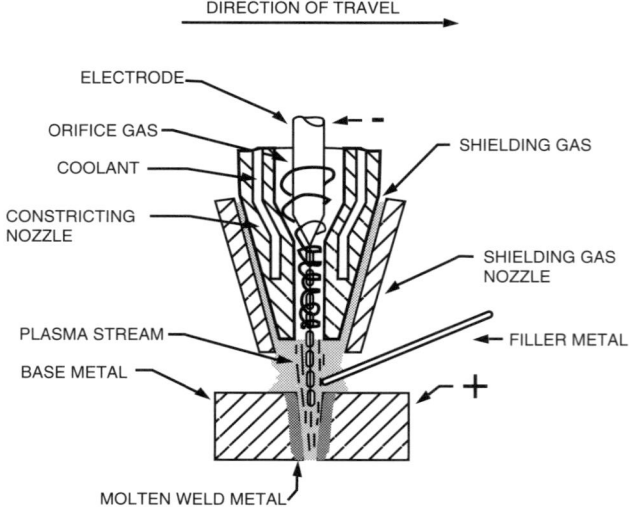

PLASMA ARC WELDING (Keyhole Mode)

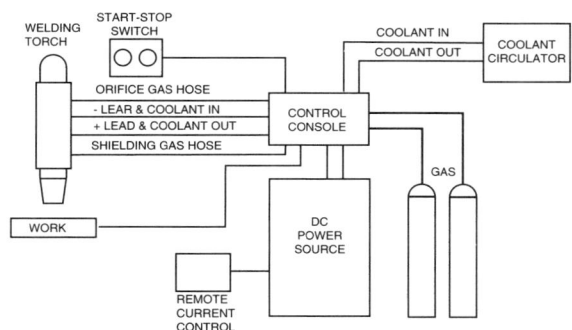

TYPICAL EQUIPMENT FOR PLASMA ARC WELDING

tungsten arc welding; however, the constricted arc provides a much more localized heat source.

Shielding for this process is obtained from the hot, ionized gas issuing from the torch which may be supplemented by an auxiliary source of shielding gas. This shielding gas may be inert gas or a mixture of gases. There is no pressure utilized and filler metal may or may not be required.

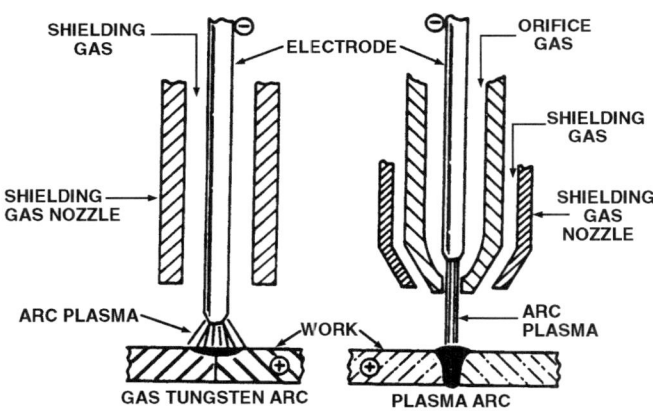

Comparison of GTAW and PAW Torches

Like the gas tungsten arc process, plasma arc welding uses a tungsten electrode, but it is recessed into the torch. In most respects, the application of this process is identical to gas tungsten arc welding, so the previous comments relative to GTAW will be applicable.

Welding Metallurgy: The constricted arc produces a more localized heat source which results in the ability to weld materials at higher travel speeds than those obtainable with gas tungsten arc welding. This will tend to reduce the heat input which will result in faster heat affected zone cooling rates. This might increase the heat affected zone hardness in some materials to the degree that cracking might result. Like gas tungsten arc welding, filler metal addition may not be necessary if the edges of the base metals are melted and fused together.

Welding Chemistry: Since the tungsten electrode is not intended to be consumed, it contributes neither deoxidation nor fluxing to the molten metal. Filler metals to be used should contain deoxidizers to reduce the tendency for the production of porosity. Base and filler metal cleanliness is also important for this process, as with gas tungsten arc welding.

Limiting Factors: All those items mentioned for gas tungsten arc welding are appropriate for plasma arc welding as well. The constricted arc is less sensitive to changes in arc length, plus the distance from the torch to the work is longer than that for gas tungsten arc welding. Both of these factors tend to make the plasma arc welding process slightly easier to control manually.

One variation of plasma arc welding used for producing complete joint penetration groove welds is referred to as the "keyhole" technique. With this method, welding is performed on a square-groove butt joint with zero root opening. The concentrated heat of the arc penetrates through the material thickness to form a small keyhole. As welding progresses, the keyhole moves along the joint melting the edges of the base metal which then flow together and solidify after the welding arc passes. This creates a high quality weld, with

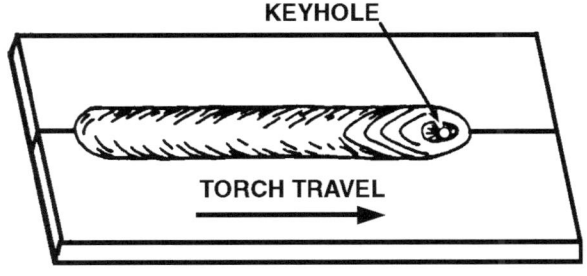

Keyhole Technique for Plasma Arc Welding

no elaborate joint edge geometry and faster welding travel speeds compared to gas tungsten arc welding. The application of this technique is limited to materials 1 inch and less in thickness.

The faster welding speeds obtainable with this method will tend to reduce the amount of distortion present after welding.

Discontinuities: All of the discontinuities mentioned in the previous discussion of gas tungsten arc welding are also applicable for plasma arc welding.

There is a reduced tendency for tungsten inclusions with plasma arc welding due to the fact that the tungsten is recessed in the torch. Use of excessive welding currents may cause overheating of the copper constricting nozzle, resulting in copper inclusions in the weld.

Since the arc is constricted, mis-tracking of the joint is quite possible, resulting in incomplete fusion, especially when the keyhole technique is being used.

Another discontinuity unique to plasma arc welding with the keyhole technique is commonly referred to as "tunneling." This is simply a pore oriented through the weld throat at the termination of a keyhole weld. It results when the welder fails to properly fill the keyhole at the weld termination.

Inspection Processes: All inspection processes are used for examination of plasma arc welds.

Submerged Arc Welding (SAW)

Submerged arc welding uses the heat of an electric arc or arcs between the electrode or electrodes and the work, all buried beneath a shielding blanket of granular flux.

Welding under a granular flux is a semiautomatic, mechanized or automatic process in which electrode feed and arc length are controlled by the wire feeder and power supply. In automatic welding, a travel mechanism moves either the torch or the work, and normally a flux recovery system recirculates the unfused granular flux back to the flux hopper for reuse.

The arc is hidden in submerged arc welding, which frees the welder or operator from his helmet but hides the path he must follow. For machine or automatic welding, the path is prealigned or a seam-tracking apparatus controls the orientation of the torch relative to the joint centerline. In semiautomatic submerged arc welding, the gun is actually moved along the joint in contact with the faces of the work (usually a T-joint or groove weld) to control the location of the weld. This process may produce deep penetration by the arc. Straight butt joints may be welded in metal 1 inch thick in one pass from each side with complete joint penetration if the joint is accurately followed and the current density is high.

Welding Metallurgy: The high heat input of this process produces rapid heating and cooling.

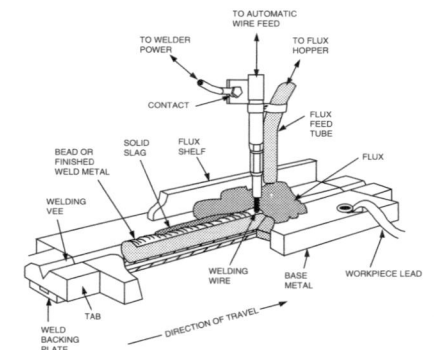
SCHEMATIC VIEW OF SUBMERGED ARC WELDING PROCESS

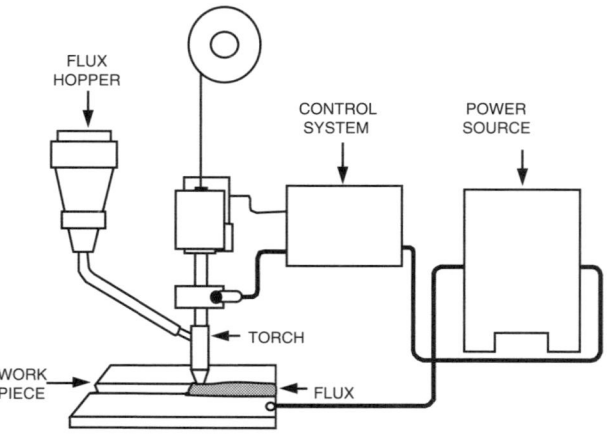
SUBMERGED ARC WELDING EQUIPMENT

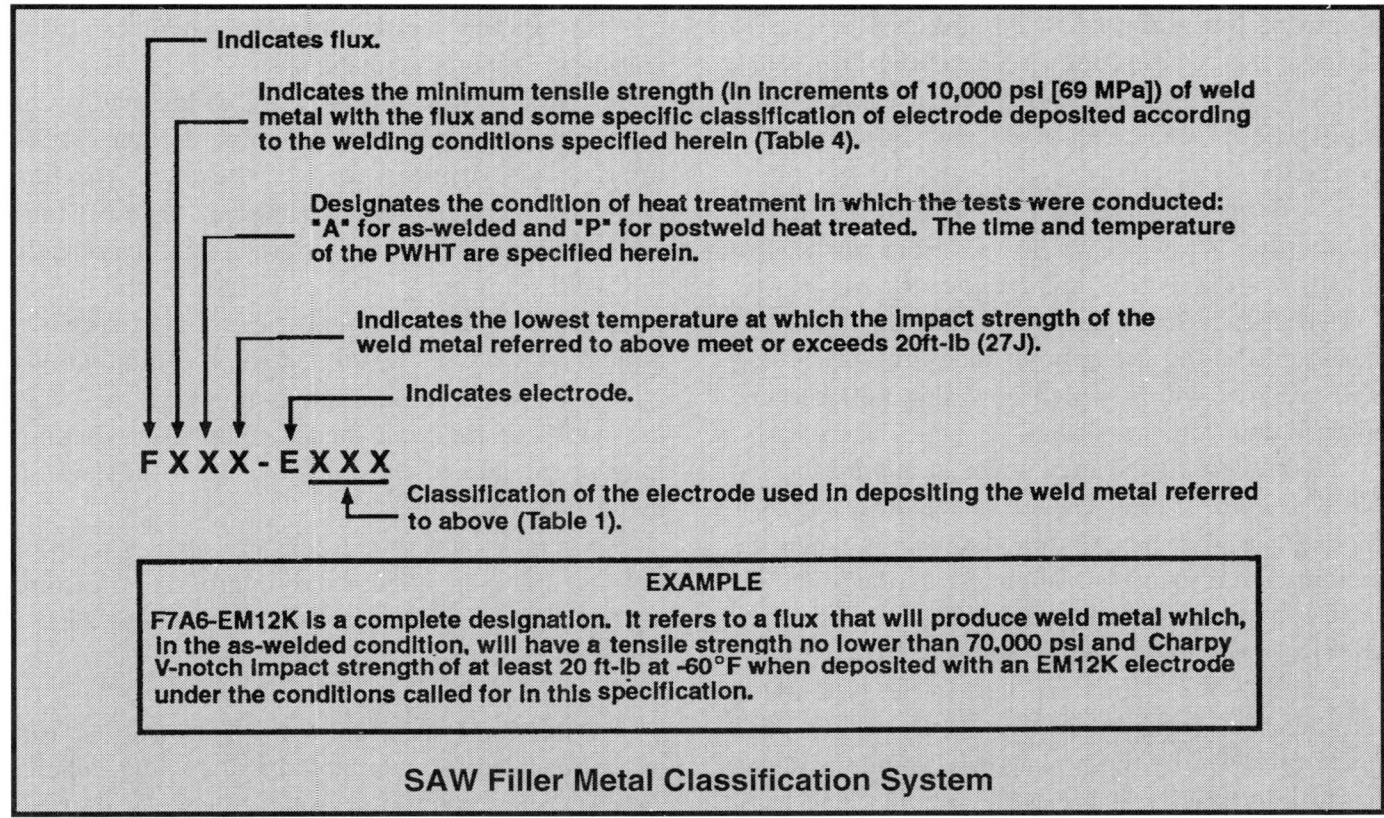

SAW Filler Metal Classification System

Sound welds are deposited rapidly and economically, with excellent metallurgical properties.

Welding Chemistry: The weld composition results from contributions from the melted base metal and the electrode, modified by chemical reactions with the flux and alloys added through the flux. Since the flux and filler wire are independently dispensed in this process, great flexibility in obtaining weld properties is possible. The inspector must make sure that the proper wire and flux combination is used to obtain the correct weld composition. Shop dirt, grease or moisture can contaminate the flux, resulting in cracks. Some fluxes require heated storage containers and hoppers to assure that the flux is dry when used.

Limiting Factors: Cleaning of the work surfaces and aligning of the machine travel with the joint are particularly important in submerged arc welding. Improper alignment will result in offset beads with incomplete joint penetration. In a highly restrained joint, joint misalignment may also cause cracks.

Welding operators are needed who are trained in controlling the variables of the process. For example, a change in the proportion of flux or arc voltage will change the weld composition.

For safety, goggles are required, but filter plates are not necessary because the arc is not visible.

Discontinuities: Welds may exhibit all of the common discontinuities. Porosity is often encountered in submerged arc welds because of the high travel speeds and rapid freezing which results. Related discontinuities, perhaps of only minor importance but hard to overlook because they are so visible, are gas dents on the bead surface. They are caused by large bubbles of gas trapped beneath the solidified slag while the metal is still molten. Changes in flux granulation (fewer fines), changes in composition, or reduction in the height of the granular flux blanket will tend to reduce this occurrence. Cleaning the work surfaces, slowing the travel speed or using a more highly deoxidized electrode may cure internal porosity.

Slag inclusions are found in many submerged arc welds. Convex bead profiles in multiple pass welds are frequently the cause. Convex beads leave narrow pockets against prior beads and along the groove faces in which slag will cling or within which new slag will become trapped under the next bead.

Incomplete fusion may occur in attempting too large a weld in one pass or in welding too fast. Incomplete joint penetration also occurs when the welding arc is carelessly aligned with the joint.

Undercut is not uncommon in submerged arc welding when high welding currents are used.

Laminations, seams and laps in the base metal may lead to cracks in submerged arc welds. The notch provided by these plate discontinuities tends to initiate cracks in the weld metal. The high stresses in the weld may also cause lamellar tearing in susceptible materials.

Cracks in submerged arc welds may occur when the material is hot or cold. Crater cracks may be anticipated unless the operator has perfected a crater-filling technique. Run-off tabs are commonly used to put starts and stops outside the weld joint area. Throat cracks in small root pass weld beads between heavy sections are typical in a highly restrained joint. In situations where the weld penetration is much greater than the width of the weld bead, centerline cracking may occur due to the undesirable solidification pattern which results.

Toe cracks and root cracks are often a form of delayed cracking attributed to hydrogen. Moisture in the flux is the prime suspect in these occurrences.

Inspection Processes: The inspection processes used for submerged arc welding include visual, radiographic, ultrasonic, magnetic particle and penetrant testing.

Electroslag Welding (ESW)

Electroslag welding uses an electrically melted metallurgical flux that melts the filler metal and the surfaces of the work. The heat is created by the electrical resistance of the flux. There is no arc except at the start of the weld before the granular flux melts and becomes conductive. The slag is then kept molten by its resistance to the flow of electric current passing between the electrode and the work.

Electroslag welding is purely a mechanized or automatic process. The melted base metal, electrode and guide tube (in consumable guide welding) collect at the bottom of the flux pool. Welding is done essentially in the flat position with welding progressing from the bottom to top of a weld joint positioned vertically. Water-cooled backing shoes in contact with the joint sides contain the

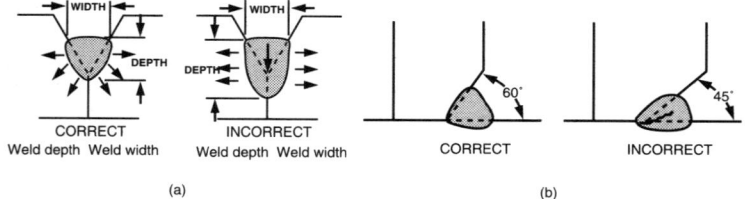
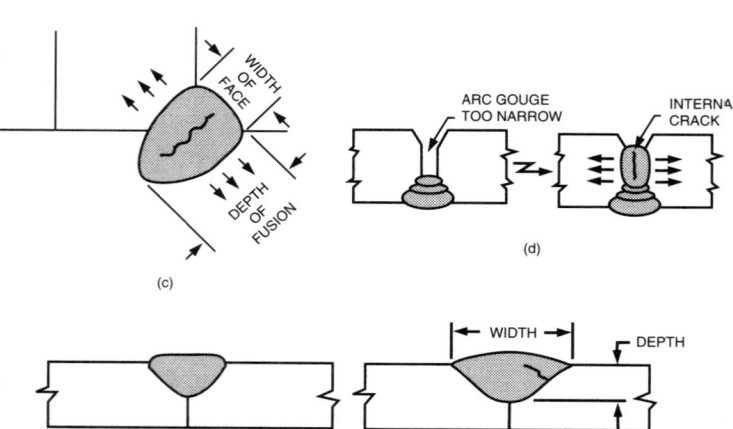

SOLIDIFICATION CRACKING BECAUSE OF WELD PROFILE

molten weld metal and the molten flux. The weld surface is molded by the contour of the shoes. The vertical progression of welding provides directional solidification of the molten pool and ready flotation of all nonmetallic impurities into the floating slag layer. As the molten metal slowly solidifies in upward progression, it joins the plates together. This process has a high deposition rate, therefore a substantial amount of base metal is melted to form part of the weld in this process.

Welding Metallurgy: Two metallurgical factors are important in electroslag welding: symmetrical stress patterns which result in no angular distortion and large weld metal grain sizes resulting from slow cooling. The stresses induced during the slow cooling of the weld is symmetrical with no distortion, which is different from the result when groove welds are welded from one side of a joint using some other welding processes. The large grains associated with electroslag welding tend to reduce the mechanical properties of the weld metal and cracking may result, especially if the weld joint is highly restrained.

Welding Chemistry: The welding composition is controlled by a combination of base metal, electrode and consumable guide tubes, if used, and the metal is protected from oxidation and refined by the molten slag. The slag used in this process is a metallurgical flux that refines both the fused base metal and the filler metal. As welding progresses, the slag also shields the weld pool along the full cross section of the joint.

Limiting Factors: Electroslag welding now has the reputation for producing the cleanest weld metal, being comparable to vacuum arc remelted steel. The only limiting factor of note is the skill level required to set up the machine. Due to the extensive time required for this setup, only thicker sections can be economically joined using this process.

Preweld preparation is simple. Faces of the joint must be aligned, but this is not as critical as with some other processes. The plate edges must be smooth and fairly uniform in thickness for proper positioning of the backing shoes.

For safety, the white-hot electroslag pool should not be watched without No. 5 filter lens protection (or darker). Safety goggles are needed when slag is being removed from weld deposits.

Electroslag Welding Equipment

Discontinuities: All of the common discontinuities may appear in electroslag welds.

Porosity is usually gross when it occurs, and in the form of piping porosity. This can be caused by moisture in the flux or by the "sweating" of the water-cooled backing shoes.

Slag inclusions are uncommon but may exist. They normally result from improper operating parameters that may allow solidified slag to become mechanically trapped. They may also result if the weld is nearly lost or is lost and restarted. The welding process depends on building up a pool of molten slag heated to about 3100° F. An inadequate starting sump (joint extension) will leave poorly melted metal and slag in the weld. Shallow run-off tabs at the top of the joint may fail to allow the joint to be completely filled, resulting in underfill.

Incomplete fusion often occurs in welds with poor shape of the pool. Welds in thick plates in which the heat is distributed by oscillating the electrode(s) may have incomplete fusion in the central portion, or near the shoes. The cooling effect of the shoe may prevent melting of the base metal out to the surface. The resulting indication resembles an undercut.

Overlap may occur if the shoes do not ride smoothly and closely on the plates, allowing the pool to grow outward from the surface plane. Ultimately, this condition leads to a run-out of the pool unless the gap is patched with insulating material or another dam. Again, this does not occur as readily in consumable guide welding because the shoes are fixed.

Laminations and delaminations in the plate are not as great a concern in an electroslag weld. The metallurgical slag floats out any inclusions in the lamination and seals such discontinuities alongside the weld.

Lamellar tearing has not been observed in electroslag welding because no through-thickness tensile stress is imposed on the base metal, at least not in butt joints. Tearing might occur in T-joints, but its occurrence is minimal.

Cracking, often severe, may occur in electroslag welds made in a deep, narrow pool at high welding speeds, due to the undesirable solidification patterns which result.

Fissures are commonly found. This is a term used to describe a discontinuity that particularly occurs in these welds. It is a small or moderately-sized separation along grain boundaries. It is readily formed here because of the large size of the grains. High restraint stresses caused by solidification of the weld metal cause the separation.

Inspection Processes: Inspection processes commonly used for electroslag welding are visual and radiographic testing. Ultrasonic testing is difficult because of the large grains typically present in electroslag welds.

Oxyacetylene Welding (OAW)

Oxyacetylene welding is a chemical welding process which relies on the chemical reaction between the oxyacetylene flame and the base metal to produce the necessary heat for melting the base and filler metals, if filler metal is used.

The oxyacetylene flame is produced by burning acetylene gas with pure oxygen fed through a torch, completing the combustion with oxygen in the air. Oxyacetylene welding is usually a manual process, although it may be mechanized for machine or automatic operation. While other fuel gases are effective for cutting operations, only acetylene provides sufficient heat to produce a satisfactory weld.

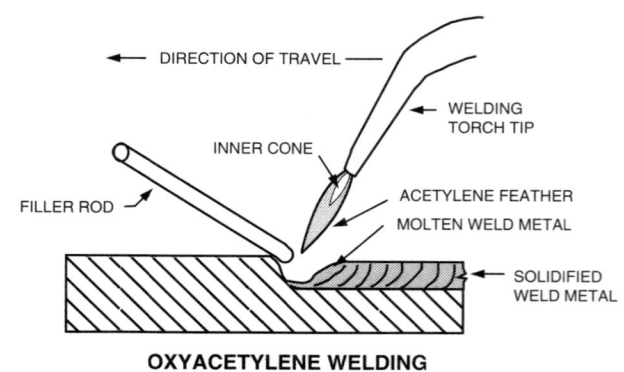

OXYACETYLENE WELDING

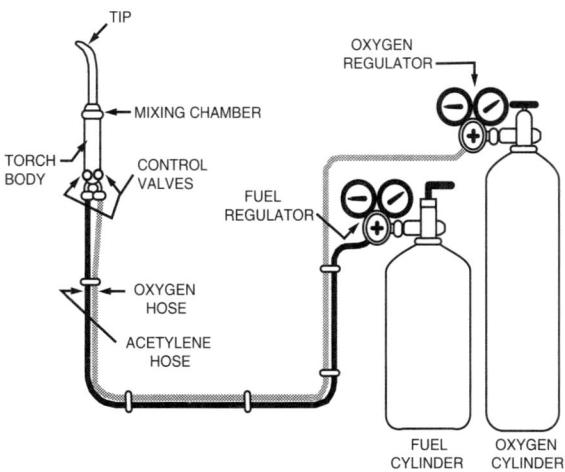

OXYFUEL WELDING EQUIPMENT

Welding Metallurgy: The temperature of the oxyacetylene flame is about 5600° F, which is sufficient to melt the commonly used metals. The flame provides both the heat and the necessary shielding of the molten metal for most applications; however, fluxes are available to improve the cleaning action on some materials. Both the heat and shielding are provided by the chemical reaction of the flame with the base metal.

An important variable is the size of the torch tip, which must be large enough to preheat the weldment and maintain a molten pool.

Welding Chemistry: The main welding chemistry consideration is what effect the nitrogen, oxygen and hydrogen in the air will have with the base metal and weld metal during the welding process.

The welder must regulate the torch flame to burn under neutral conditions, with the primary reaction exactly balanced, yielding only carbon monoxide and hydrogen. The flame atmosphere is then neither carburizing nor oxidizing, although it is chemically reducing. This flame adjustment must be learned from the appearance of the inner flame cone and the noise of the flame. The hot metal is then protected from the atmosphere by the combustion products in the neutral flame, and fluxes, if used.

Limiting Factors: The primary limitation of oxyacetylene welding is the level of skill required to minimize discontinuities. Adjustment of the heat source is totally controlled by the welder. He must set it for neutral conditions. There are no absolute guides; to secure the right flame, the welder must read the color and shape of the inner cone. He has pressure gages to set, but the size of the flame and the fuel ratio are set by judgment only.

The primary combustion area at the center of the flame has an intense white color, while the secondary combustion area varies from light orange to blue.

The welder must also control the application of heat to the work, the position and direction of the flame, and the manipulation of the filler rod (if needed). For safety, the eyes must be protected from the brilliance of the hot weld pool with No. 4 or 5 filter lenses in goggles.

In pre-weld preparation, the joint opening must be sufficiently large for oxyacetylene weld penetration. Also, joints in the vertical and overhead positions require appropriate technique to overcome the effects of gravity. The joint must be cleaned properly to avoid contaminating the weld.

Discontinuities: The most commonly found discontinuities are porosity, incomplete fusion, incomplete joint penetration, undercut, underfill, overlap, and various forms of cracks. Proper welding technique will eliminate most of these problems.

Uniformly scattered porosity generally betrays faulty welding technique, improper filler metal or contaminated base metal.

Incomplete fusion of the base metal edges frequently occurs when they are inadvertently oxidized. It occurs even with the best flame adjustment if improper manipulation is used.

Undercut, underfill and overlap are other weld faults attributed directly to the welder.

Cracks are generally hot cracks in oxyacetylene welds. The process does not lead to cold cracking because the slow application of heat provides adequate preheat and makes the weld cool slowly. Throat cracks may result if the the weld deposit is too thin to resist weld shrinkage stresses.

Inspection Processes: The inspection processes for oxyacetylene welding include visual, penetrant, magnetic particle, radiographic and ultrasonic testing.

Stud Welding (SW)

Stud welding is a two-step process used to join attachments to metal surfaces. It heats a metal stud or similar part and a workpiece by drawing an arc between them, and then rapidly forces the heated surfaces together under pressure until they have united.

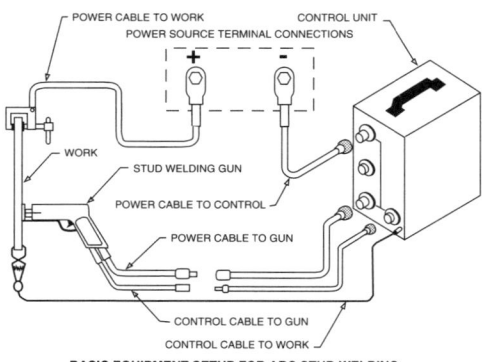

BASIC EQUIPMENT SETUP FOR ARC STUD WELDING

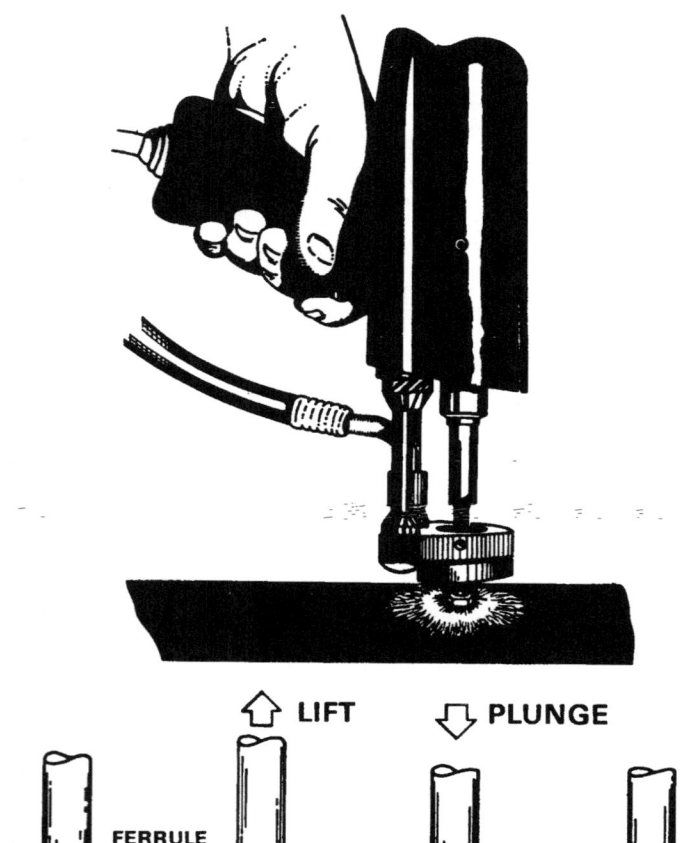

The process may be fully automatic or semiautomatic. A stud gun holds the tip of the stud against the workpiece. It then applies heat by creating an electrical arc between the stud and the underlying spot on the work surface. A timer in the control circuit then cuts off the current, and the stud-holding mechanism in the gun plunges the stud into the molten pool to make a weld. Freezing is almost instantaneous, as the weld is chilled by the surrounding base metal. When the weld is complete, the gun is removed and the ceramic ferrule is removed.

Studs may be welded in all positions. Stud welding is widely used in shipbuilding, bridges, automobiles, railroads, boilers, and industrial equipment.

Welding Chemistry: Partial shielding may be provided by a ceramic ferrule surrounding the stud or by a shielding gas. A flux may also be present on the stud tip.

Limiting Factors: The welding operator must correctly adjust the stud arc equipment. Defective gun action, improper current levels, and arc blow may result in incomplete fusion or undercut.

For safety, goggles must be worn whenever welding operators are knocking ferrules from completed stud welds and when testing the finished stud welds.

Discontinuities: There are only three main discontinuities which may be encountered with this process. Incomplete fusion could occur if the stud is not joined to the base metal across its entire cross section. Undercut may also occur if excessive welding current is used. Finally, the weld is not considered complete unless there is a small reinforcing fillet (or flash) present around the entire circumference of the stud base.

Inspection Processes: The examinations given to stud welds are visual examination for uniform build-up, and mechanical testing by hammer blows to bend the stud and restraighten it. Studs with defective welds break off or give a "dead" sound.

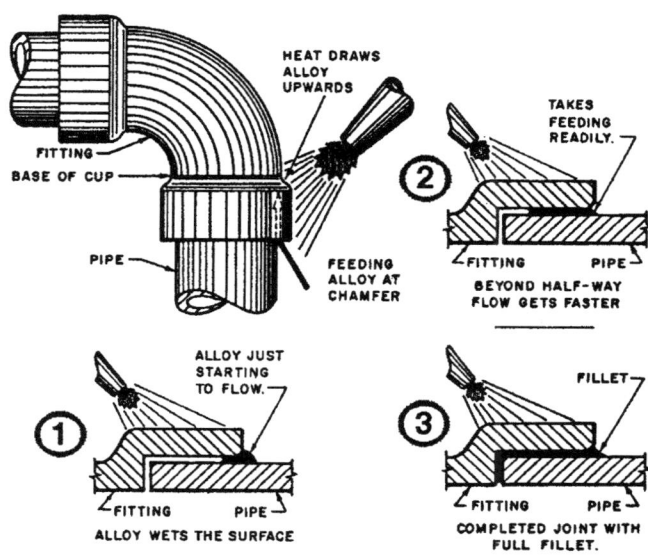

Brazing Process

Brazing (B)

Brazing is a group of joining processes that produces coalescence of materials by heating them to the brazing temperature in the presence of a filler metal having a melting point above 840° F and below the melting point of the base metals. The filler metal is distributed between the closely fitted faying surfaces of the joint by capillary action. The major difference between the various brazing methods is the manner in which the heat is applied to the joint. Brazing methods currently of industrial significance include the following:

- Torch brazing (TB)
- Furnace brazing (FB)
- Induction brazing (IB)
- Resistance brazing (RB)
- Dip brazing (DB)
- Infrared brazing (IRB)
- Diffusion brazing (DFB)

Torch Brazing (TB)

Torch brazing is accomplished by heating with a gas torch or torches. The fuel gas (acetylene, propane, natural gas, etc.) may be burned with air, compressed air or oxygen. Brazing filler metal may be preplaced in the forms of rings, washers, strips, slugs, powder, etc., or it may be face-fed, that is, fed from hand-held filler metal, usually in the form of wire or rod. Proper cleaning and fluxing before brazing are essential. This process may be performed manually, by machine or automatically.

Manual torch brazing is particularly useful on assemblies involving sections of unequal mass. Machine operations are set up when the rate of production warrants, using one or more torches equipped with single- or multiple-flame tips. The machine may move either the work or the torches, or both.

Torch brazing should be a reliable process. The flux, by its melting, indicates when the work is at a temperature suitable for brazing. The filler metal then flows into the joint by capillary action and should be visible at all exposed edges of the joint.

Furnace Brazing (FB)

Furnace brazing is preferred when the parts to be brazed can be assembled with the filler metal preplaced near or in the joint. The preplaced brazing filler metal may be in the form of wire, foil, powder, paste, or tape. Most high production brazing is done in a reducing gas atmosphere, such as hydrogen, but much furnace brazing is done in a vacuum, which prevents oxidation. Brazing in a vacuum assures the clean surfaces needed for good wetting and flow of filler metals without the use of fluxes. However, several base metals and filler metals are harmed by vacuum brazing because the constituents may evaporate.

Induction Brazing (IB)

Induction brazing is an automatic process that uses an induction coil to heat the metal. The parts are placed in or near a coil carrying alternating current. They become heated by an electric current induced in the parts to be joined. The brazing filler metal is usually preplaced. Careful design of the joint and the coil setup are necessary to assure that all surfaces reach brazing temperature at the same time.

Resistance Brazing (RB)

Resistance brazing is an automatic or semiautomatic process that uses the electrical resistance in the joint being welded as the heating device. The brazing filler metal is preplaced or added in some convenient form. Fluxing must take into account the conductivity of the flux. Most fluxes are electrical insulators when solid. The parts being brazed are held between two electrodes, and proper pressure and current are applied. The pressure is maintained until the joint solidifies.

Dip Brazing (DB)

Dip brazing is performed in a chemical bath or a molten metal bath. In chemical bath dip brazing, the brazing filler metal is preplaced and the assemblies are then immersed in a molten salt. The salt bath furnishes the heat necessary for brazing and usually protects against oxidation; if not, suitable flux should be used. In molten metal dip brazing, the parts are immersed in a bath of molten brazing filler metal. A cover of flux is maintained over the molten bath.

Infrared Brazing (IRB)

Infrared brazing uses the high intensity quartz lamp to provide radiant heat to bring the work to brazing temperature. Assemblies to be brazed are supported so that the energy impinges on the joints to be brazed.

Diffusion Brazing (DFB)

Diffusion brazing uses a filler metal which will diffuse into the base metal to create joint properties approaching those of the base metal. Migration of atoms (in the solid state) away from their home positions in the crystal lattice results in interdiffusion of the filler metal and base metal. This results in partially or completely eliminating any trace of filler metal as a layer in the joint. Such a braze develops increased mechanical properties and a higher remelt temperature. In some joints, a diffusion of atoms is planned that will create a liquid phase, which may be distributed by capillary action as in other brazing methods.

Brazing Chemistry: Brazing compounds are complex alloys with carefully adjusted melting points and flow characteristics. The work must be cleaned of oxides and films, which explains the reason for fluxes used in every type of brazing except furnace brazing. Furnace brazing cleans the surfaces by using reducing type gases or a vacuum

Limiting Factors: Preparation for brazing requires proper cleaning of each joint and jigging to hold correct joint alignment during capillary flow of the brazing filler metal.

For safety, all personnel must guard against toxic fumes of volatilized cadmium (from parts and filler metals) and fluorides (from fluxes). Eye protection is essential.

Discontinuities: Typical discontinuities found in brazed joints include voids, erosion and incomplete joint penetration. Voids and

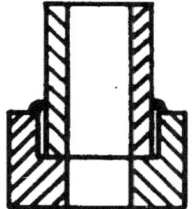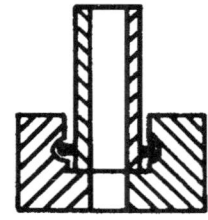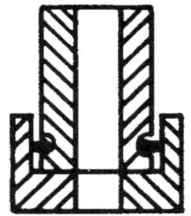

Methods of Preplacing Brazing Filler Metal in Wire Form

incomplete joint penetration are commonly caused by underheating or by improper fluxing (or both). Erosion is a discontinuity resembling undercut. It results from overheating during brazing.

Inspection Processes: Inspection processes for detecting internal defects in brazed joints include torsion or mechanical testing, radiography and leak testing.

Cutting Processes

So far this discussion has dealt exclusively with those methods used to join metals together. However, in the welding environment, there are also needs for metal removal or severing. There are numerous methods available, including both thermal and mechanical types.

Thermal cutting is the standard method of preparing base metal joints for welding, of cutting out defects for weld repair, and of backgouging. The area cut is confined to a narrow, well-defined zone of controlled width called the kerf. Metal is removed by combustion (oxidation) in oxyfuel gas cutting, by simple melting in the plasma jet, or by arc melting in a blast of air.

The cutting process may be manual, semiautomatic or automatic. Standards for quality of cut surfaces vary with the requirements of the structure being welded. If the resulting irregularities are excessive, they can be chipped with a blunt tool or ground back with an abrasive wheel.

For safety, filter lenses are required for observing thermal cutting operations, and safety goggles with side shields are needed for protection from sparks and spatter.

The three primary thermal cutting processes are the following:
- Oxyfuel gas cutting (OFC)
- Air carbon arc cutting (CAC-A)
- Plasma arc cutting (PAC)

Oxyfuel Gas Cutting (OFC)

Oxyfuel gas cutting was at one time exclusively oxyacetylene cutting (OFC-A), but today oxyfuel cutting may be with oxygen and natural gas (OFC-N), propane (OFC-P), hydrogen (OFC-H), a proprietary mixture such as stabilized methylacetylene and propadiene, or metal powder

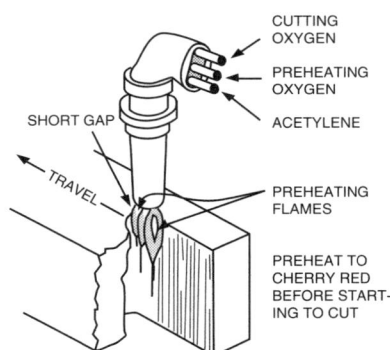

PROCESS DIAGRAM OXYGEN CUTTING

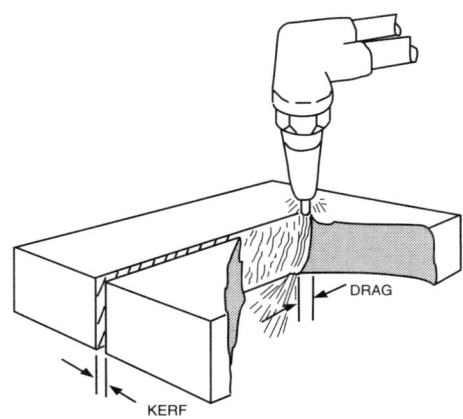

KERF AND DRAG IN OXYFUEL GAS CUTTING

cutting (POC). Metal powder cutting is used for stainless steel, aluminum and copper alloys. Torch modifications to suit each fuel gas are required.

Oxyfuel gas cutting severs ferrous metals by oxidizing the iron in oxygen to form iron oxide. Above the oxidation temperature of about 1700° F, the familiar oxidation of iron (rusting) becomes combustion which is confined to a narrow zone. The metal to be cut is heated to the oxidation temperature by preheat flames disposed around the oxygen cutting jet.

The quality of the cut surface varies over wide limits. The skill of the cutter or operator affects all operations, since the cutting flame must be manually adjusted even for automatic cutting.

The major limitation of this method is the fact that it will only effectively cut those materials which oxidize at a temperature below their melting point. Consequently, it is difficult to produce a quality cut in stainless steels using this method.

Air Carbon Arc Cutting (CAC-A)

Air carbon arc cutting melts the metal with an electric carbon arc and then blows it out with a high velocity air jet, traveling parallel to the electrode and striking the weld pool just behind the arc.

Plasma Arc Cutting (PAC)

Plasma arc cutting uses the greater heat of the plasma arc (27,000° F) to cut through any metal, ferrous or nonferrous. It removes the molten material with a high velocity jet of hot ionized gas. The process uses a constricted arc between a water-cooled electrode (direct current electrode negative) and the workpiece. The orifice which constricts the arc is also water-cooled.

The quality of plasma arc cutting is superior to other types of thermal cutting because of the high temperature of the jet. The kerf is normally slightly unsquare; however, use of water-injected torches will tend to minimize this effect.

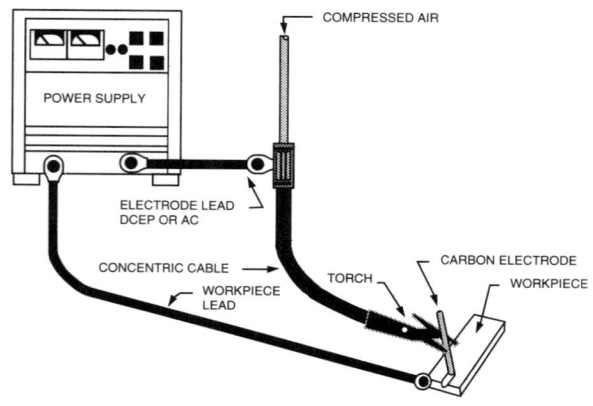

TYPICAL OPERATING PROCEDURES FOR AIR CARBON ARC GOUGING

TYPICAL AIR CARBON ARC GOUGING EQUIPMENT

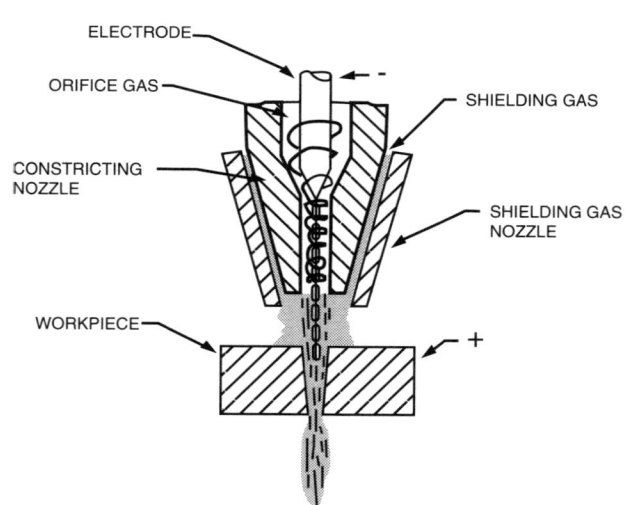

TYPICAL PLASMA ARC CUTTING TORCH

Mechanical Cutting

Joints are also prepared for welding by mechanical means such as milling, grinding, shaping, sawing, shearing, and chipping. The major concern after mechanical cutting is the removal of sulfurized cutting oils used to lubricate the cutting tools. Sulfur may cause cracking in welds, and all oils are a source of hydrogen.

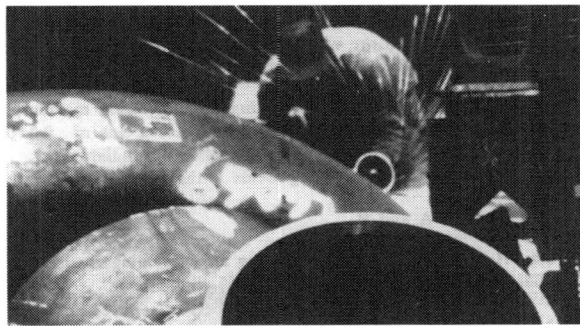
Mechanical Cutting

Summary

The welding inspector is primarily responsible for judging the quality of welds produced by numerous processes. Knowledge of the various processes will be extremely helpful in deciding what types of welding problems might be anticipated. Actual welding experience is also quite helpful for the welding inspector; however, this is not a requirement for qualification as a CWI.

REVIEW - CHAPTER 9
WELDING, BRAZING AND CUTTING PROCESSES

Q9-1 Of the following, which is not a necessary requirement for a welding process?
a. source of energy
b. electricity
c. means of shielding molten metal
d. base material
e. none of the above

Q9-2 Which of the following are functions of the flux coating of a SMAW electrode?
a. insulating
b. alloying
c. deoxidation
d. shielding
e. all of the above

Q9-3 In the AWS system of SMAW electrode designations, the next to the last number refers to:
a. usability
b. electrode coating
c. position
d. strength
e. none of the above

Q9-4 Which of the following is an incorrect statement about a SMAW electrode designated as an E7024?
a. It is a low hydrogen type.
b. The weld deposit has a minimum tensile strength of 70,000 psi.
c. It is suitable for use in the flat and horizontal fillet positions only.
d. all of the above
e. none of the above

Q9-5 Which of the following is not an essential part of a typical SMAW system?
  a. constant current power supply
  b. wire feeder
  c. covered electrode
  d. electrode lead
  e. work lead

Q9-6 Shielding of the molten metal in GMAW is accomplished through the use of:
  a. granular flux
  b. slag
  c. fuel gas and oxygen
  d. a and b above
  e. inert and reactive gases

Q9-7 Which of the following is not considered a type of metal transfer for GMAW?
  a. short circuiting
  b. spray
  c. globular
  d. droplet
  e. pulsed arc

Q9-8 Which of the following types of metal transfer in GMAW is considered to be the lowest energy, and therefore prone to incomplete fusion?
  a. short circuiting
  b. spray
  c. globular
  d. droplet
  e. pulsed arc

Q9-9 What type of welding process is pictured below?

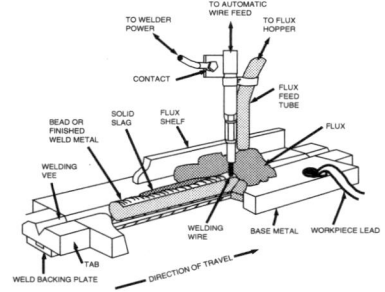

a. SMAW
b. GMAW
c. FCAW
d. SAW
e. ESW

Q9-10 Which of the following is not considered an arc welding process?
a. SMAW
b. GMAW
c. FCAW
d. ESW
e. none of the above

Q9-11 In the electrode designation system for FCAW, the second number refers to:
a. strength
b. position
c. chemical composition
d. usability
e. none of the above

Q9-12 Which of the following is not always an essential element of a FCAW system?
a. constant voltage power supply
b. tubular electrode
c. wire feeder
d. shielding gas
e. work (ground) lead

Q9-13 What aspect of the GTAW and PAW processes is different from the other arc welding processes?
   a. nonconsumable electrode
   b. power supply
   c. shielding
   d. all of the above
   e. none of the above

Q9-14 Shielding for the GTAW and PAW processes is accomplished through the use of:
   a. granular flux
   b. slag
   c. inert gas
   d. reactive gas
   e. none of the above

Q9-15 A green stripe on a tungsten electrode designates:
   a. pure tungsten
   b. 1% thoriated tungsten
   c. 2% thoriated tungsten
   d. zirconated tungsten
   e. none of the above

Q9-16 When welding aluminum with the GTAW process, what type of welding current is most commonly used?
   a. DCEP
   b. DCEN
   c. AC
   d. a and b above
   e. b and c above

Q9-17 SAW and ESW are similar in that:
   a. both are arc welding processes
   b. both use shielding gases
   c. both use a granular flux
   d. a and b above
   e. a and c above

Q9-18 The diagram below depicts what welding process?

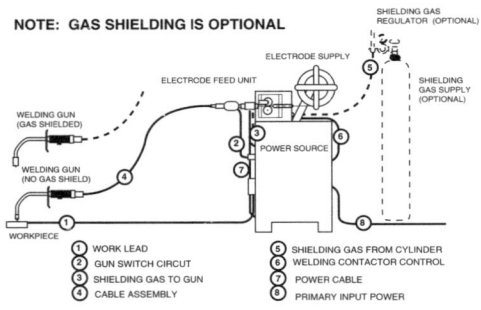

- a. SMAW
- b. ESW
- c. FCAW
- d. SAW
- e. PAW

Q9-19 Solidification cracking due to improper width-to-depth ratio of the weld nugget may be a problem with which welding process?
- a. OFW
- b. SW
- c. SAW
- d. all of the above
- e. none of the above

Q9-20 A welding process done essentially in the flat position with welding progressing from the bottom to top of the weld joint positioned vertically identifies:
- a. GMAW
- b. SAW
- c. ESW
- d. a and b above
- e. b and c above

Q9-21 Which of the following are not common to both GTAW and PAW?
- a. nonconsumable tungsten electrode
- b. copper constricting nozzle
- c. shielding gas nozzle
- d. externally-applied filler metal
- e. none of the above

Q9-22 What technique is employed with PAW to produce full penetration welds?
   a. stringer beads
   b. weave beads
   c. keyhole
   d. backstep
   e. none of the above

Q9-23 What welding process produces welds in the flat position, in a single pass, with the progression vertically upward along the joint?
   a. SAW
   b. ESW
   c. FCAW
   d. a and b above
   e. b and c above

Q9-24 Which of the following is not an advantage of the ESW process?
   a. high deposition rate
   b. ease of setup
   c. capable of joining thick sections
   d. no tendency for angular distortion
   e. none of the above

Q9-25 Which welding process is considered to be a chemical welding process?
   a. SMAW
   b. ESW
   c. SAW
   d. OAW
   e. none of the above

Q9-26 Which arc welding process provides an efficient means of joining attachments to some planar surface?
   a. OAW
   b. SW
   c. GMAW
   d. GTAW
   e. SMAW

Q9-27 Brazing differs from welding in that:
a. no filler metal is used.
b. an oxyfuel flame is used.
c. the base metal is not melted.
d. all of the above
e. none of the above

Q9-28 For satisfactory results, a braze joint should have:
a. a large surface area.
b. a small gap between pieces to be joined.
c. a precise bevel.
d. a and b above
e. b and c above

Q9-29 Which of the following is not an advantage of brazing?
a. ease of joining thick sections
b. ability to join dissimilar metals
c. ability to join thin sections
d. a and b above
e. b and c above

Q9-30 Of the following metals, which cannot be effectively cut using OFC?
a. high carbon steel
b. low carbon steel
c. medium carbon steel
d. stainless steel
e. none of the above

Q9-31 Which of the following gases can be used to perform OFC?
a. MAPP
b. propane
c. acetylene
d. natural
e. all of the above

Q9-32 Which of the following cutting processes can be used to cut any metal?
   a. OFC
   b. CAC-A
   c. PAC
   d. a and b above
   e. b and c above

Q9-33 The width of a cut is technically referred to as the:
   a. gap
   b. dross
   c. kerf
   d. drag
   e. none of the above

Q9-34 Which process is illustrated below?

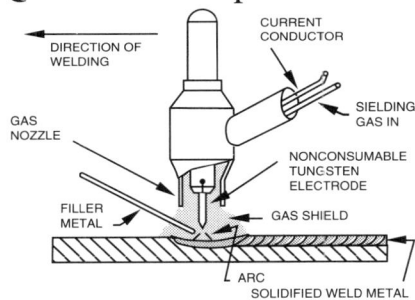

   a. GMAW
   b. PAW
   c. GTAW
   d. BMAW
   e. CAW

Q9-35 ESW designates which process?
   a. electric slag arc welding
   b. electroslag arc welding
   c. electric slag welding
   d. electroslag welding
   e. electric stud welding

Q9-36 Which process is illustrated below?

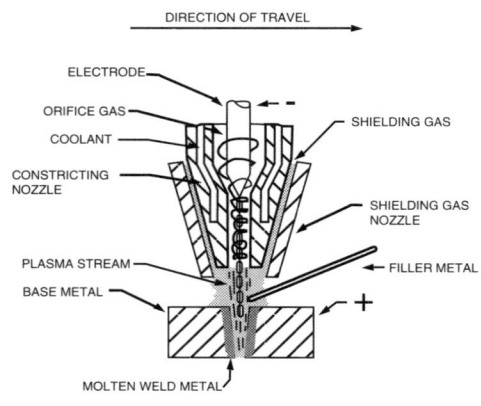

a. GMAW
b. PAW
c. GTAW
d. TIG
e. CAW

Q9-37 Which process is illustrated below?

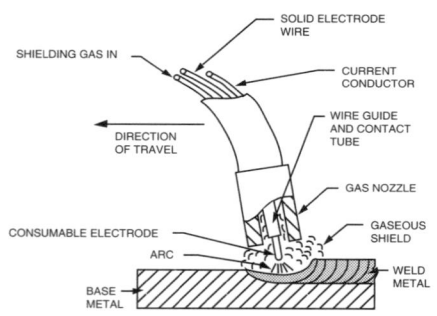

a. GMAW
b. PAW
c. GTAW
d. TIG
e. CAW

Q9-38 Which process is illustrated below?

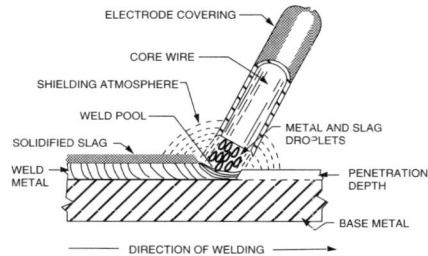

- a. GMAW
- b. SAW
- c. GTAW
- d. SW
- e. SMAW

Q9-39 SMAW designates which process?
- a. stick metal arc welding
- b. shielded metal arc welding
- c. submerged arc welding
- d. seam metal arc welding
- e. short circuiting metal arc welding

Q9-40 SW designates which process?
- a. stud welding
- b. stud arc welding
- c. submerged welding
- d. stick welding
- e. submerged arc welding

Q9-41 SAW designates which process?
- a. stud welding
- b. stud arc welding
- c. submerged welding
- d. stick welding
- e. submerged arc welding

Q9-42 FCAW designates which process?
- a. flux cored arc welding
- b. flux centered arc welding
- c. furnace controlled arc welding
- d. friction arc welding
- e. flow arc welding

Q9-43 GMAW designates which process?
- a. gas machine arc welding
- b. gas method arc welding
- c. gas material arc welding
- d. gas metal arc welding
- e. general material arc welding

Q9-44 GTAW designates which process?
- a. gas tungsten arc welding
- b. general tungsten arc welding
- c. globular transfer arc welding
- d. gas torch arc welding
- e. none of the above

Q9-45 PAW designates which process?
- a. plasma arc welding
- b. pressure arc welding
- c. plate arc welding
- d. percussion arc welding
- e. none of the above

Q9-46 Which of the following could result in the creation of porosity in the GTAW of 6061-T6 aluminum structural members for an aircraft application?
- a. insufficient cleaning of the weld joint
- b. contaminated filler metal
- c. leak in the shielding gas hose
- d. presence of drafts during the welding operation
- e. all of the above

Q9-47 An ER70S-6 electrode can be used with which of the processes?
   a. GTAW
   b. GMAW
   c. PAW
   d. all of the above
   e. none of the above

Q9-48 A granular flux is a characteristic of which of the following?
   a. ESW
   b. SAW
   c. SMAW
   d. both a and b
   e. both b and c

Q9-49 Which of the classifications listed below produces the strongest weld metal?
   a. ER70S-6
   b. E70T-5
   c. E7018
   d. F7A2-EM12K
   e. no difference

Q9-50 Which process is classified as a chemical welding method?
   a. GTAW
   b. GMAW
   c. ESW
   d. PAW
   e. OAW

Q9-51 Which cutting methods use electricity?
   a. PAC
   b. CAC-A
   c. OAC
   d. a and b above
   e. b and c above

Q9-52 A tubular electrode is a significant characteristic of which process?
   a. SAW
   b. ESW
   c. FCAW
   d. SMAW
   e. GMAW

Q9-53 Which arc welding process is used very effectively for the welding of various types of attachments to surfaces of plates and structural members?
   a. SMAW
   b. GMAW
   c. FCAW
   d. SW
   e. ESW

Q9-54 Molding shoes is a term associated with which process?
   a. SAW
   b. GTAW
   c. ESW
   d. FCAW
   e. GMAW

Q9-55 Short circuiting metal transfer is a mode of operation for:
   a. GTAW
   b. FCAW
   c. SMAW
   d. none of the above
   e. all of the above

Q9-56 The need for electrode holding ovens for some types of its filler metal is a disadvantage of which process below?
   a. SMAW
   b. FCAW
   c. SAW
   d. all of the above
   e. a and b

Q9-57 Which process uses a carbon electrode?
- a. SMAW
- b. GMAW
- c. GTAW
- d. CAC-A
- e. PAC

Q9-58 EWTh-1 is an electrode designation for which of those processes listed below?
- a. GTAW
- b. PAW
- c. GMAW
- d. a and b above
- e. b and c above

Q9-59 Of the following which brazing process is preferred when the parts to be brazed can be assembled with the filler metal preplaced near or in the joint?
- a. torch
- b. induction
- c. furnace
- d. diffusion
- e. none of the above

Q9-60 F7P6-EM12 is a filler metal designation for:
- a. SMAW
- b. GMAW
- c. FCAW
- d. SAW
- e. PAW

Q9-61 The ability to perform keyhole welding is a primary advantage of:
- a. GTAW
- b. PAW
- c. SMAW
- d. FCAW
- e. SAW

Q9-62 Which gases can be used for OFW?
   a. MAPP
   b. acetylene
   c. natural gas
   d. propane
   e. all of the above

Q9-63 Which gases below can be used for TB?
   a. acetylene
   b. oxygen
   c. natural gas
   d. propane
   e. all of the above

Q9-64 A ferrule is an item used for shielding in which process below?
   a. ESW
   b. PAW
   c. PAC
   d. SW
   e. FB

Q9-65 When GTAW is used, what type of current results in the greatest amount of penetration?
   a. dcen
   b. dcep
   c. ac
   d. hwac
   e. no difference

Q9-66 A constricting nozzle is one of the components for which welding process?
   a. PAW
   b. GTAW
   c. SAW
   d. GMAW
   e. SW

Q9-67 What gases can be used for GMAW?
a. carbon dioxide
b. argon
c. 75% argon-25% carbon dioxide
d. 98% argon-2% oxygen
e. all of the above

Q9-68 The process which can be used either with or without an external shielding gas is:
a. GMAW
b. SMAW
c. FCAW
d. GTAW
e. PAW.

Q9-69 Which of the welding processes below is generally considered to provide the highest deposition rate?
a. SAW
b. ESW
c. FCAW
d. SMAW
e. GMAW

Q9-70 When welding carbon steel with the OAW process, the torch should be adjusted to provide:
a. an oxidizing flame
b. a carburizing flame
c. a neutral flame
d. a heating flame
e. none of the above

Q9-71 Of the following which of the processes make use of water-injected torches to minimize the effect of irregular kerf?
a. PAC
b. CAC-A
c. GTAW
d. a and b above
e. b and c above

Q9-72 The use of a constricting orifice, is the distinguishing feature of which of the following?
- a. GTAW
- b. GMAW
- c. FCAW
- d. PAW
- e. none of the above

Q9-73 Which of the following processes utilize a flux to provide necessary shielding?
- a. SMAW
- b. SAW
- c. GMAW
- d. a and b above
- e. all of the above

CHAPTER 10: WELD AND BASE METAL DISCONTINUITIES

Introduction

Discontinuities are imperfections in welds or base metals. Ideally, a sound weld should have no discontinuities at all, but welds are not perfect; imperfections do exist in varying degrees.

There is a temptation to call discontinuities defects, but as a matter of terminology the terms discontinuity and defect should be carefully distinguished by all inspectors. A defect is rejectable. Some discontinuities are acceptable. A discontinuity becomes a defect when it exceeds acceptable limits imposed by acceptance standards. An imperfection of lesser magnitude than that is still a discontinuity, but it is not a defect.

The welding inspector's primary job is to inspect the fabricator's work to see that it meets the requirements of the contract. To be able to do this, one must be familiar with acceptance standards, which spell out the acceptable limits for discontinuities. If a particular type of discontinuity is permissible in the welds to be inspected, the acceptance standards, code or specification must specify the criteria used to discriminate between acceptable imperfections and defects.

The criteria used to discriminate between acceptable imperfections and defects are described in the following terms:

- Type of discontinuity
- Size of discontinuity
- Location of discontinuity

All three criteria must be considered to judge whether a discontinuity is severe enough to be considered a defect.

Types of Discontinuities

Discontinuities have been categorized under the following twelve types:

- Porosity
- Inclusions, both metallic and nonmetallic
- Underfill
- Incomplete fusion
- Incomplete joint penetration
- Overlap
- Undercut
- Lamination and delamination
- Seams and laps
- Lamellar tearing
- Crack
- Arc strike

TABLE 1
COMMON TYPES OF DISCONTINUITIES

Type of Discontinuity	Section	Location*	Remarks
1 Porosity	2.3	WM	Porosity is also commonly found in the heat-affected zone if base metal is a casting.
(a) Uniformly scattered	2.3.1		
(b) Cluster	2.3.2		
(c) Linear	2.3.3		
(d) Piping	2.3.4		
2 Inclusions	2.4	WM	
(a) Slag	2.4.1		
(b) Tungsten	2.4.2		
3 Incomplete fusion	2.5	WM, BM/WM	Also between passes.
4 Incomplete joint penetration	2.6	BM	Weld root.
5 Undercut	2.7	BM/WM	Adjacent to weld toe or weld root in base metal.
6 Underfill	2.8	WM	Weld face or root surface.
7 Overlap	2.9	BM/WM	Weld toe or root surface.
8 Laminations	2.10	BM	Base metal, generally near midthickness of section.
9 Delamination	2.11	BM	Base metal, generally near midthickness of section.
10 Seam and laps	2.12		Base metal surface almost always aligned with rolling direction.
11 Lamellar tears	2.13	BM	Base metal, near HAZ.
12 Cracks (includes hot cracks and cold cracks described in text)	2.14 2.14.1		
(a) Longitudinal	2.14.2, 2.14.3	WM, HAZ	Weld metal or base metal adjacent to weld interface.
(b) Transverse	2.14.2, 2.14.4	WM, HAZ, BM	Weld metal (may propagate into HAZ and base metal).
(c) Crater	2.14.5	WM	Weld metal at point where arc is terminated.
(d) Throat	2.14.6	WM	Parallel to weld axis.
(e) Toe	2.14.7	BM/WM	
(f) Root	2.14.8	WM	Root surface.
(g) Underbead and heat-affected zone	2.14.9	BM/WM	
13 Insufficient throat	2.15	WM	Weld face.
14 Convexity or weld reinforcement	2.16	WM	Weld face.
15 Insufficient leg	2.17	WM	Fillet weld.

*WM—weld metal
BM—base metal
HAZ—heat affected zone
BM/WM—weld interface

SOURCE: ANSI/AWS B1.10-86

TABLE 2
DISCONTINUITIES COMMONLY ENCOUNTERED WITH WELDING PROCESSES

Welding Process	Type of Discontinuity						
Arc	Porosity	Slag	Incomplete Fusion	Incomplete Penetration	Undercut	Overlap	Cracks
SW—Stud welding			X		X		X
PAW—Plasma arc welding	X		X	X			X
SAW—Submerged arc welding	X	X	X	X	X	X	X
GTAW—Gas arc tungsten welding	X		X	X			X
GMAW—Gas metal arc welding	X	X	X	X	X	X	X
FCAW—Flux cored arc welding	X	X	X	X	X	X	X
SMAW—Shielded metal arc welding	X	X	X	X	X	X	X
CAW—Carbon arc welding	X	X	X	X	X	X	X
Resistance							
RSW—Resistance spot welding			X				X
RSEW—Resistance seam welding			X				X
PW—Projection welding			X				X
FW—Flash welding			X				X
UW—Upset welding			X				X
Oxyfuel Gas							
OAW—Oxyacetylene welding	X		X	X	X	X	X
OHW—Oxyhydrogen welding	X		X	X			X
PGW—Pressure gas welding	X		X				X
Solid state*							
CW—Cold welding			X				X
DFW—Diffusion welding			X				X
EXW—Explosion welding			X				
FOW—Forge welding			X				
FRW—Friction welding			X				
USW—Ultrasonic welding			X				
Other							
EBW—Electron beam welding	X		X	X			X
ESW—Electroslag welding	X	X	X	X	X	X	X
IW—Induction welding			X				X
LBW—Laser beam welding	X		X				X
PEW—Percussion welding			X				X
TW—Thermit welding	X	X	X				X

*Solid State is not a fusion process, so incomplete joining is incomplete welding rather than incomplete fusion.

SOURCE: ANSI/AWS B1.10-86

Porosity

Porosity results when gas is entrapped in solidifying metal. This will be discussed only with respect to welds (although porosity is also commonly seen in castings). The entrapped gas comes from either the gas used in the welding process or the gas released from chemical reactions occurring during the welding process. Proper welding technique avoids gas formation and entrapment. Faulty or dirty materials may produce gas. The gas becomes trapped in the form of porous discontinuities in the weld.

Porosity usually occurs in the form of rounded discontinuities, but in a severe case the porosity is cylindrical. These large cylindrical pores are called piping porosity (or "wormholes"). The presence of porosity is a sign that the welding process is not being properly controlled or that the base metal and welding fluxes are contaminated

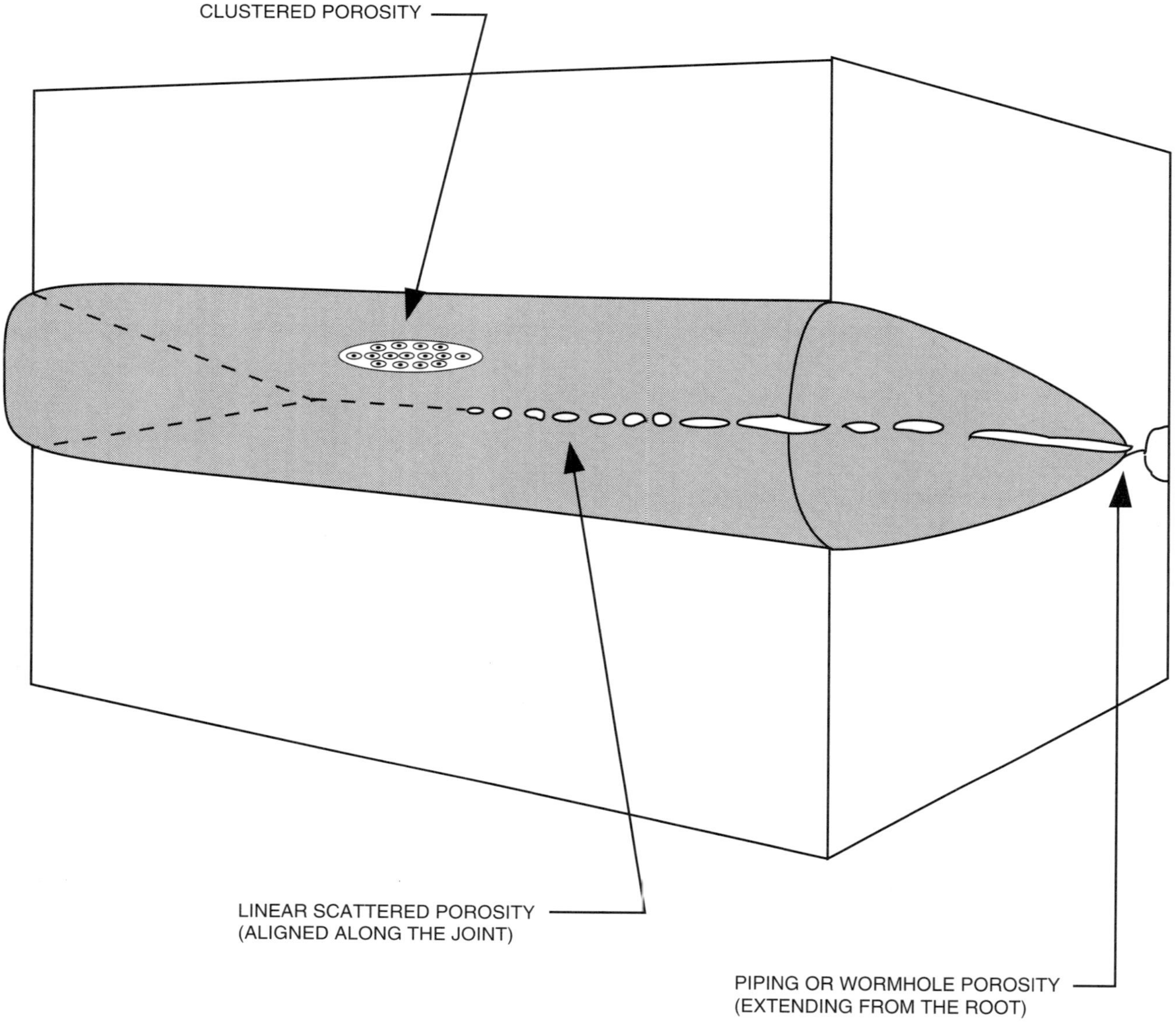

CLUSTERED POROSITY

LINEAR SCATTERED POROSITY
(ALIGNED ALONG THE JOINT)

PIPING OR WORMHOLE POROSITY
(EXTENDING FROM THE ROOT)

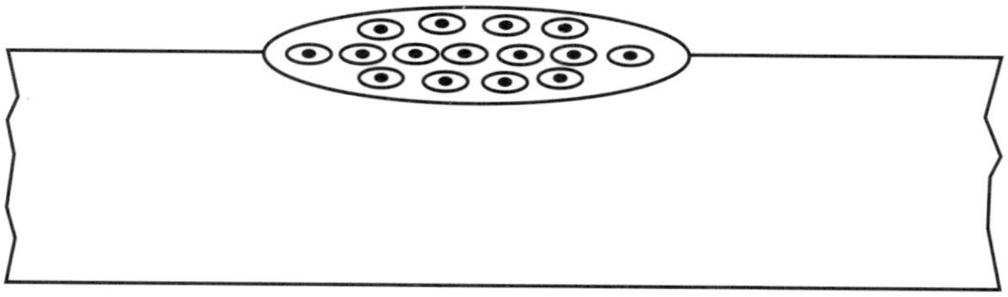

UNIFORMLY SCATTERED POROSITY

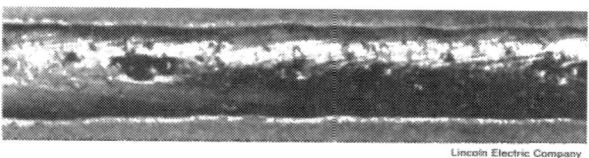

Lincoln Electric Company

POROSITY

Porosity Illustrated

with gas producing elements. Generally speaking, porosity in small amounts does not significantly intensify stress, making it a less critical discontinuity compared to those with sharp ends.

The distribution of porosity can be helpful in determining what type of fault caused the porosity. If the porosity is uniformly scattered, the cause could either be faulty materials or poor technique used throughout the weld.

A cluster of porosity is likely to result from improper initiation or termination of the weld.

Linear porosity aligned along a joint boundary would suggest that contamination triggered a chemical reaction, which produced unwanted gas. Such contamination could have been eliminated by preparing the joint properly.

Piping porosity is an elongated gas discontinuity extending from the weld root toward the surface and is also evidence of the presence of surface contamination.

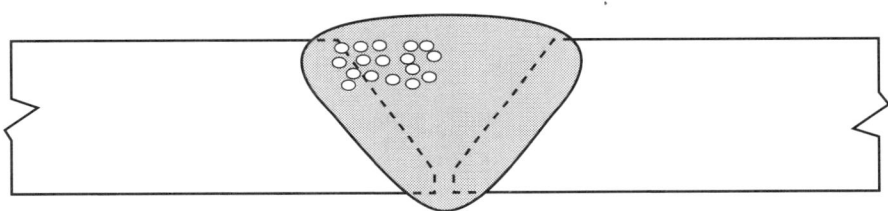

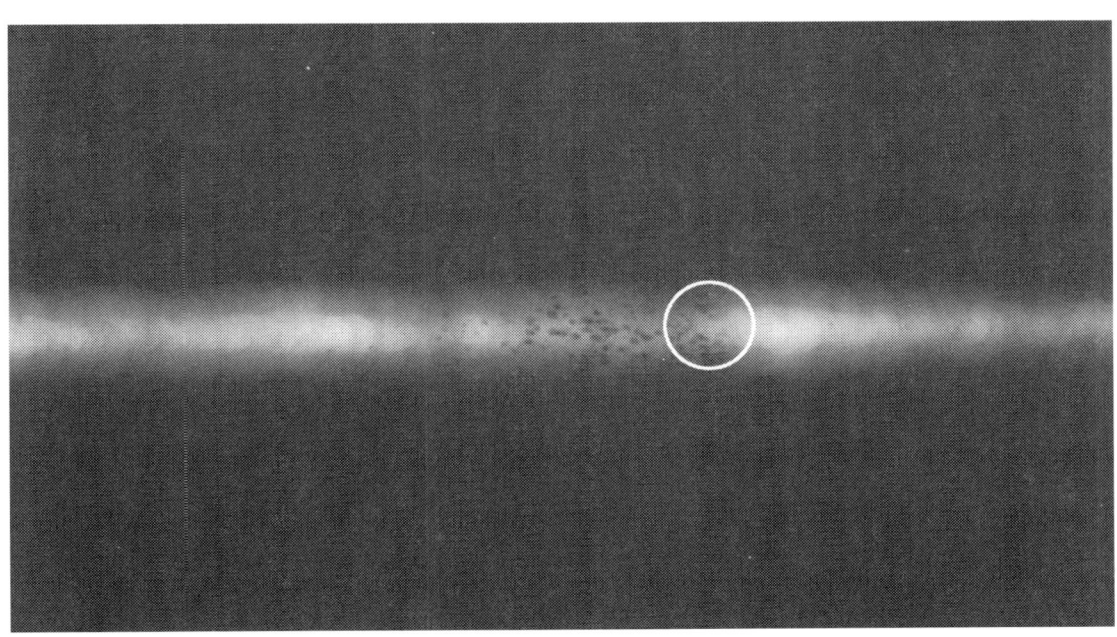

RADIOGRAPH OF CLUSTER POROSITY

10-6

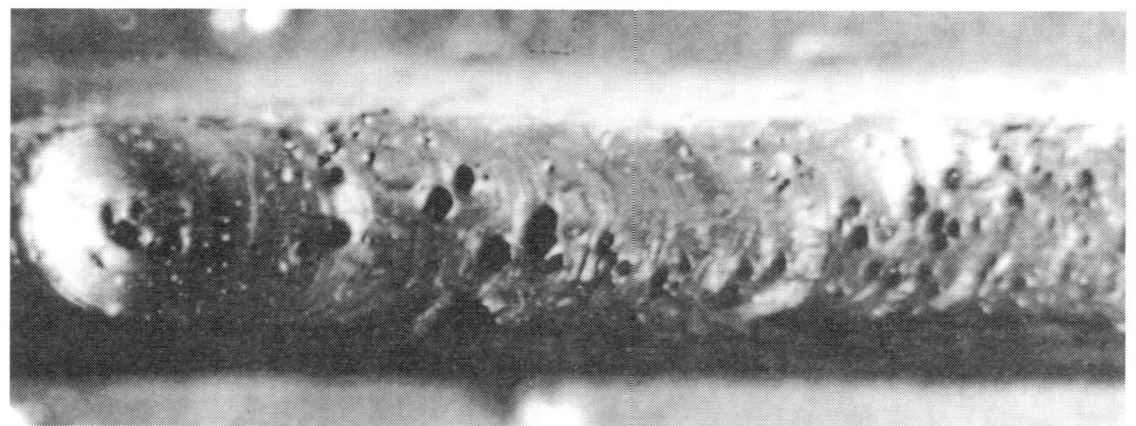

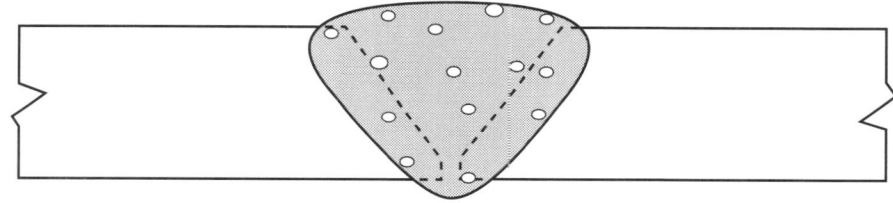

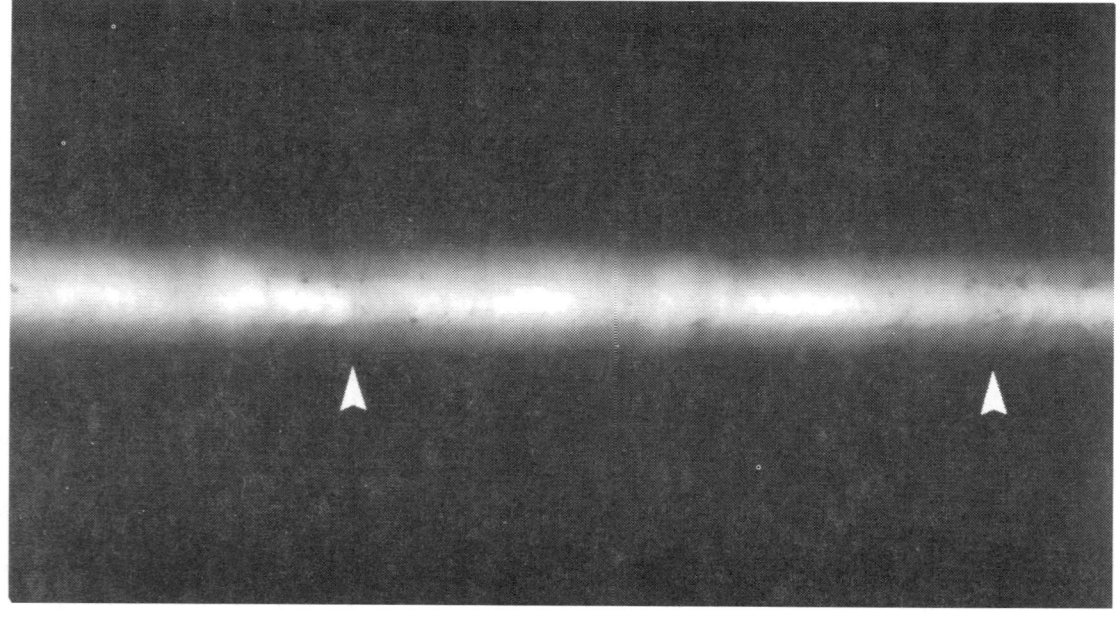

RADIOGRAPH OF SCATTERED POROSITY

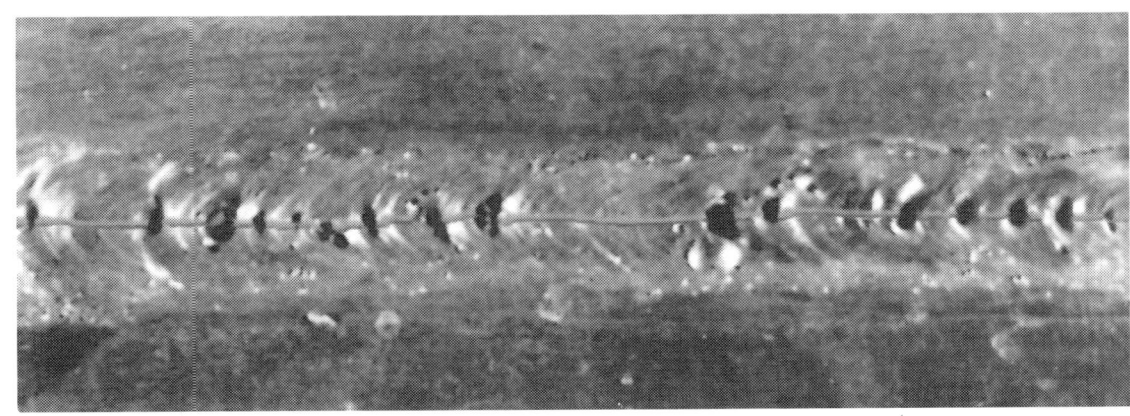

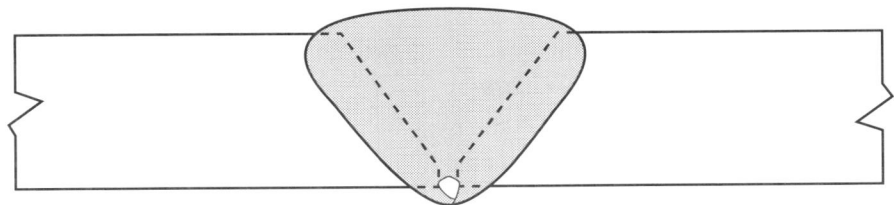

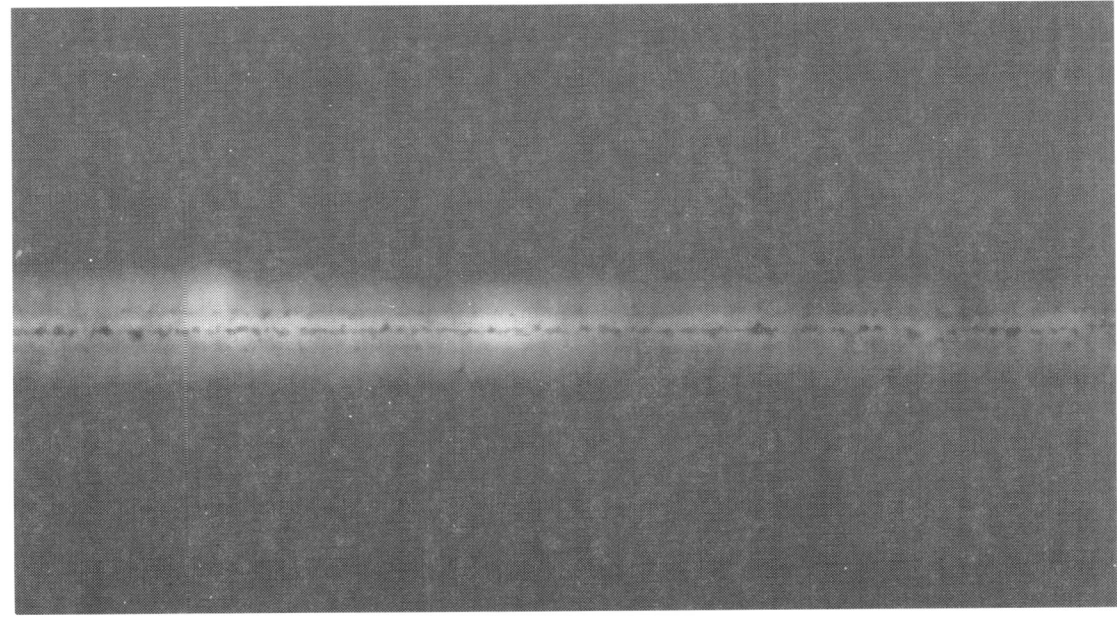

RADIOGRAPH OF LINEAR (ALIGNED) POROSITY

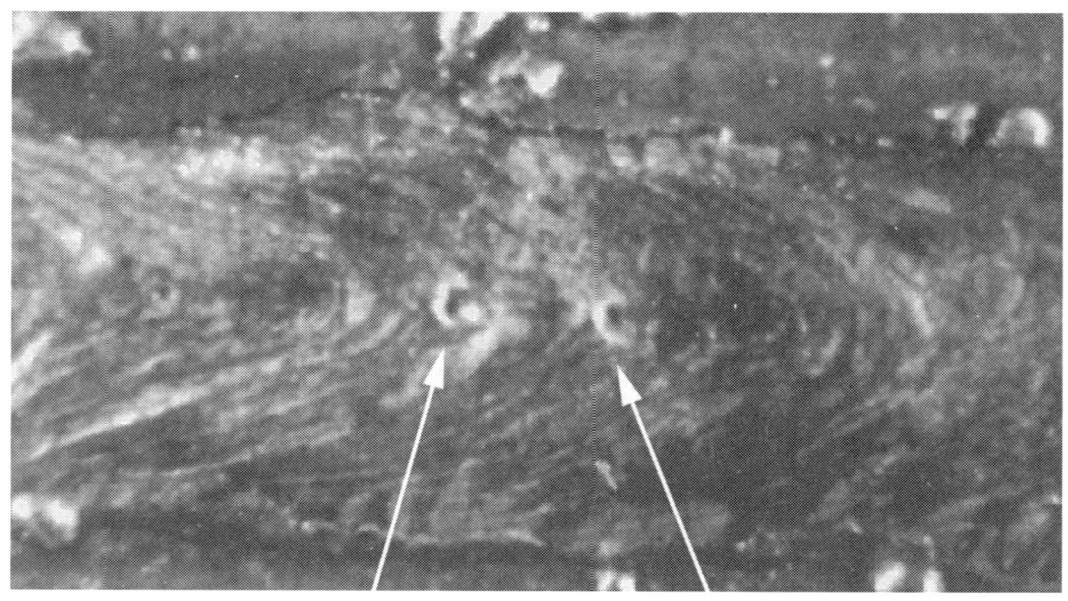

SURFACE APPEARANCE OF PIPING POROSITY

Inclusions

Inclusions result when solid materials are entrapped in solidifying metal. Inclusions interrupt the continuity of the weld, and some loss in structural integrity will result where they are present.

Nonmetallic (Slag and Oxides) Inclusions: These types of inclusions result from faulty welding or cleaning techniques and/or the failure of the designer to provide proper access for welding within the joint. Molten slag and oxides will flow to the top of the weld if allowed. Sharp notches in joint boundaries or between passes often cause slag to become entrapped under the molten weld metal. Parallel lines of elongated slag inclusions, sometimes called "wagon tracks" because of their radiographic appearance, often result if the welder produces a convex root pass in an open root pipe joint and fails to adequately clean the slag on either side of this weld pass.

Metallic Inclusions: These inclusions are usually tungsten particles trapped in weld metal. They most often occur in gas tungsten arc welding but may also result if the plasma arc welding process is improperly applied. These tungsten inclusions appear as light areas on radiographs because tungsten is highly opaque to radiation. This is just the opposite from most other discontinuities, which will show up as dark regions on the radiographic film.

Copper inclusions result if copper backing bars or backing shoes are used, such as in electroslag welding. Improper application of the plasma arc welding process could result in overheating of the copper constricting nozzle which could also cause copper inclusions in the weld.

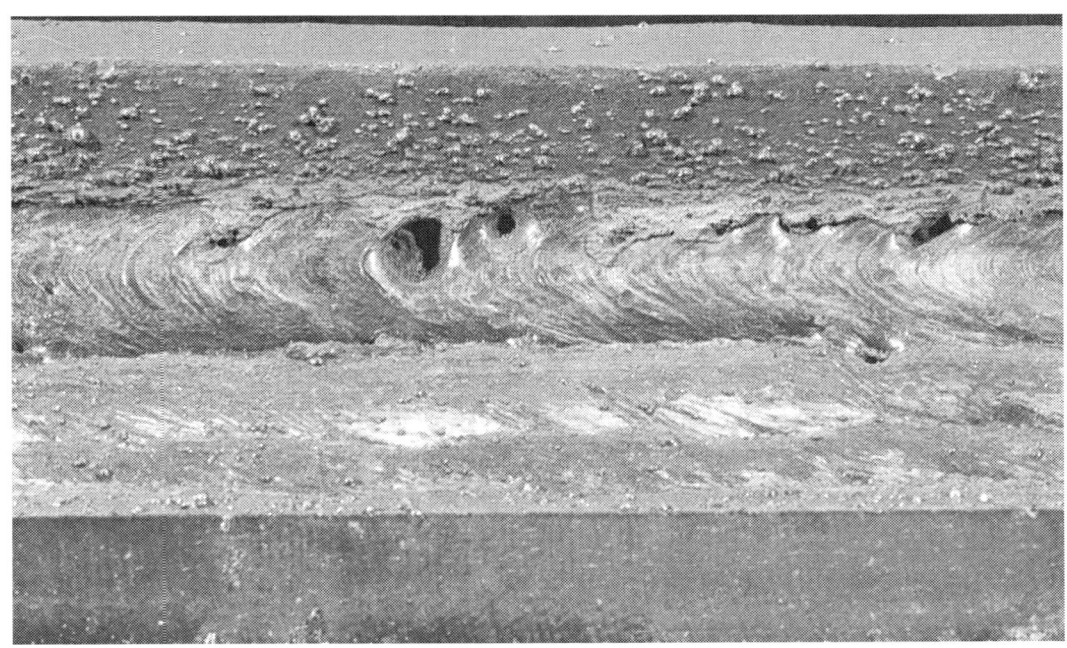

SURFACE SLAG INCLUSIONS

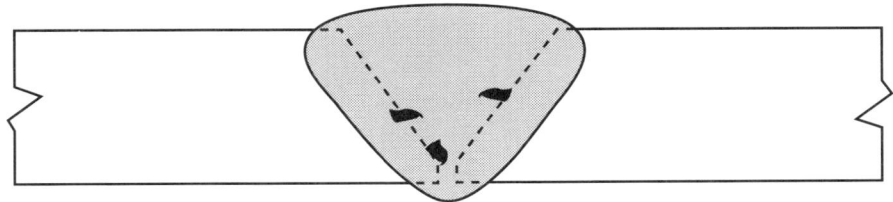

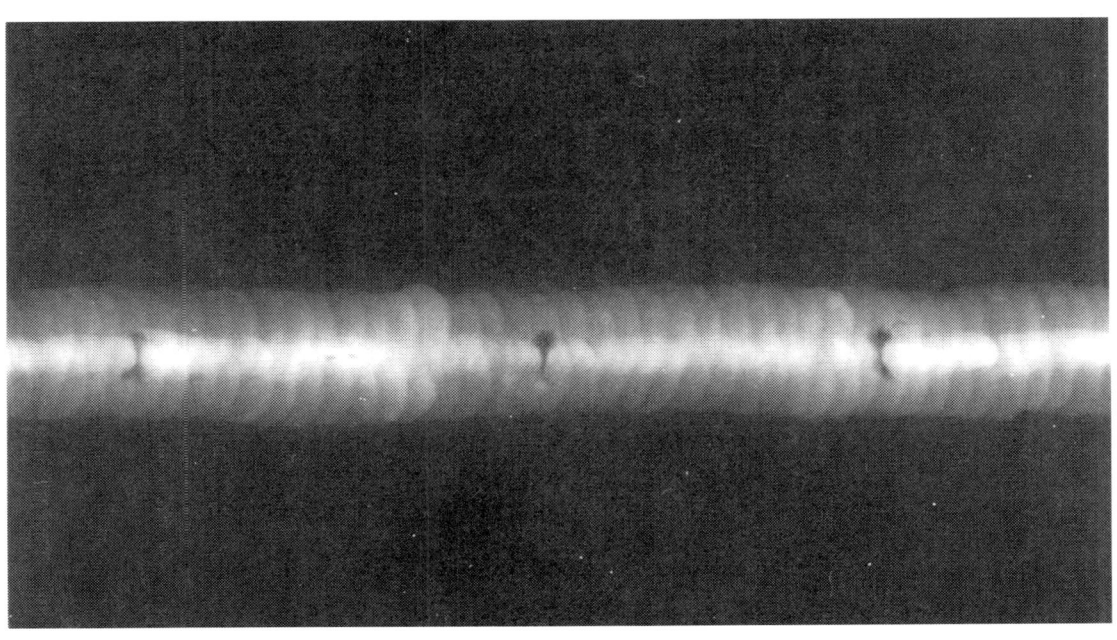

RADIOGRAPH OF INTERPASS SLAG INCLUSIONS

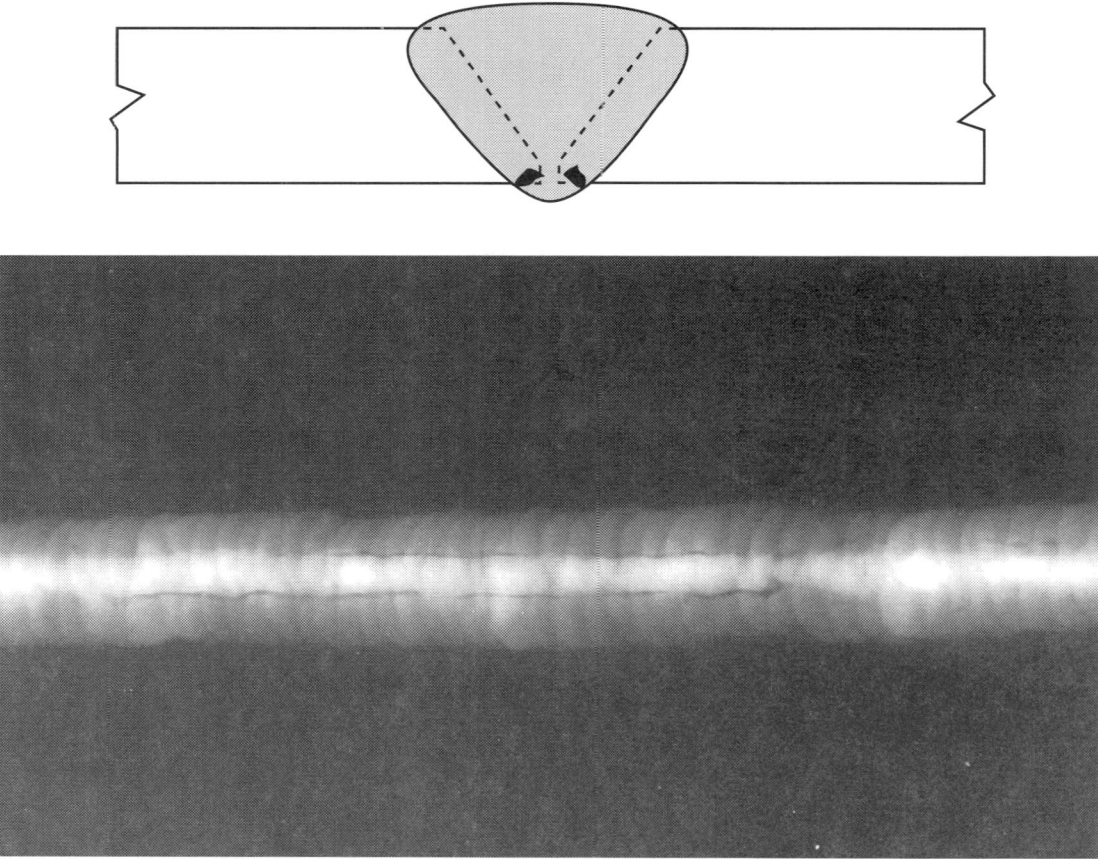

RADIOGRAPH OF ELONGATED SLAG LINES (WAGON TRACKS)

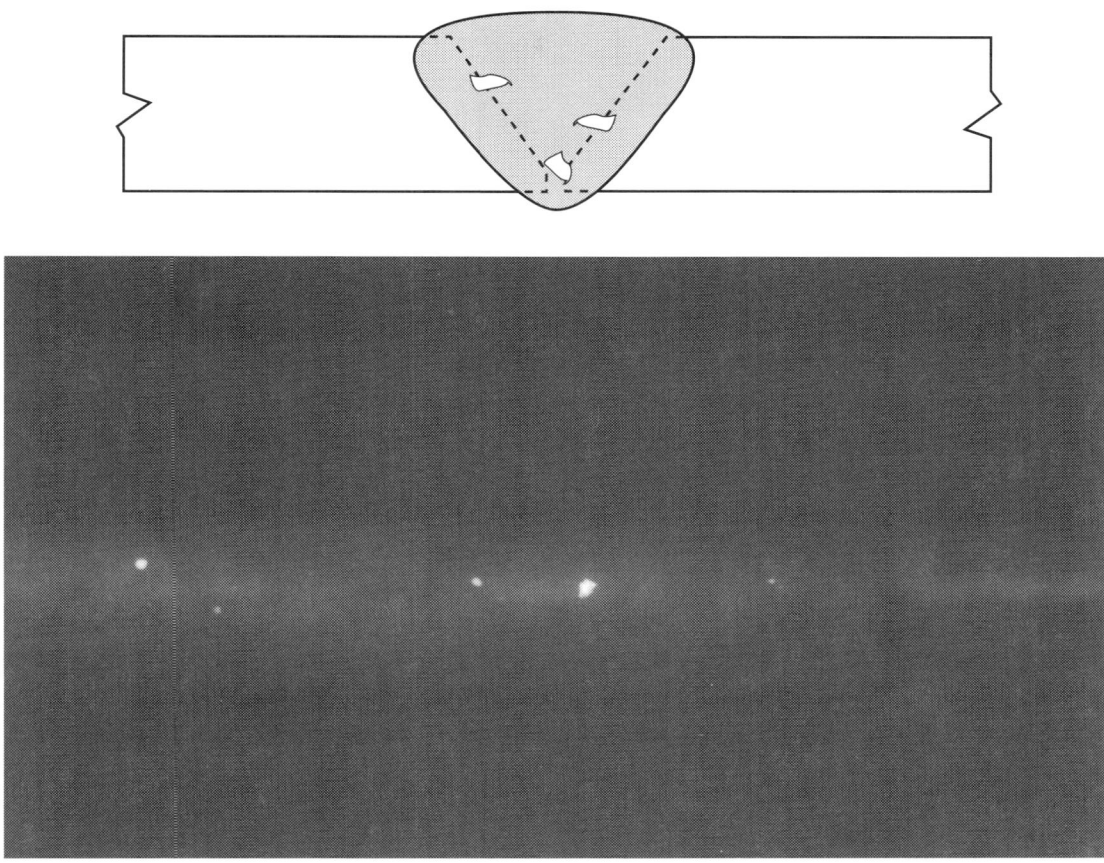

RADIOGRAPH OF TUNGSTEN INCLUSIONS

Underfill - is a depression on the face or root surface of a groove weld below the surface plane of the adjacent base metal. In other words, the welder or welding operator has failed to completely fill the groove, resulting in an undersize weld. On pipe welds, underfill at the weld root may also be referred to as "internal concavity" or "suck-back".

Incomplete Fusion - is the failure of the liquid weld metal to fuse into the entire groove face of the joint or to adjacent weld beads. Incomplete fusion is usually caused by insufficient application of heat to all faces of the joint. However, incomplete fusion can also be caused by the presence of oxides which inhibit fusion by remaining tightly secured to the base metal.

Incomplete Joint Penetration - results when the weld metal fails to extend completely through the joint thickness. The amount of joint penetration required in any joint should be specified on drawings. Whether that amount of joint penetration can be obtained depends upon the accessibility of the heat source and filler rod to the face area. This discontinuity can also result from improper joint designs.

The presence of incomplete joint penetration in a joint requiring complete joint penetration could also be referred to as inadequate joint penetration, or joint penetration which is less than that specified. Many codes require the use of joint backing for single-groove welds or backgouging of double-groove welds to assure that complete joint penetration can be attained.

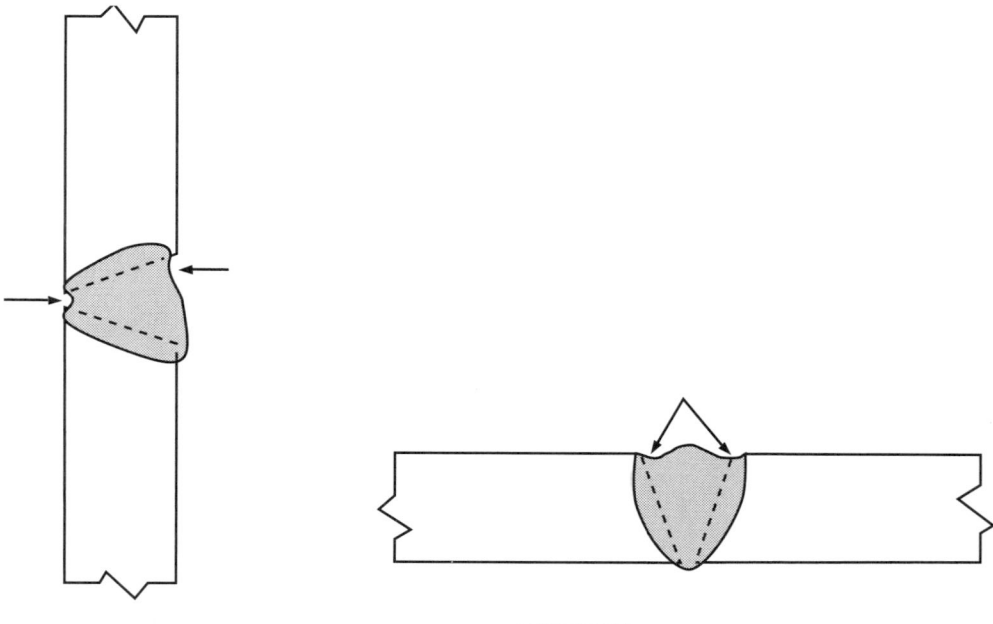

UNDERFILL

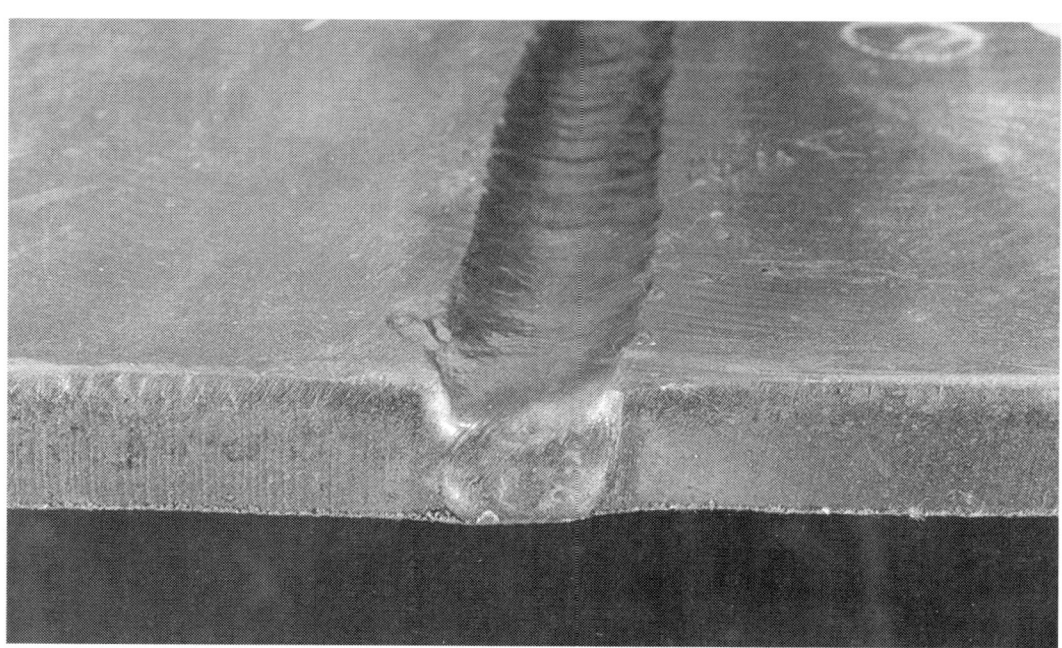

UNDERFILL USING FLUX CORED ARC WELDING IN STEEL

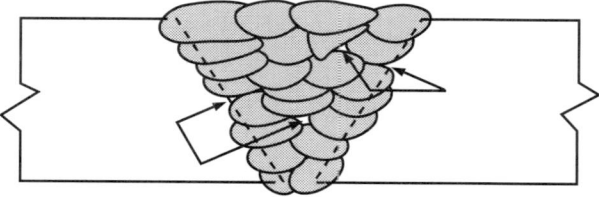

VARIOUS LOCATIONS OF INCOMPLETE FUSION

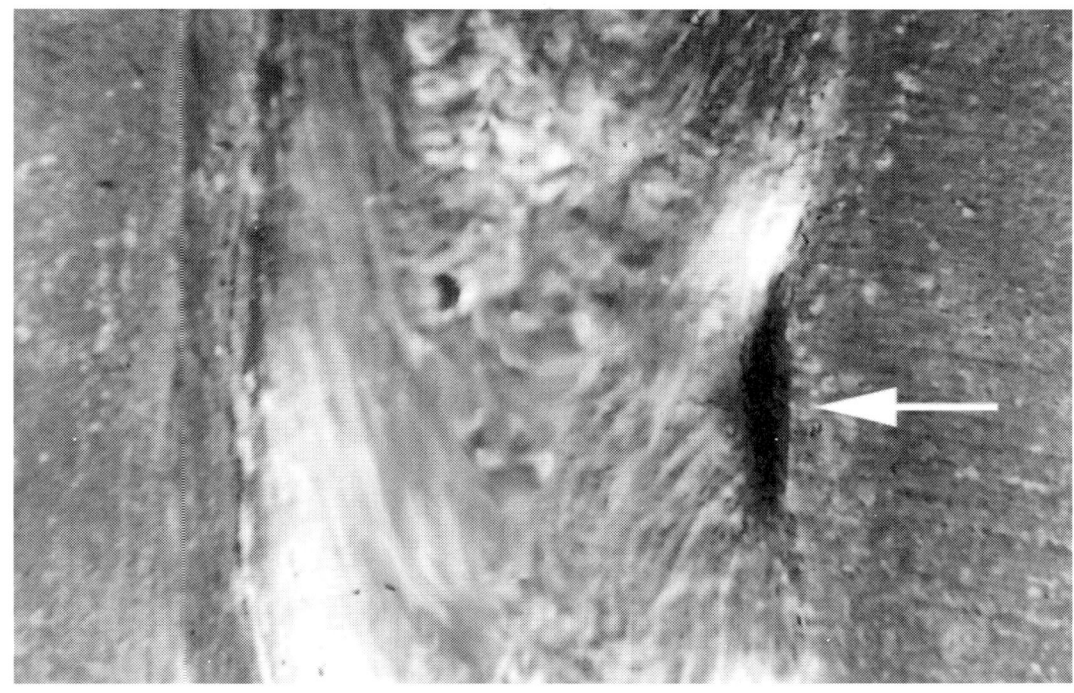

INCOMPLETE FUSION AT WELD FACE

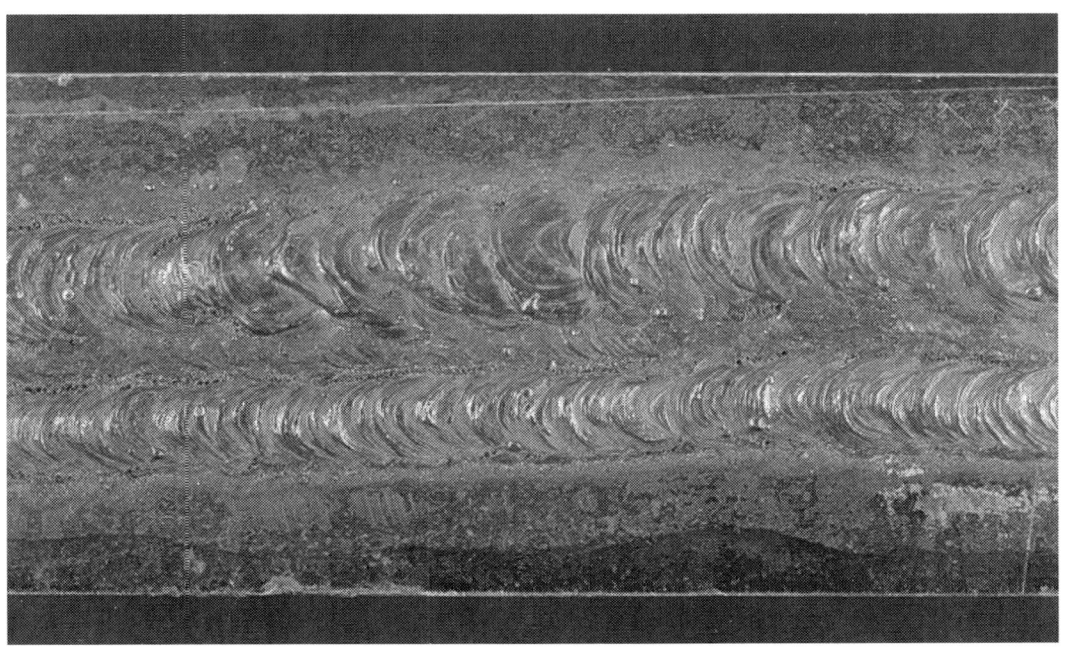

INCOMPLETE FUSION BETWEEN INDIVIDUAL WELD BEADS

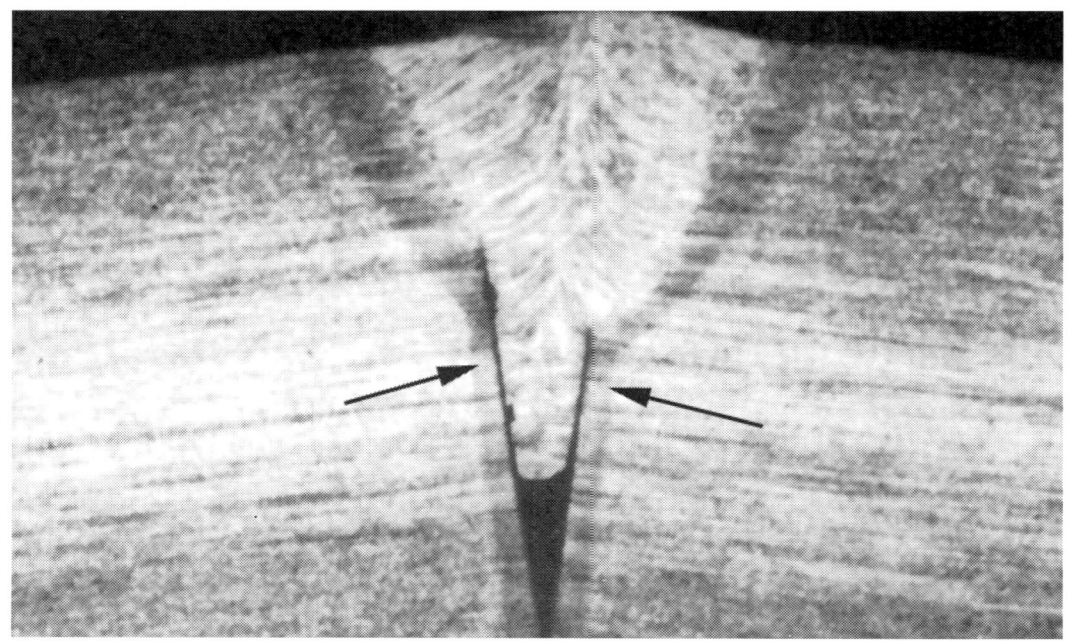

INCOMPLETE FUSION BETWEEN THE WELD AND BASE METAL

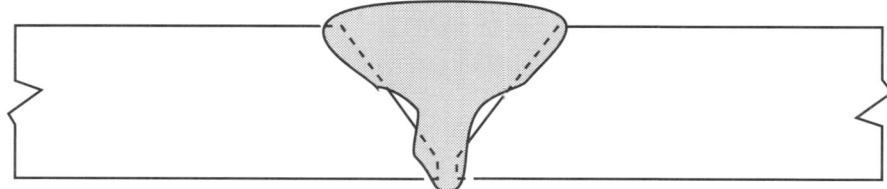

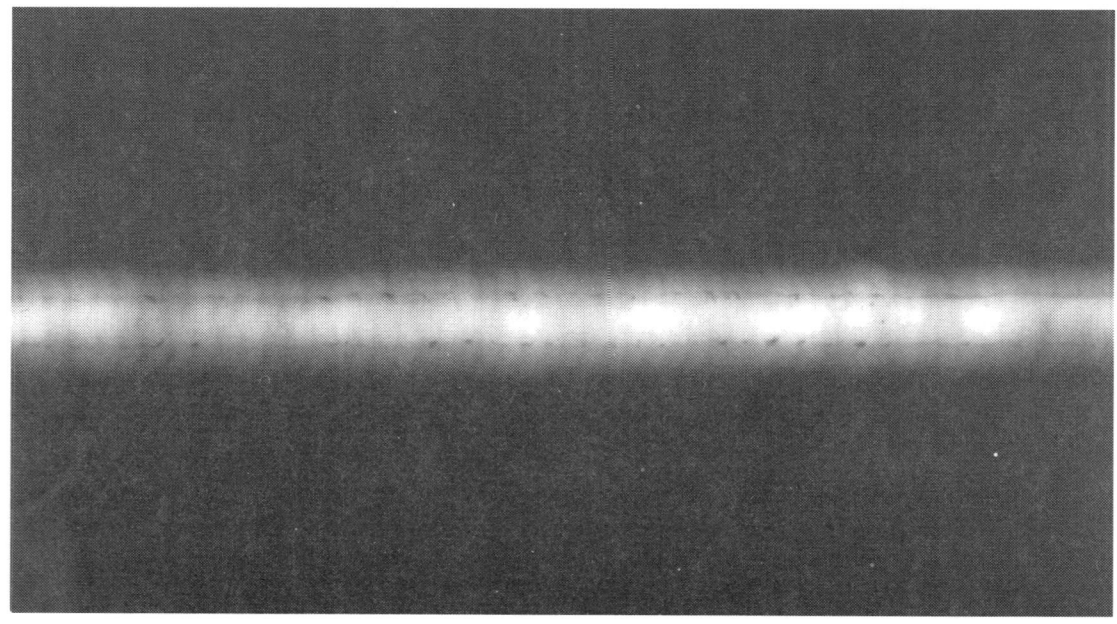

RADIOGRAPH OF LACK OF SIDE WALL FUSION (LOF)

10-15

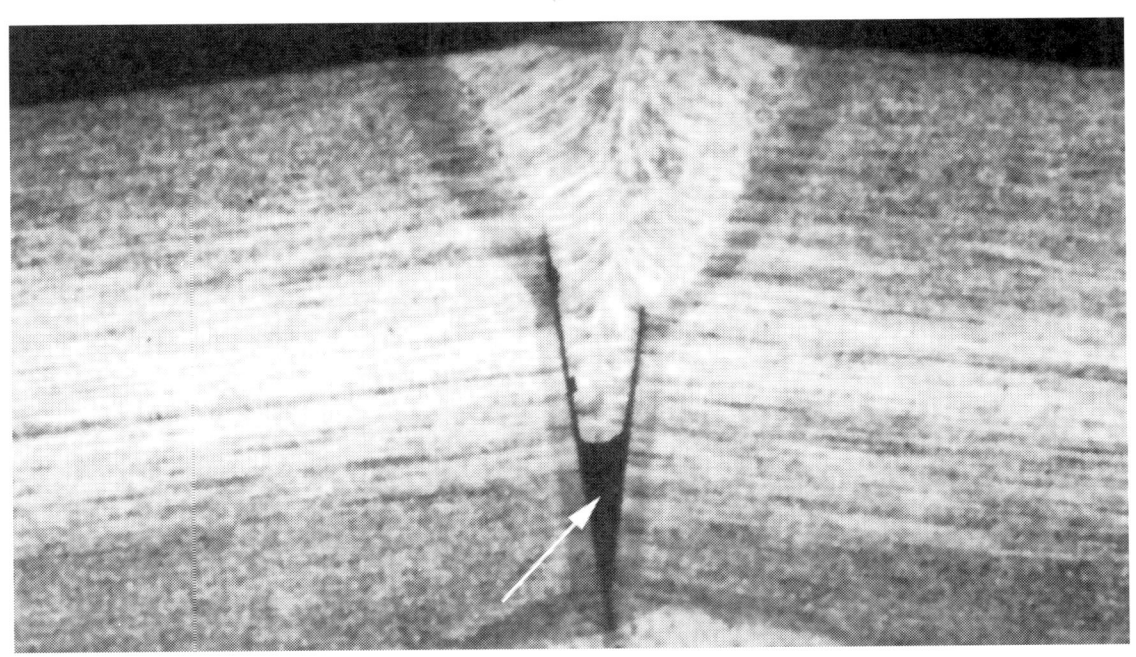

INCOMPLETE JOINT PENETRATION

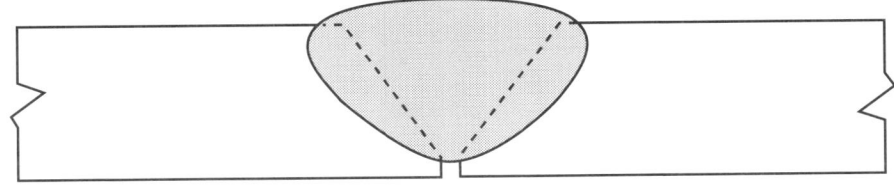

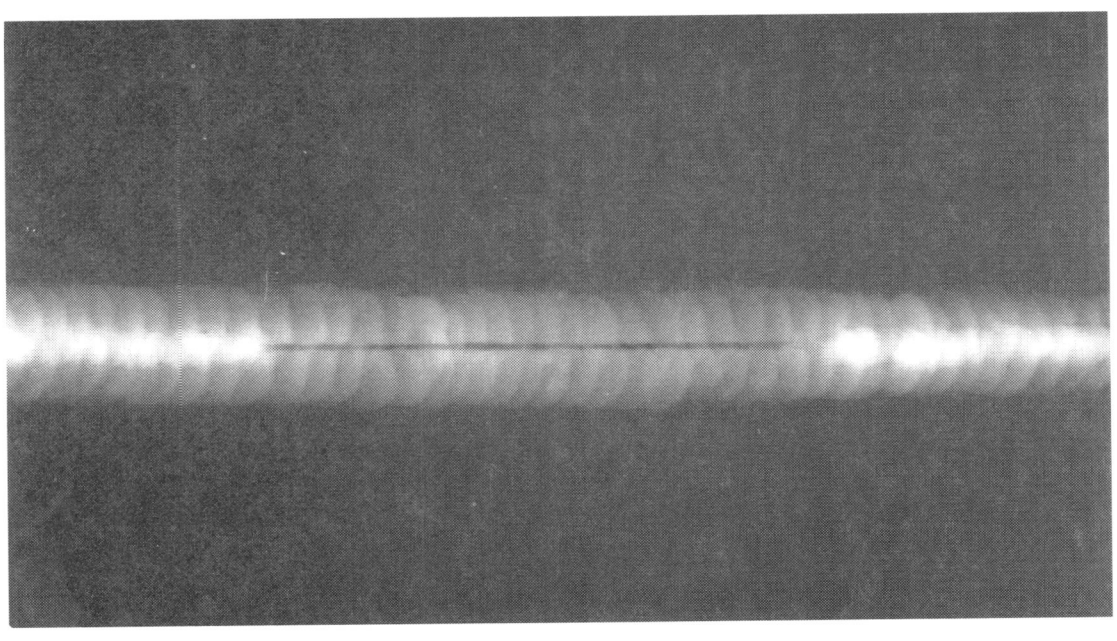

RADIOGRAPH OF INCOMPLETE JOINT PENETRATION

Overlap

Overlap is the protrusion of weld metal beyond the toe or weld root of the weld joint without fusion. The resulting discontinuity is a severe mechanical notch on the surface. This discontinuity is similar to incomplete fusion with the difference being the location where the fusion failed to take place.

Overlap is caused by the inability of the weld metal to fuse with the surface, especially when tightly adhering oxides cover the base metal. Overlap results from lack of control of the welding process in the form of insufficient heat (current too low), inadequate travel speed, improper selection of welding materials (lack of deoxidizers), or improper preparation of the joint (failure to remove mill scale or other surface coatings). Excessive weld metal buildup on a groove weld is referred to as "excess weld reinforcement".

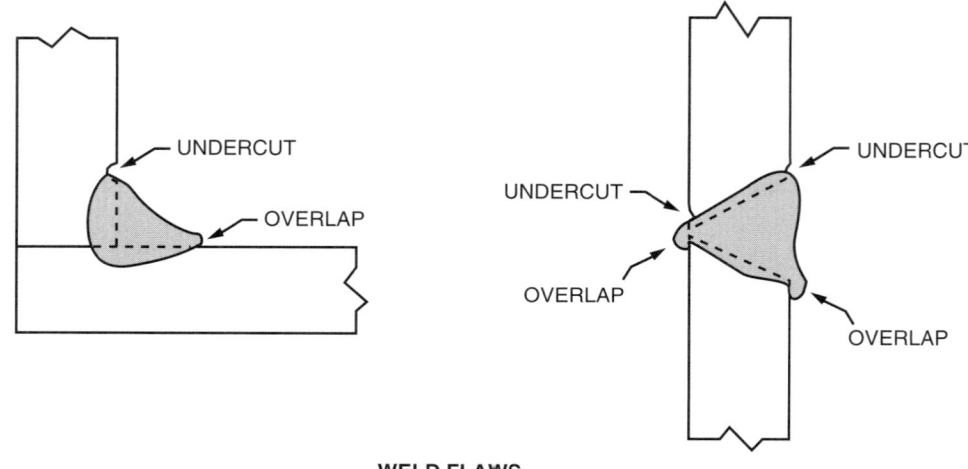

WELD FLAWS

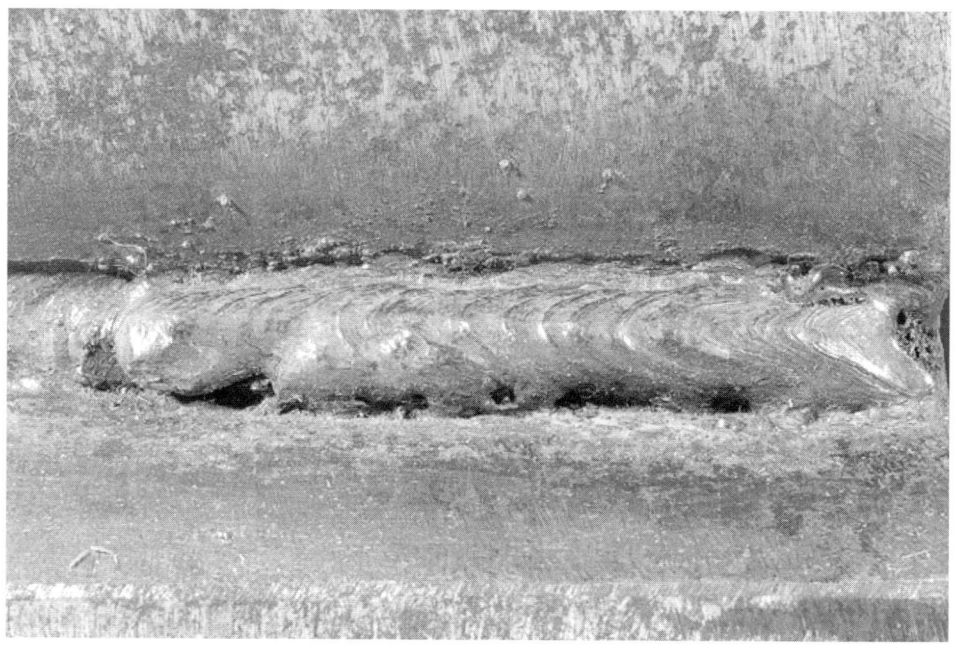

OVERLAP

Undercut - is a surface discontinuity resulting from melting of the base metal at either the weld toe or weld root. It takes the form of a mechanical notch at the these locations. Undercut is caused by the application of excessive heat (excessive weld current) or improper electrode manipulation, which melts away the base metal. Use of excessive travel speeds will also cause undercut.

UNDERCUT AT FILLET WELD TOE

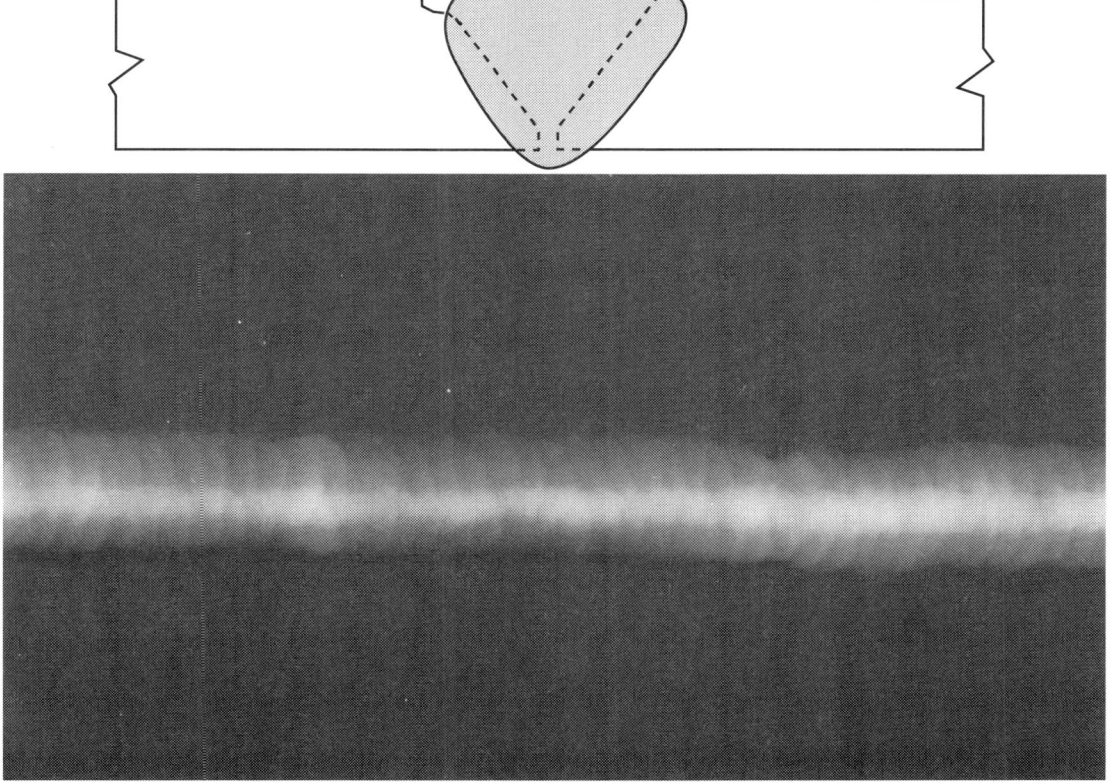

RADIOGRAPH OF EXTERNAL (SURFACE UNDERCUT)

Lamination and Delamination

Laminations are flat, generally elongated, planar base metal discontinuities found near the center of rolled products. Laminations are formed when gas voids in the shrinkage cavity in the ingot are rolled flat but are not subsequently welded under the pressure of hot rolling. They generally run parallel to the surface of the rolled product and are most commonly found in structural shapes and plates.

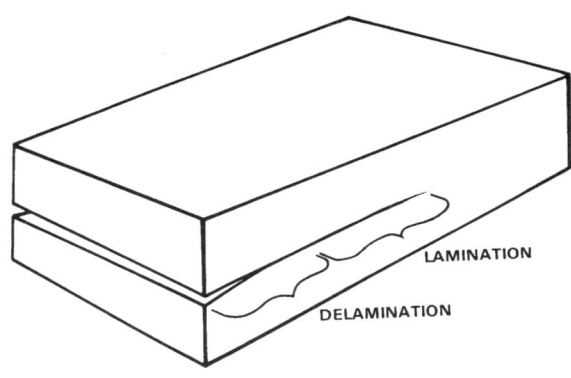

Lamination and Delamination

They most often appear near the centerline of the material thickness. Since it would open as a sandwich, metal containing laminations cannot reliably carry stress in the through-thickness direction.

A delamination is the separation of a lamination under stress. The stress may be a result of distortion during flame cutting; it may be residual stress from welding, or it may be applied stress.

Ultrasonic testing is the only effective means of locating laminations unless they extend to an exposed edge of the material. They will not be revealed by radiographic testing.

Seams and Laps

Seams and laps are linear base metal discontinuities found in rolled products which result from improper steelmaking practices. Seams and laps differ from laminations in that they always appear on the rolled surfaces. When they are parallel to the principal stress, they are generally not considered to be a critical defect. When perpendicular to the applied or residual stress, they will often propagate as a crack. Welding over seams and laps can cause cracking.

Lamellar Tearing

Lamellar Tearing is a fracture separation in heavy weldments, found within or just beneath the heat affected zone of thick plates which were not adequately refined by the steel mill. Heavy plates and structural shapes receive limited working from their ingot stage to the final thickness, which may not remove all traces of ingotism. Rolling and forging impart good properties in the direction of metal flow (the "X" direction) but the strength and ductility perpendicular to the rolled surface (the through-thickness or "Z" direction) remain poor.

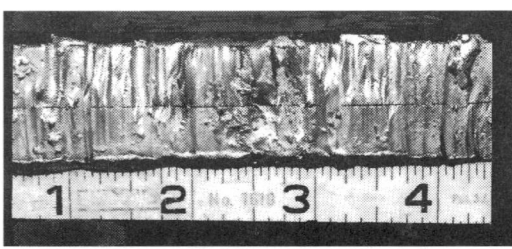

Laminations

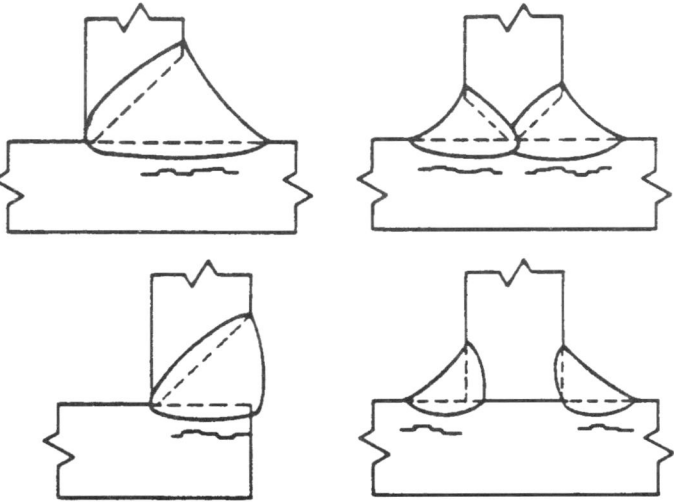

Weld configurations which may cause lamellar tearing

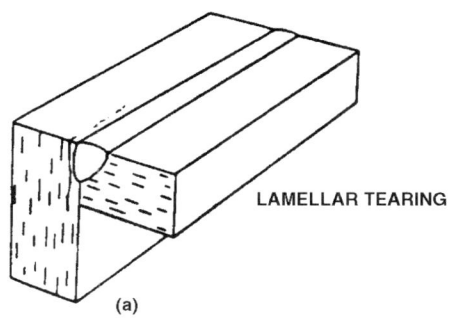 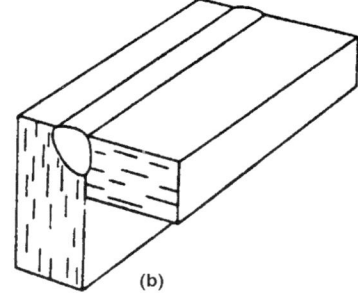

Redesigned Corner Joint to Prevent Lamellar Tearing

Massive welds poorly located adjacent to a thick plate will transmit weld shrinkage stresses into the plate in its weakest direction, creating tears parallel to the surface which then are linked together by shear fractures, forming steps connected by risers perpendicular to the surface. The phenomenon is called lamellar tearing, because the plate opens up as though it were made of stacked sheets or lamellae. The engineer should change the joint design to bring the shrinkage stresses more in line with the rolling direction.

A reduction in the amount of weld required will also reduce the tendency for this type of discontintuity.

Lamellar tears may extend over long distances and are located more deeply than underbead cracks, which differ in shape, cause and location.

Cracks

Cracks may occur in the weld or base metal, or both, when localized stresses exceed the strength of the material. Cracking is generally associated with discontinuities in welds and base metals, with notches, with high residual stresses, and often with hydrogen embrittlement. Welding-related cracks often look as though the metal were brittle. There is little evidence at the crack boundaries that the metal deformed before it cracked. Cracks can be classified as either *hot cracks* or *cold cracks*.

Hot cracks develop at at high temperatures. They commonly form on preferential solidification of alloys of the metal near the melting point. Hot cracks propagate between the grains when the preferential solidification occurs. Cold cracks develop after solidification is complete and are often service-related. Delayed cracks are commonly caused by the presence of hydrogen in a crack-susceptible microstructure subjected to some applied stress. Cold cracks may propagate either through or between grains.

Longitudinal cracks lie parallel to the weld axis. They are called longitudinal cracks, whether they are centerline cracks in the weld metal or toe cracks in the heat affected zone of the base metal.

Transverse cracks are perpendicular to the weld axis. They may remain within the weld metal or extend from the weld metal into the adjacent heat affected zone and the remainder of the base metal. In some weldments, transverse cracks will form in the heat affected zone of the base metal and not in the weld.

Crater cracks occur in the crater formed by improper termination of a weld pass. They are considered hot cracks and are sometimes referred to as "star cracks" because they often radiate in several directions from the center of the crater. However, they also have other shapes. Crater cracks are usually shallow, allowing for their removal with minimal grinding.

A **throat crack** is a longitudinal crack in the weld face of either a groove or fillet weld.

Toe cracks are generally cold cracks. They begin and grow from the weld toe where residual

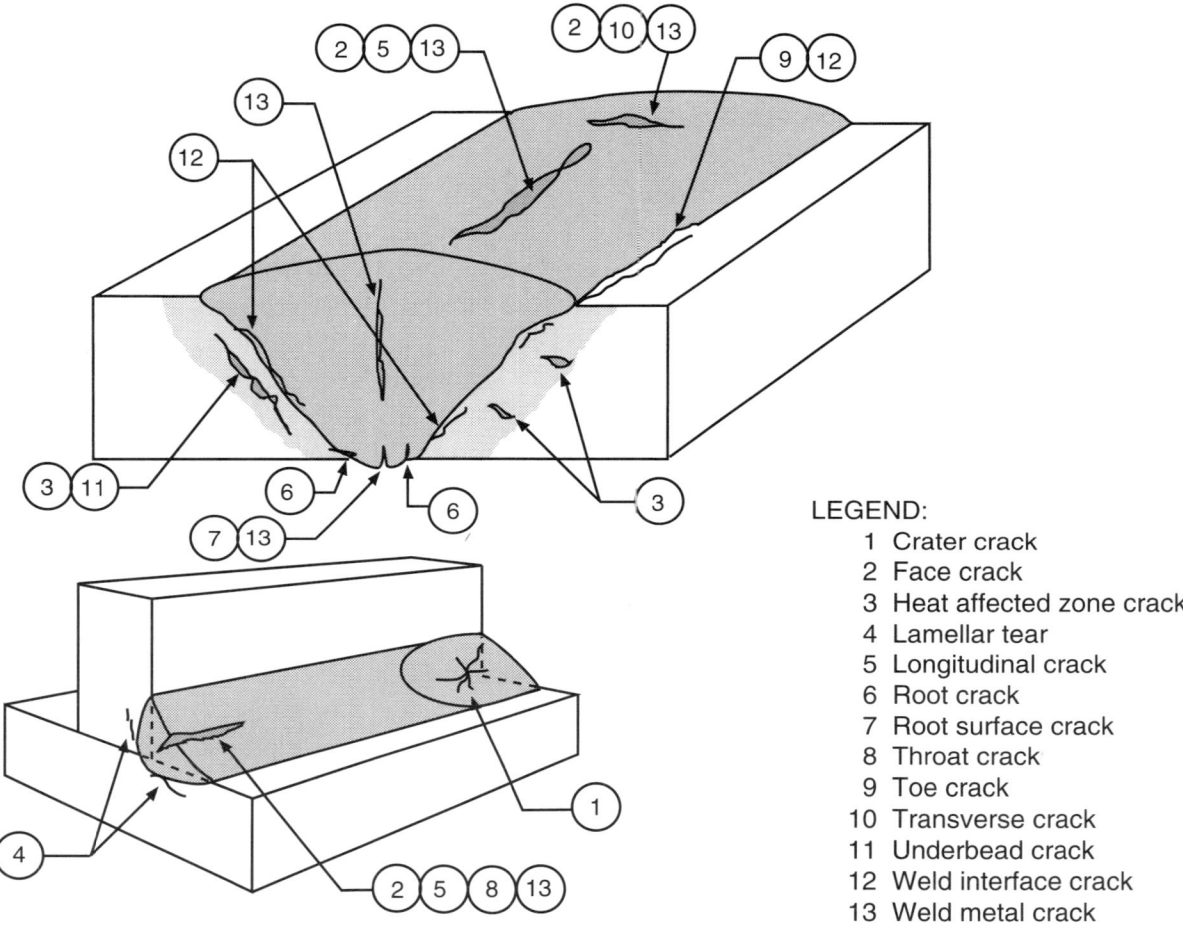

LEGEND:
1 Crater crack
2 Face crack
3 Heat affected zone crack
4 Lamellar tear
5 Longitudinal crack
6 Root crack
7 Root surface crack
8 Throat crack
9 Toe crack
10 Transverse crack
11 Underbead crack
12 Weld interface crack
13 Weld metal crack

CRACK TYPES

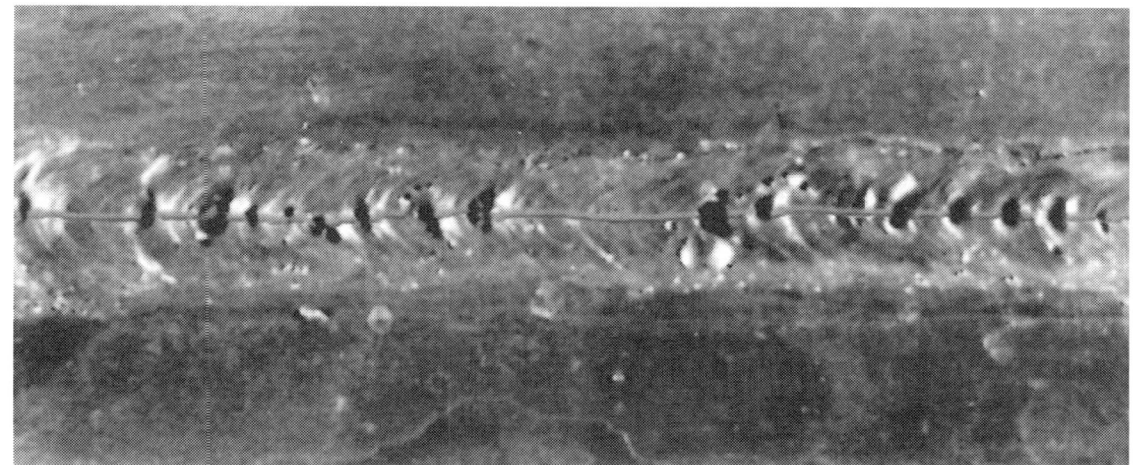

LONGITUDINAL CRACK AND LINEAR POROSITY

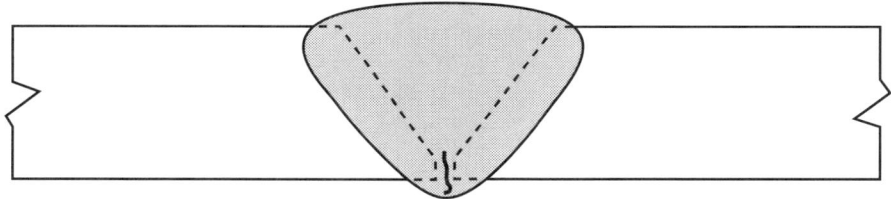

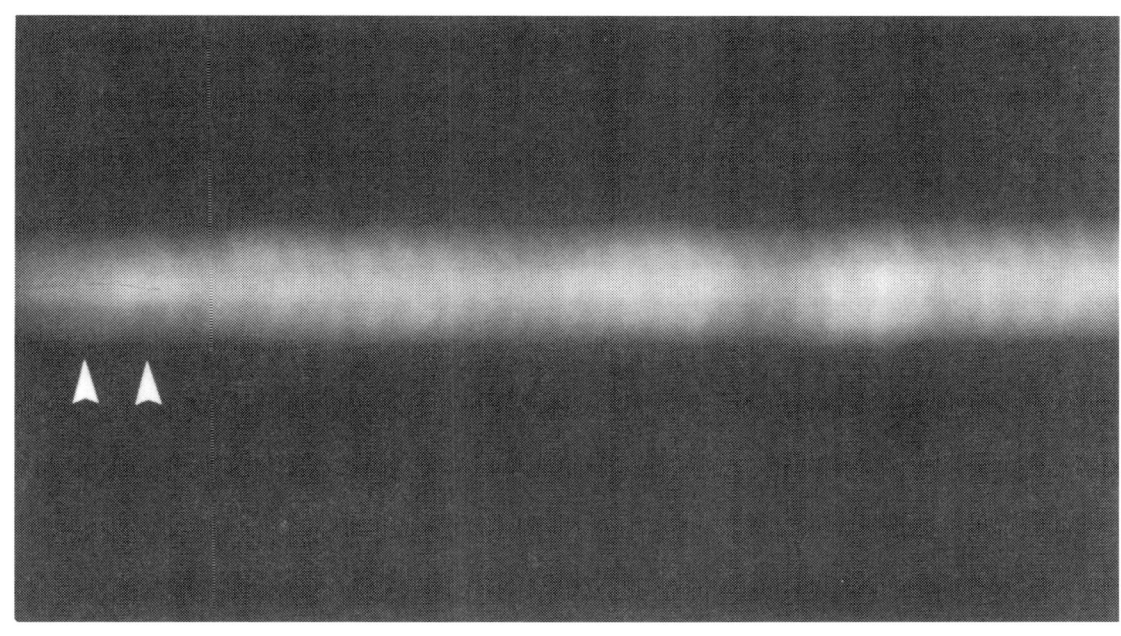

RADIOGRAPH OF LONGITUDINAL CRACK

10-22

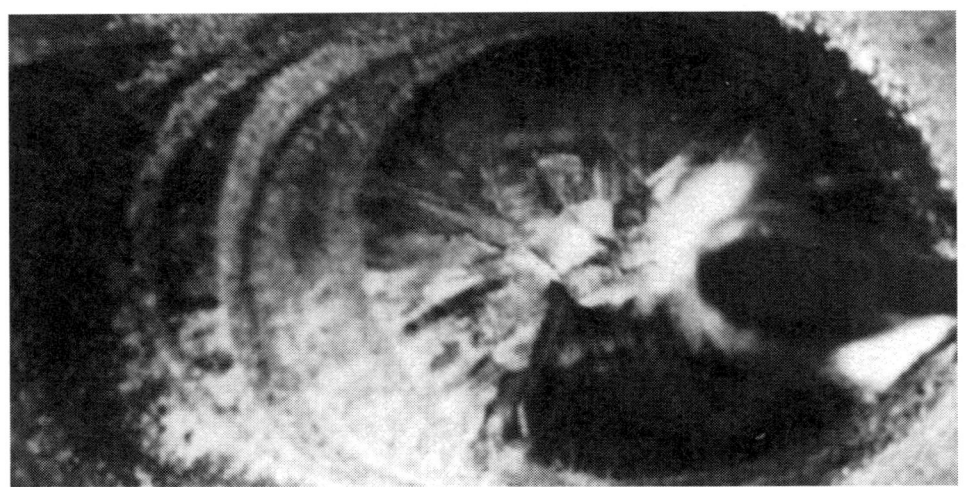

CRATER CRACK

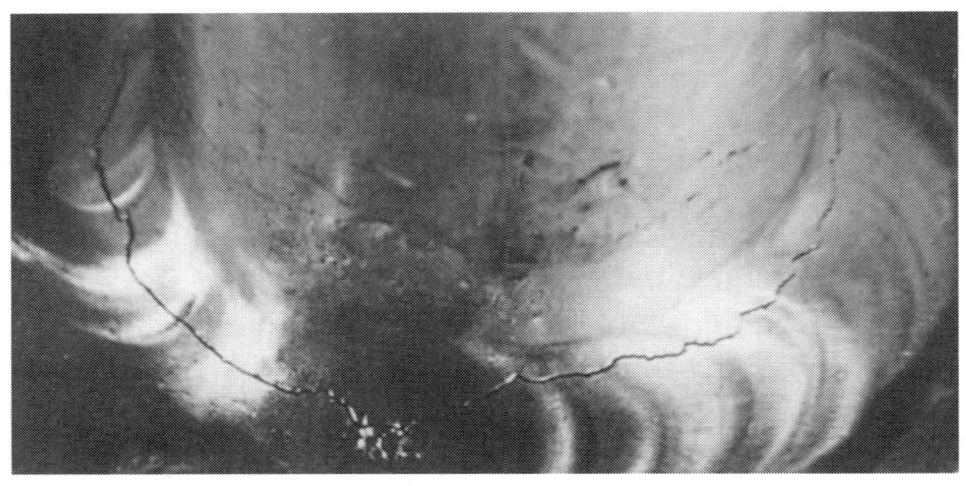

LONGITUDINAL CRACK PROPAGATING FROM CRATER CRACK

10-23

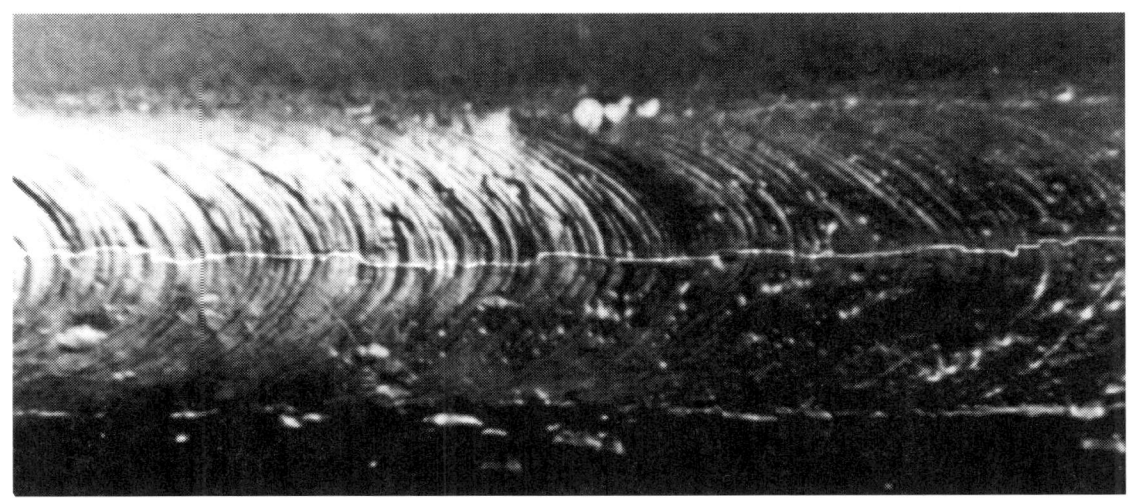

THROAT CRACK

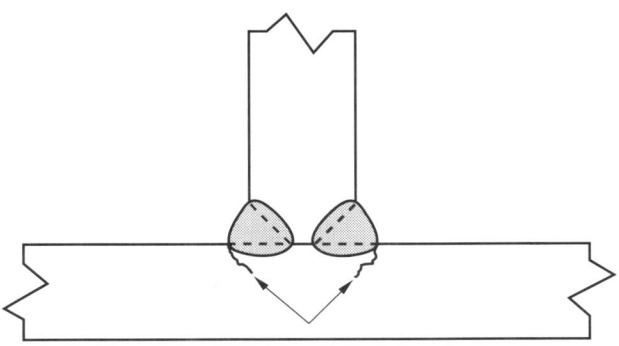

TOE CRACKS

stresses are high, especially when the weld exhibits excessive reinforcement or convexity.

Toe cracks initiate approximately perpendicular to the metal surface, but may tend to curve and follow the weld heat affected zone.

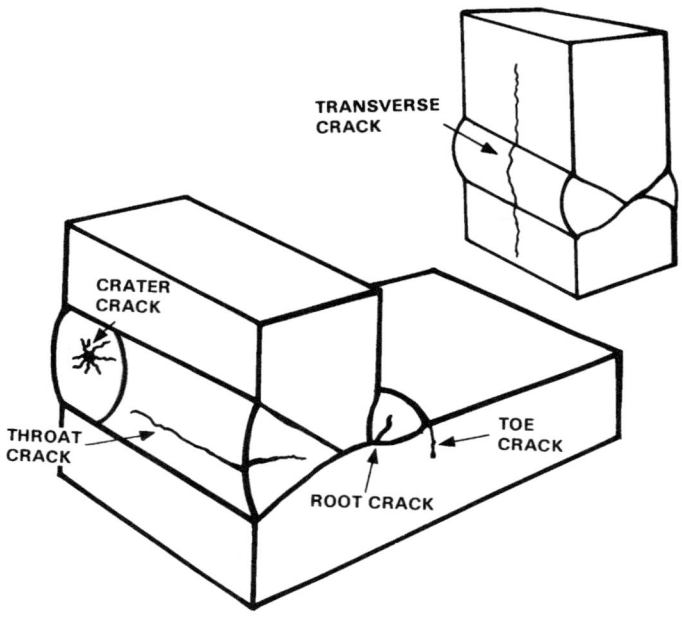

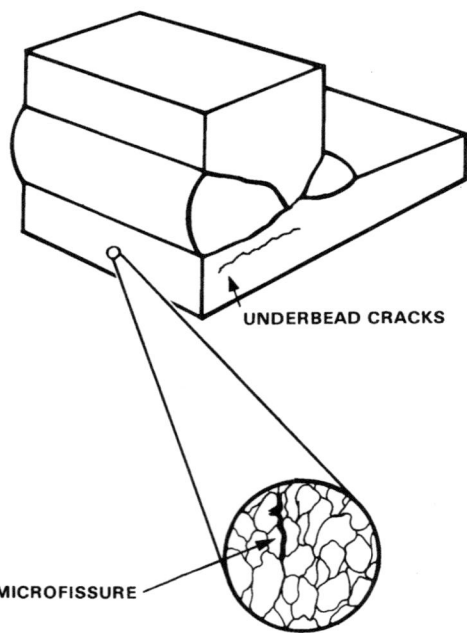

Root cracks are longitudinal cracks in the weld root. The are generally a form of hot cracking.

Underbead and heat-affected zone cracks are usually cold cracks that form in the heat affected zone of the base metal. They are most often short, but they may join to form a continuous crack, especially when three simultaneous conditions are present: (1) hydrogen, (2) high-strength material (Rockwell "C" hardness of 30 or higher), and (3) high residual stress. Underbead and heat affected zone cracks can be either longitudinal or transverse.

Fissures are small or moderate size separations along grain boundaries. This discontinuity is easiest to see in electroslag welds because of the large grains commonly present. The separations may be either hot or cold cracks. The term *microfissure* is used if they are so small that magnification must be used to detect the separation. They are termed *macrofissure* if the separation is large enough to be seen with the unaided eye.

Arc Strike

Arc strikes represent unintentional melting or heating outside the intended weld deposit area. They are usually caused by the welding arc but can be produced beneath an improperly secured work connection during welding. They can also result from improper contact of the prods used for magnetic particle testing.

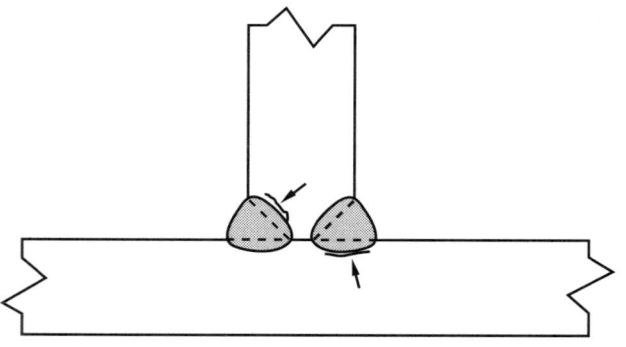

Underbead Cracks

The result is a small, remelted area that can be the source of undercutting, hardening or localized cracking, depending upon the metal composition. For that reason, they represent a dangerous condition which could result in catastrophic failure of the weldment.

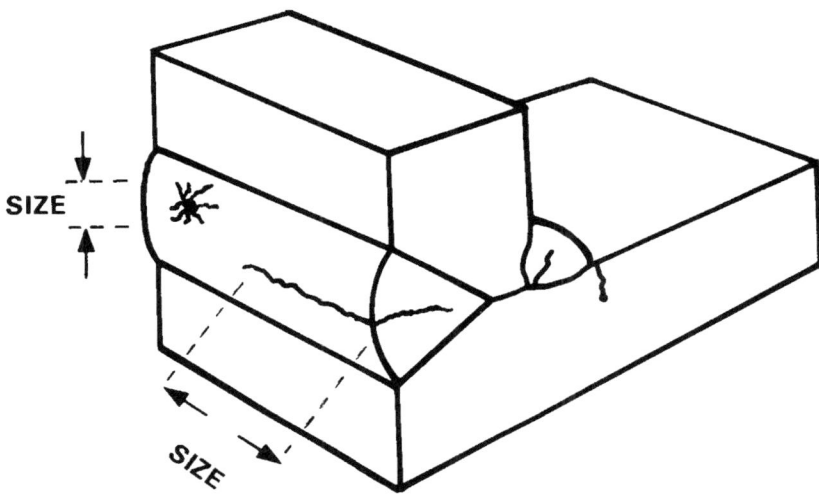

Size of Discontinuity

The size of the discontinuity must be considered when evaluating the structural integrity of the entire weld. Acceptance standards specify the allowable size of discontinuity in terms of its linear dimensions.

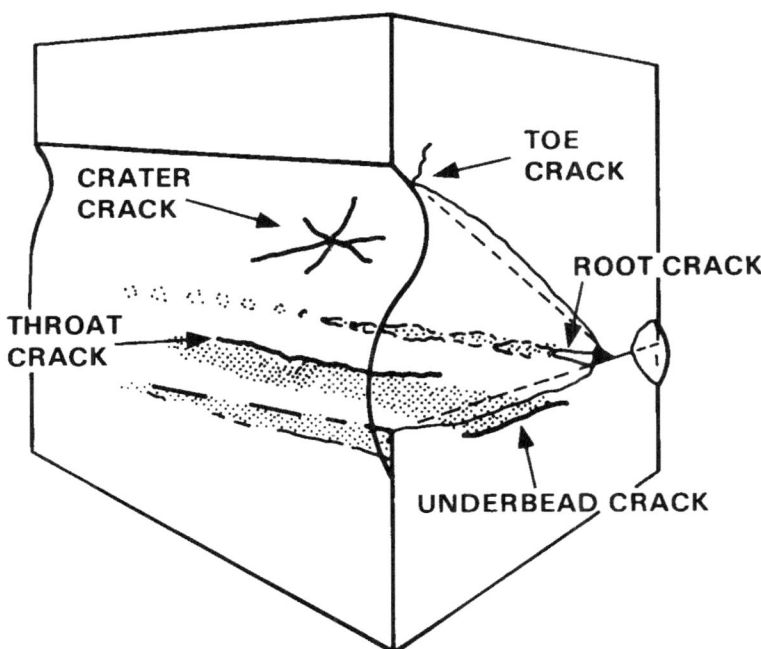

Some discontinuities are acceptable as long as their size does not exceed specified limits. However, other types, such as cracks, are normally unacceptable regardless of length. In general, nonlinear discontinuities are permitted to be larger than linear types.

Location of Discontinuity

The location of a discontinuity may suggest the cause of the problem as well as its seriousness. The location of porosity can identify where contamination exists. The welding inspector must consider the location and orientation of some discontinuities to determine how much the load-carrying capacity of the structure will be degraded.

For structures which will be subjected to fatigue, or cyclic, types of loads, those discontinuities exposed to the surface are generally considered to be more severe. In fact, small surface discontinuities may be more damaging than subsurface ones, even though the size of those subsurface discontinuities is much greater.

Summary

The welding inspector will be asked to examine welds to determine their acceptability in accordance with various codes and specifications. One of the aspects of this activity will be the visual identification of weld discontinuities. The inspector must be capable of identifying the type of discontinuity, because different types of discontinuities have different levels of permissibility.

In addition to identifying the types of weld discontinuities present, it will also be helpful to know what conditions might have lead to the creation of those discontinuities. That way, corrective action can be taken to prevent further occurrence.

REVIEW - CHAPTER 10 WELD AND BASE METAL DISCONTINUITIES

Q10-1 A discontinuity is:
   a. always a defect.
   b. always a reject.
   c. always acceptable.
   d. rejectable if it exceeds code limits.
   e. none of the above

Q10-2 Of the following, which is commonly caused by the presence of hydrogen in a crack susceptable microstructure subjected to applied stress?
   a. lamellar tearing.
   b. delamination.
   c. porosity.
   d. delayed cracking.
   e. none of the above

Q10-3 Porosity, occuring in the form of large cylindrical pores is called:
   a. clustered porosity
   b. linear scattered porosity
   c. uniformly scattered porosity
   d. elongated porosity
   e. none of the above

Q10-4 Which of the following discontinuities is least likely to be seen visually?
   a. toe crack
   b. undercut
   c. lamellar tear
   d. overlap
   e. none of the above

Q10-5 Underbead cracks can result from which of the following welding practices?
   a. use of wet electrodes
   b. welding on contaminated steels
   c. welding over paint
   d. all of the above
   e. none of the above

Q10-6 The weld discontinuity which results from improper termination of the welding arc is referred to as:
   a. undercut
   b. overlap
   c. crater crack
   d. incomplete fusion
   e. all of the above

Q10-7 All but which of the following processes may result in the presence of slag inclusions in the completed weld?
   a. SMAW
   b. PAW
   c. FCAW
   d. SAW
   e. none of the above

Q10-8 That discontinuity which results from the entrapment of gas within the weld cross section is referred to as:
   a. crack
   b. slag inclusion
   c. incomplete fusion
   d. porosity
   e. none of the above

Q10-9 What base metal discontinuity, located at the weld toe, is caused by the welder traveling too rapidly?
a. underfill
b. undercut
c. incomplete fusion
d. overlap
e. none of the above

Q10-10 What weld discontinuity results when the welder travels too slowly causing excess weld metal to pour out of the joint and lay on the base metal surface without fusing?
a. undercut
b. underfill
c. overlap
d. incomplete fusion
e. none of the above

Q10-11 What weld metal discontinuity results when the welder fails to completely fill the weld groove?
a. underfill
b. undercut
c. overlap
d. incomplete fusion
e. none of the above

Q10-12 Excessive weld metal buildup on a groove weld is referred to as:
a. excess convexity
b. excess weld reinforcement
c. overfill
d. all of the above
e. none of the above

Q10-13 The weld discontinuity which results from the initiation of the welding arc outside the weld joint is referred to as:
a. incomplete fusion
b. undercut
c. overlap
d. scratch start
e. arc strike

Q10-14 Of the following, which weld discontinuity shows up as a light region on a radiograph?
a. porosity
b. incomplete joint penetration
c. a and b above
d. tungsten inclusion
e. none of the above

Q10-15 What base metal discontinuity results from improper steelmaking practice and is associated with the rolled surface?
a. lamination
b. delamination
c. seam
d. crack
e. none of the above

Questions **Q10-16 through Q10-20** refer to the figure on the facing page:

Q10-16 What discontinuity is shown by #12b?
    a. longitudinal crack
    b. transverse crack
    c. face crack
    d. toe crack
    e. root crack

Q10-17 What discontinuity is shown by #11?
    a. lamination
    b. base metal crack
    c. lamellar tear
    d. seam
    e. lap

Q10-18 What discontinuity is shown by #12g?
    a. toe crack
    b. incomplete fusion
    c. root crack
    d. lamellar tear
    e. underbead crack

Q10-19 What discontinuity is shown by #5?
    a. undercut
    b. underfill
    c. overlap
    d. incomplete fusion
    e. toe crack

Q10-20 What discontinuity is shown by #10?
    a. lamination
    b. seam
    c. delamination
    d. base metal crack
    e. incomplete fusion

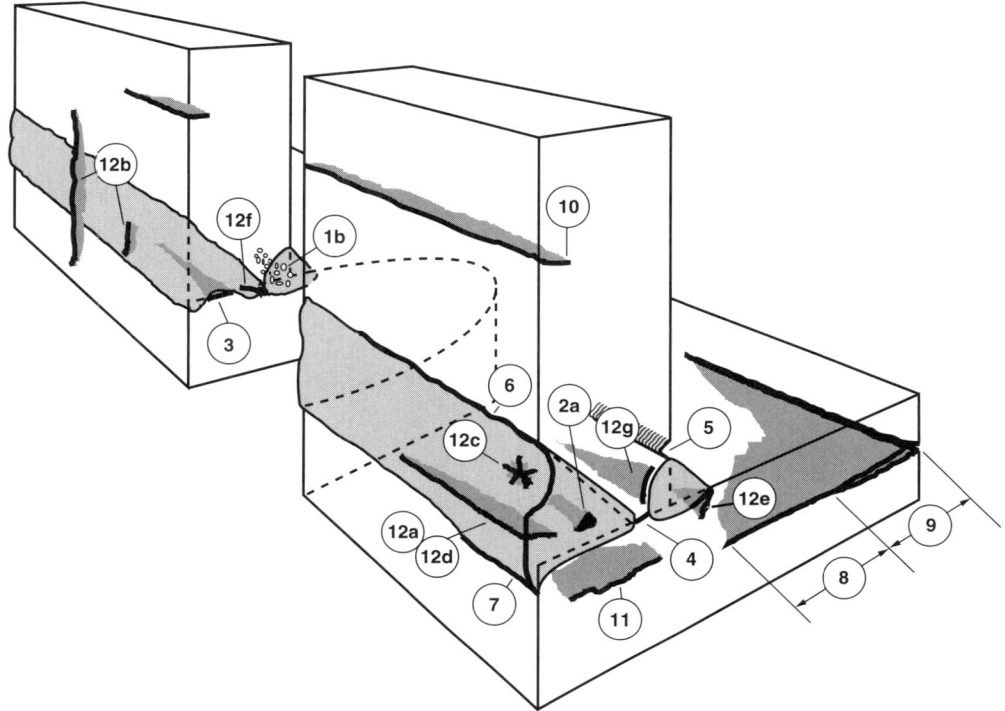

CHAPTER 11:
NONDESTRUCTIVE EXAMINATION PROCESSES

Introduction

The knowledgeable welding inspector will have the authority under some specifications to accept welds by visual examination alone, but cannot certify some of the welds in a weldment or structure so easily. The welding inspector will need help in collecting the necessary evidence of specification compliance.

To collect the necessary evidence, the welding inspector can use one or more of the following methods:

- Nondestructive examination performed by technicians qualified in accordance with the guidelines of the American Society for Nondestructive Testing Recommended Practice No. SNT-TC-1A, when required
- Chemical analysis qualified to ASTM standards
- Metallurgical analysis
- Mechanical testing

This chapter will help the welding inspector to:

- Know the principal features to be considered when choosing a nondestructive examination method.
- Understand how this examination differs in different applications of welding.
- Understand fundamental aspects of the various techniques of nondestructive examination.
- Relate the proper examination technique to a given weld joint design and welding process.
- Know the responsibilities in regard to sample selection, examination methods, retests, repairs, and evaluation of data.

Selection of Examination Method

There are three principal features to be considered when choosing an examination method:

- Limitations of the examination method
- Acceptance standards
- Economics

Limitations of the Examination Method

The limitations of the examination method are a consideration in determining which method will provide the best results for a particular test. For example, radiography can detect cracks that are aligned parallel to the radiation beam; such cracks are usually normal to the plate surfaces. However, radiography usually cannot detect laminations in the plate; ultrasonic examination can detect laminations in the plate and, with the proper choice of technique, can usually be relied upon to detect cracks normal to the plate surfaces.

Acceptance Standards

The statement, "The weld shall be radiographically examined," has no meaning unless acceptance standards are stated. The acceptance standards must state each type of discontinuity and whether or not such a discontinuity is permissible. If a particular type of discontinuity is permissible, then the acceptance standards must specify the maximum size at which the discontinuity is acceptable. Acceptance standards are an integral part of almost all of the codes and standards listed in Appendix A and are commonly used as references in purchase specifications.

Economics (Cost)

Each examination method has a different cost in a particular situation. Two basic cost factors to be considered in the selection of a nondestructive examination method are the initial equipment availability and cost, and the cost of performing the tests. Visual examination is almost always the cheapest, but is also limited to the detection of surface discontinuities. In general, costs of radiographic, ultrasonic, and eddy current examinations are greater than those of visual, magnetic particle, and liquid penetrant methods of examination.

Selection of the proper examination method can be quite complex. In many cases, a single test may not be adequate. It may be necessary to apply two or more tests to assure complete coverage. For example, the examination of a large carbon steel weldment might include radiography for subsurface evaluation and magnetic particle testing for an examination of the surface. For critical applications or large volumes of testing, it is suggested that help be obtained from a competent laboratory or consultant.

Examination Methods and Limitations

Visual Examination

As the welding inspector gains experience, it will be noted that a weld displays its pedigree to a large degree in its surface appearance. When the welding inspector has seen that the preparation of the joint was good, and knows that the welder had

HOPTON JONES INDUSTRIES

the requisite ability, the odds favor a good-looking weld being sound and within specification.

Remember: The evidence needed by the inspector is provided by the governing code, engineering standards, job specifications, or job requirements. The welding inspector cannot accept less nor ask for more than is required by these documents.

Visual examination is limited to the detection of surface discontinuities. However, visual examination, when conscientiously applied before, during and after welding, can prevent the vast majority of those discontinuities that would be detected by other examination methods. In fact, the ability to prevent many discontinuities before the weld is complete is perhaps the most important feature of visual examination.

Visual Examination Practice: Any good visual weld examination program requires that the inspector have a working knowledge of:

- Applicable codes, standards and specifications, including weld acceptance criteria
- Workmanship standards
- Welding processes being used
- Good weld fabrication practices

The welding inspector's responsibilities commence before welding begins. In fact, these

initial steps are critical enough to set the tone for the resulting quality of the entire project. If these preliminary duties are thoroughly attended to, the attainment of quality welding will be more assured. Some of the items which may be checked include:

- Review applicable documentation
- Check welding procedures
- Check individual welder qualifications
- Establish hold points
- Develop an inspection plan
- Develop a plan for recording inspection results and maintaining those records
- Develop a system for identification of rejects
- Check condition of welding equipment
- Verify base metal and filler metal identifications
- Check quality and condition of base and filler metals
- Check weld preparations
- Check joint fit
- Check adequacy of alignment devices
- Check weld joint cleanliness
- Check preheat, when required

Once welding begins, the welding inspector's duties are primarily related to the assurance that the welding is being done in accordance with the welding procedures and any applicable standards. Here, the goal is to identify problems before they occur, or at least as soon as possible after they occur, so they can be corrected in the most economic and effective way. Below are some of the items which should be checked during the actual welding operation:

- Check welding variables for compliance with welding procedure
- Check quality of individual weld passes
- Check interpass cleaning
- Check interpass temperature
- Check placement and sequencing of individual weld passes
- Check back-gouged surfaces
- Arrange for in-process NDE, when required

After the welding has been completed, the role of visual examination is simply the verification that all of the preceding steps have been successfully completed and the resulting weldment is acceptable. Some of the items to be checked include:

- Check finished weld appearance
- Check weld size
- Check weld length
- Check dimensional accuracy of weldment
- Arrange for additional NDE, when required
- Monitor postweld heat treatment, when required
- Prepare inspection reports

There are numerous advantages of visual examination, including: low cost to apply, little need for expensive equipment, and the ability to find problems when they occur to allow for economical and expedient correction.

 # Most Common Test

⭐ Penetrant Testing (PT)

Penetrant testing is a sensitive method used for locating discontinuities, such as cracks and porosity, in nonporous materials. The discontinuities must be clean and open to the surface. The method employs a penetrating liquid which is applied to the surface to be examined and which enters the discontinuity. The excess penetrant is then cleaned from the surface. Any penetrant which subsequently exudes from openings and causes discoloration of developer powder on the surface indicates the location of a discontinuity.

There are <u>two general classifications</u> of penetrants: <u>visible dye</u> and <u>fluorescent</u>. They differ in that the visible dye can be observed under normal white light, while the fluorescent type requires an ultraviolet (or black) light to produce an indication. There are several methods for effective application of penetrants, such as, dipping, brushing, flooding, or spraying. Use of fluorescent penetrants results in a more sensitive test, simply because our eyes can more readily detect a fluorescent indication. Consequently, for critical applications where even the smallest discontinuities must be detected, the fluorescent penetrant method would be desirable.

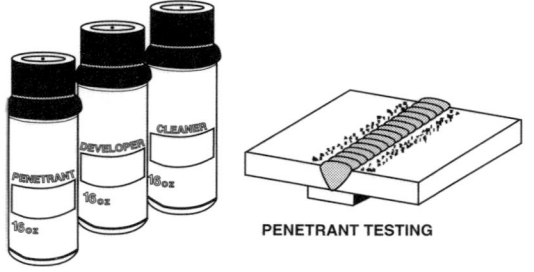
PENETRANT TESTING

In general, the penetrant test consists of <u>four simple steps</u>, including: <u>precleaning</u> the test surface, <u>application of the penetrant</u> which remains for some prescribed time (<u>called the "dwell time"</u>), <u>removal of excess penetrant</u> from the test surface, and <u>application of the developer</u>. Following the application and drying of the developer, all indications are evaluated in accordance with applicable standards. For a detailed explanation of these steps, please refer to <u>ASTM E-165</u>.

Uses

This method is applicable to magnetic materials; however, it is particularly useful on nonmagnetic materials such as <u>aluminum</u>, magnesium and austenitic <u>stainless steels</u>, which cannot be examined by magnetic particle testing.

⭐ Advantages *Simple, Portable*

Penetrant testing is relatively <u>inexpensive</u> and reasonably rapid. The process is simple and operators find little difficulty in learning to apply it properly. There are few, if any, false or nonrelevant indications on reasonably smooth surfaces.

Limitations, *Slow, Surface only, smooth surface*

The main limitation of penetrant testing is that discontinuities must be clean and open to the surface to be detected. Some substances used as penetrants can have a deleterious effect on welds and base metals and can affect the service life of the weldment or end use of the product. Penetrants are difficult to remove completely from discontinuities and if they are corrosive to the

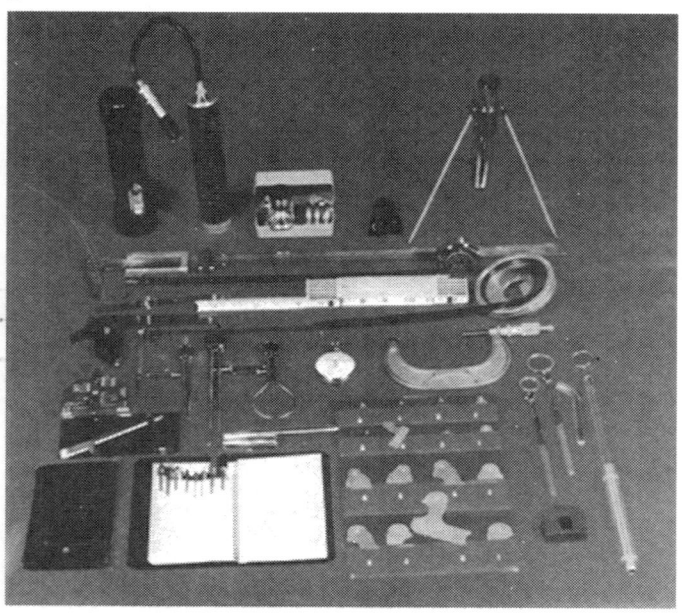
VISUAL INSPECTION TOOLS

material or otherwise not compatible with the use of the product, they should be avoided.

The contour of the surface being examined should not contain sharp depressions between beads or weld ripples that would interfere with complete cleaning and excess penetrant removal. Otherwise, you may observe false or irrelevant indications. If these conditions do exist, the weld surface should be ground smooth before inspection.

Interpretations *Results, Sketches, Photo / Lift off Tape (Nobody uses This)*

It is important that the technician not reduce the recommended dwell times for penetrant or developer. Large cracks and voids will reveal themselves quickly, but those might be found without this aid. Tightly closed cracks absorb the penetrant slowly and respond to the developer equally slowly.

Inadequate removal of the superficial penetrant will give many false indications, such as a general glow under the blacklight for fluorescent penetrant testing, or a pink coloration of the developer with visible dye penetrant testing. Precleaning the surface without adding more penetrant will give a clearer indication, showing only seepage from surface discontinuities. Indications which repeat at the same location throughout several cleanings undoubtedly reveal open discontinuities.

Care must also be taken to <u>avoid spraying cleaner directly on the test surface between application of the penetrant and developer.</u>

✱ Magnetic Particle Testing (MT)

Magnetic particle testing is used for locating surface or near surface discontinuities in ferromagnetic materials. This method involves the establishment of a <u>magnetic field</u> within the material and may be examined with the aid of prods, yokes, or coils. The pattern of discontinuities is revealed by the <u>buildup of iron powder particles</u> which are applied to the surface, either as a <u>dry powder</u> or <u>suspended in a liquid</u>.

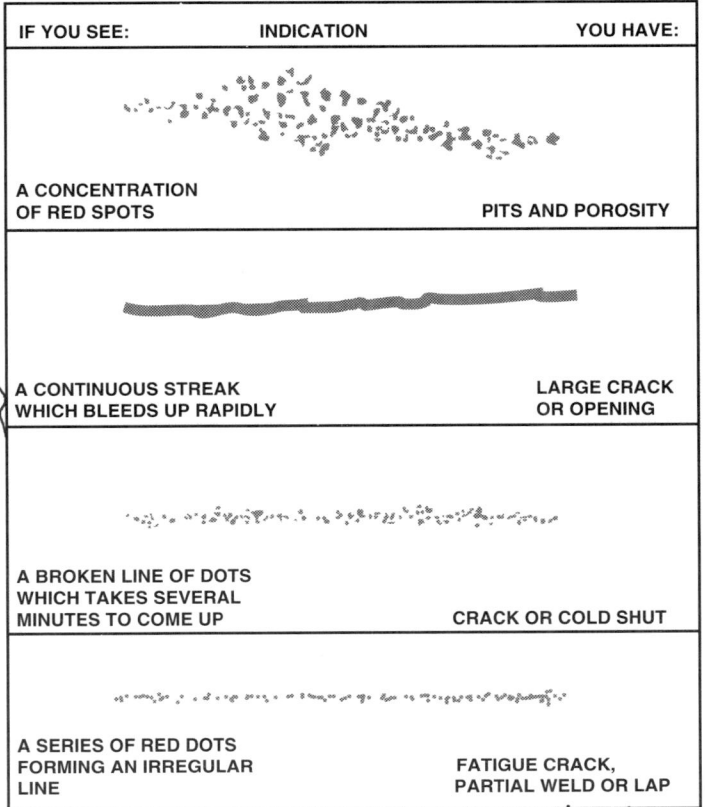

The particles may be <u>dyed</u> *All colors* for greater visibility or coated with <u>fluorescent</u> dye to be viewed under ultraviolet light with the same improved sensitivity associated with fluorescent penetrant testing. The type of surface and the type of discontinuity suspected should determine the material selected.

A test material or component can be magnetized either by passing electric current through it or by placing it in a magnetic field. The electric current used to generate the magnetic field may be alternating current (ac), direct current (dc) or half-wave rectified direct current (hwdc). <u>Alternating current</u> is generally limited to the detection of surface discontinuities; however, its characteristically changing magnetic field tends to increase the particle mobility on the surface to result in an improvement of the test sensitivity when examining rough surfaces.

A dc <u>magnetic field</u> differs from the ac field in that it tends to penetrate more deeply into the test piece, which results in the ability to detect

Permanet Mag

11-6

discontinuities slightly below the surface. However, the magnetic particles do not tend to move as readily on the surface as with ac. Half-wave rectified dc combines the benefits of both types of magnetizing current: the enhanced particle mobility of ac and the deeper penetration of dc.

Uses

Magnetic particle testing can provide a great deal of information about the quality of weldments. This method may be used to inspect welds, plate edges prior to welding and weld repairs. This method is capable of detecting:

- Surface cracks of all kinds in the weld or base metal
- Laminations or other discontinuities on the prepared edge of the base metal
- Incomplete fusion (if at or near the surface)
- Undercut
- Subsurface cracks (if it occurs near enough to the surface to cause an interruption of the magnetic field at the surface)

Advantages *Low cost*

The basic equipment for magnetic particle testing is relatively <u>simple</u>. It includes devices for creating a magnetic field of the proper strength and direction. There might also be controls for adjusting the current and an ammeter to indicate the amount of magnetizing current for each examination.

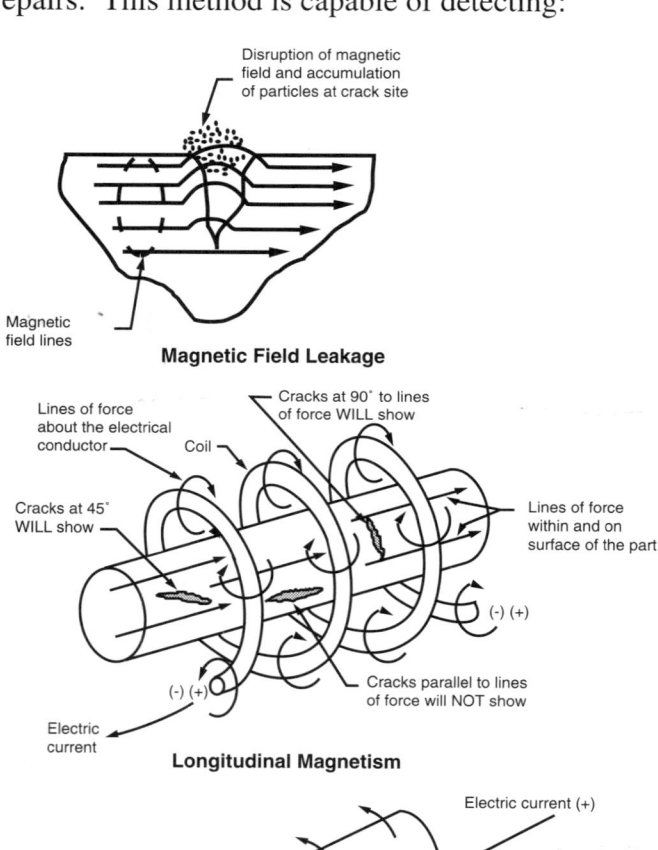

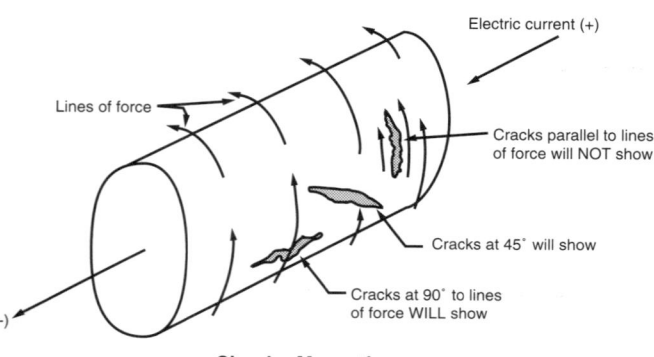

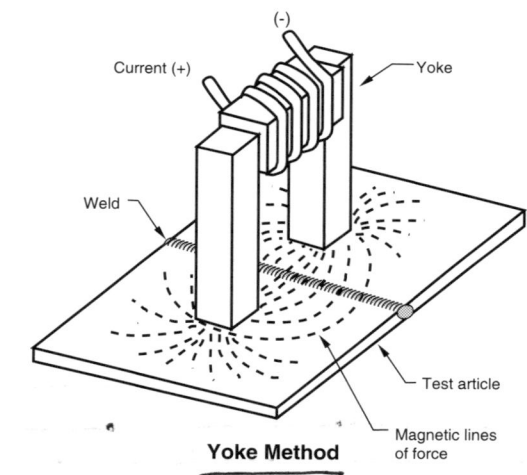

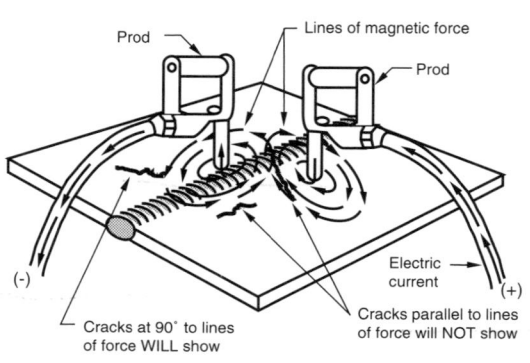

Results:
sketches
Photo
Lift off Tape

Compared to penetrant testing, this method has certain other advantages. It reveals discontinuities that are not open to the surface or filled with some substance (cracks filled with carbon, slag or other contaminants may not be detectable using penetrant testing). Magnetic particle testing is also generally faster and more economical than penetrant testing, unless extremely large test surfaces are involved. Another advantage is the fact that there is less cleaning required prior to examining the part with magnetic particle testing.

Limitations *metal only*

The magnetic particle method of examination is applicable only to ferromagnetic materials in which the deposited weld metal is also ferromagnetic. This method cannot be used to examine nonferromagnetic materials such as aluminum, magnesium or the austenitic stainless steels.

Difficulties may arise when examining weldments where the magnetic characteristics of the deposited weld metal differ appreciably from those of the base metal or where the magnetic field is not properly oriented. Joints between metals of dissimilar magnetic characteristics may create magnetic particle indications even though the joints themselves are sound.

It's important that the technician not expect the magnetic particle method to find deep-seated discontinuities. They are more readily discovered using other examination methods such as radiography or ultrasonic testing. These methods, and the visual examination of sectioned test samples, can be used to qualify the magnetic particle procedure for subsurface examination.

For discovery of discontinuities lying in all orientations the magnetic particle test must be applied in at least two directions, approximately 90° apart. Only then can the tester be assured that all discontinuities will be revealed.

※ Demagnetize Part when completed.

★ Radiographic Testing (RT)

Radiographic testing is suitable for all materials. However, the applicability of radiography for weld examination depends a great deal upon the weld joint location, joint configuration and material thickness. Almost any

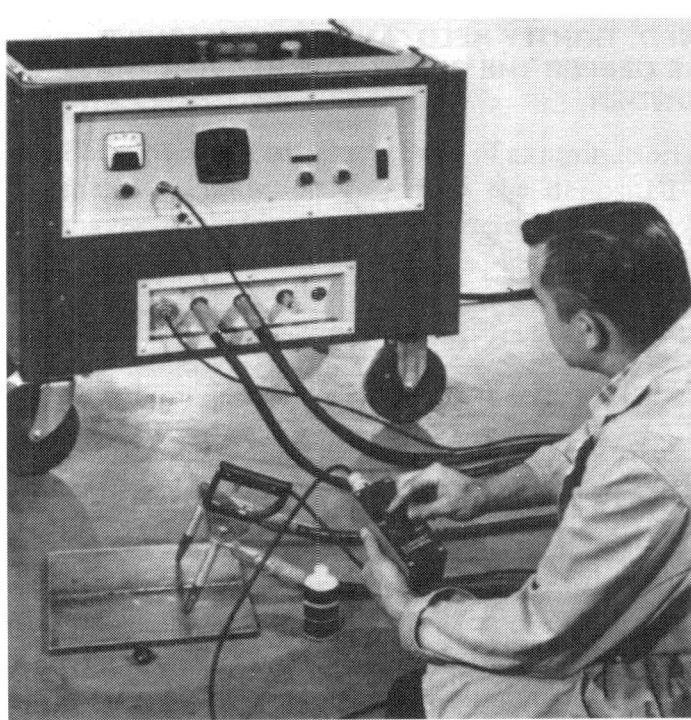

MAGNAFLUX CORP.

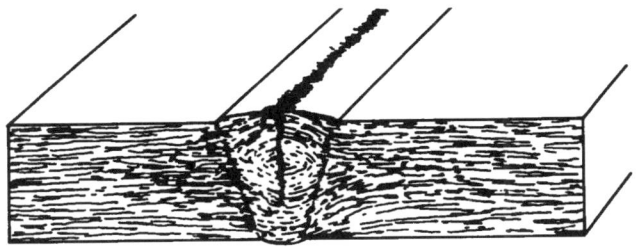

SURFACE CRACKS GIVE POWDER PATTERNS THAT ARE SHARPLY DEFINED, TIGHTLY HELD, AND USUALLY BUILT UP HEAVILY. THE DEEPER THE CRACK, THE HEAVIER THE BUILD-UP OF POWDER.

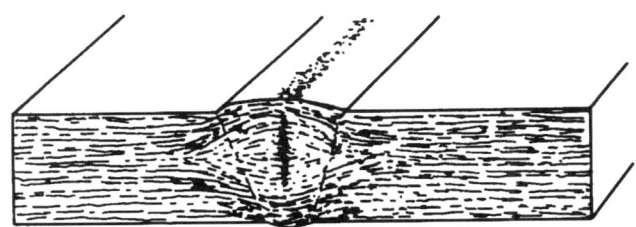

SUBSURFACE CRACKS PRODUCE A LESS SHARPLY DEFINED, FUZZY PATTERN. THE POWDER IS ALSO LESS TIGHTLY ADHERENT.

[Handwritten notes at top:]
11-8
GAMMA RADIATION — Iridium 192, COBALT 60, Cesium 137 — STAYS ON All Times
X-RAY MACHINE when off its off
1. SOURCE
2. FILM BEHIND OBJECT
3. EXPOSE RADIATION
4. DEVELOPE FILM
5. EVALUATE

[Handwritten near Advantages:] All MATERIALS. FILM is PERMANENT RECORDS

---

weld thickness can be radiographed, but insufficient joint access may prevent best use of the method. The welding inspector should keep this limitation in mind when asking for radiographic examinations.

Radiography uses X- or gamma radiation that penetrates through the part and produces an image on a film or plate. The density of the material in a discontinuity (air in the case of a crack, incomplete fusion or porosity) is less than that of the solid metal. Different density materials attenuate the radiation in different amounts and consequently produce optical density differences on the film or plate.

Material density can be affected by the material itself (For example, tungsten is much denser than steel or aluminum, so tungsten is more effective at preventing the radiation from passing through, resulting in a low density indication on the film.) or by the thickness of a given material (The thicker the material, the more effective it is at stopping the radiation and thus producing a lighter film image.). The selection of the radiation source (energy of the emitted rays) for a particular thickness of weld is a critical factor. If the energy of the source is too high or too low for a given thickness of material, then low contrast and poor radiographic sensitivity result. The use of a variable light intensity viewer is helpful when viewing and analyzing radiographs.

There must also be some means of determining the film density at various points. This is usually accomplished through the use of a device referred to as a densitometer. To indicate the acquired sensitivity of a given radiograph, a device referred to as an image quality indicator, or penetrameter, is placed adjacent to the area of interest. This penetrameter is a particular thickness, based on the thickness of the test part, and contains a series of holes having different diameters. The sensitivity of the radiograph is therefore determined by which of the three holes is visible on the radiograph.

Advantages

Radiography can detect:
- Surface discontinuities such as undercut, incomplete joint penetration, excessive weld reinforcement, underfill, etc.
- Subsurface discontinuities that cannot be detected with visual, magnetic particle or penetrant methods and may not be detected by the ultrasonic method.

Limitations

The cost of radiography usually goes up as the joint becomes more complex, and the amount of information that can be obtained becomes more limited.

Discontinuities must be more or less aligned with the radiation beam. This is not a problem for slag or porosity because these are usually round in cross section and are aligned with the beam from any direction. However, cracks, incomplete fusion and incomplete joint penetration must be aligned with the beam to be detected. Laminations and lamellar tearing are almost never detected with

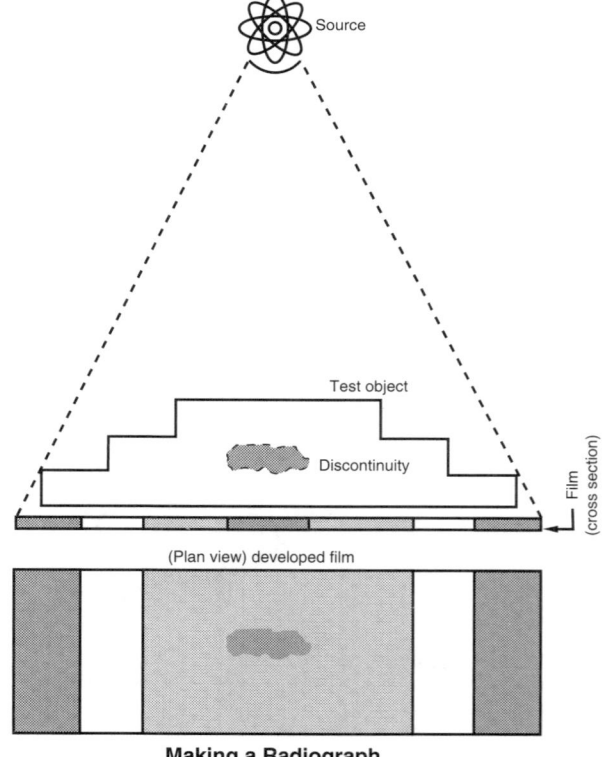

Making a Radiograph

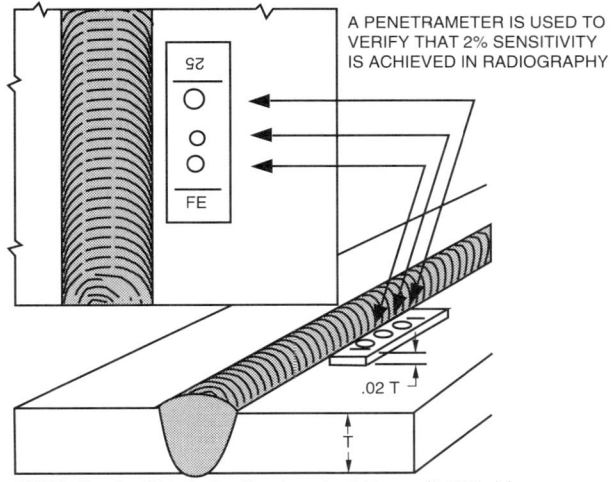

Typical Penetrameter

NOTE: Number 25 indicates Penetrameter thickness (0.025 in.) for use on 1-1/4 in. thick steel (2% sensitivity)

radiography due to their inherent orientation with respect to the radiation. hazard

The high cost of radiation sources and related equipment and facilities tends to be one of the limitations of this method. There must be a high capital expense to initially set up for this type of testing.

The use of radiography has one negative aspect not associated with the other nondestructive methods, and that is the radiation hazard. Excessive exposure to radiation from an X-ray machine or a radioactive isotope cannot be detected by any of the human senses but may cause disease, permanent injury or death. All states require that radiographers be licensed or have special training, or both. Radiography is, however, a safe operation when conducted in accordance with established procedures.

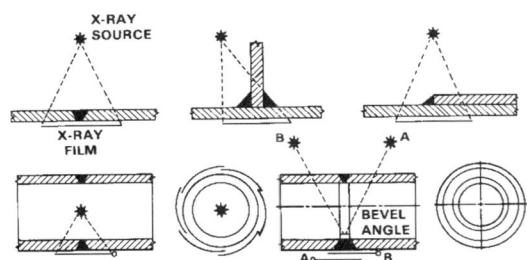

Shown here are the typical arrangements of the x-ray source and film. The angle of exposure and the geometry of the weld influence interpretation of the negative.

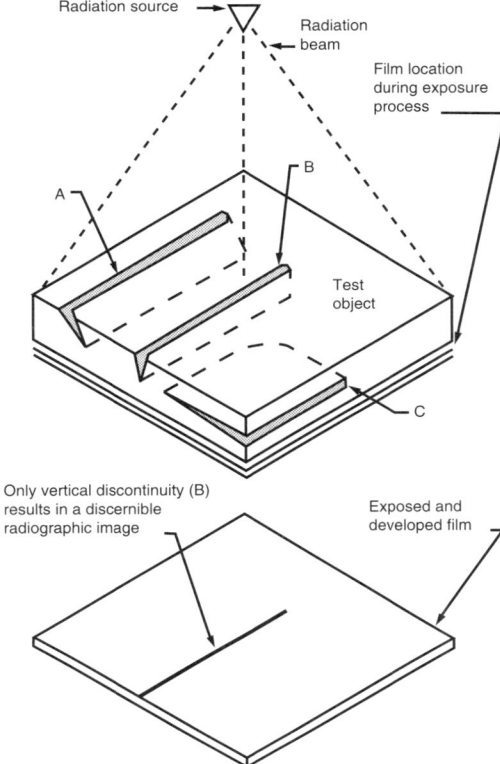

Detection of planar defects at various orientations by radiography

Ultrasonic Testing (UT) Best pay for the tech.

The ultrasonic method is applicable to almost all materials. The ultrasonic method uses the transmission of mechanical energy in wave form at frequencies above the audible range. Reflections of this energy by discontinuities in metals are detected in a manner somewhat similar to the detection of reflected light waves in transparent media.

Uses

In the pulse-echo technique, a transducer transmits a pulse of high frequency sound into and through the material and the reflected sound is received from a discontinuity, the opposite surface or other surfaces of the part. The reflected sound is received as an echo which, together with the original pulse, is displayed on the screen of a cathode ray tube (CRT).

Material can convert

11-10

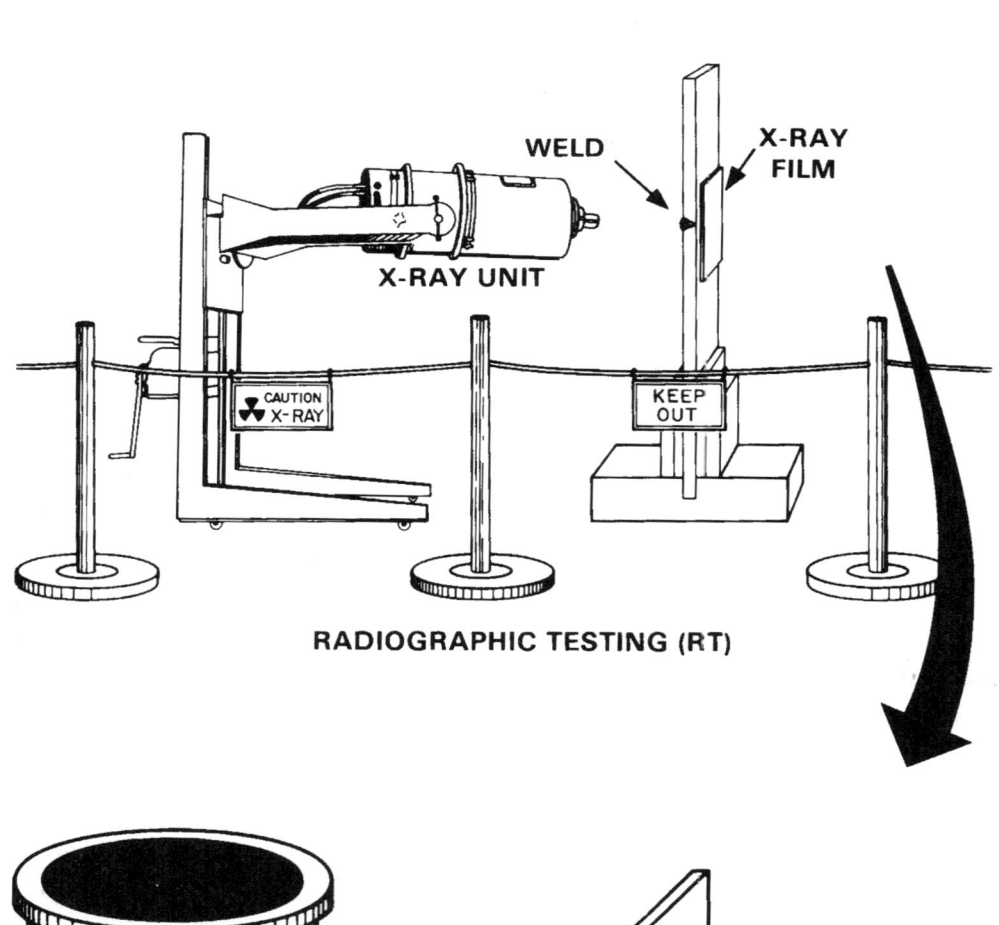

RADIOGRAPHIC TESTING (RT)

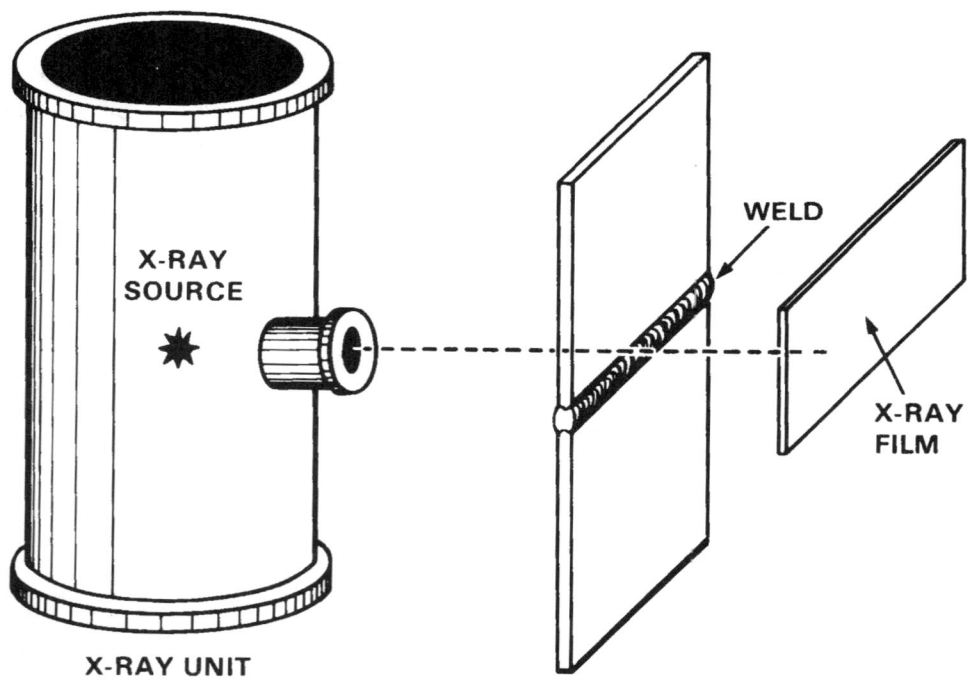

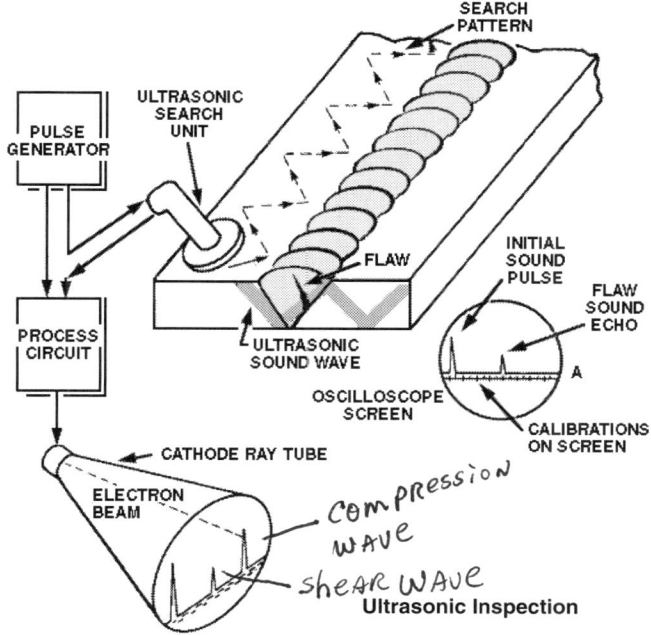
compression wave
shear wave
Ultrasonic Inspection

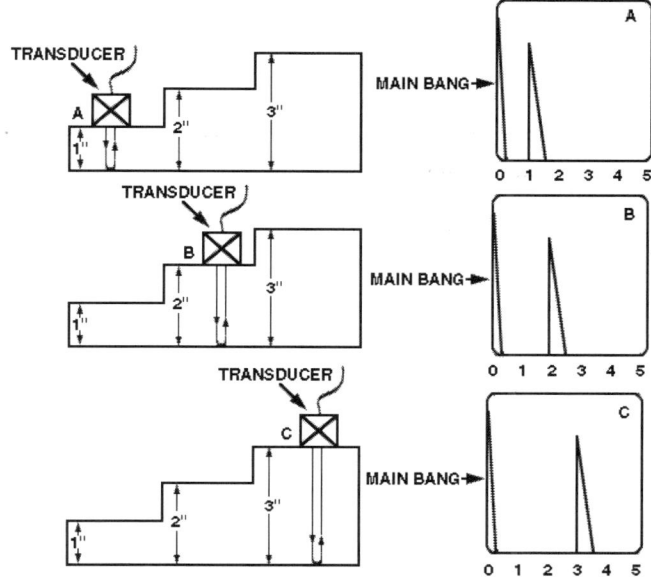
Calibration Sequence for Longitudinal Beam Transducer

Before testing is begun, the instrument is calibrated against a reference standard. When the sound beam intercepts the plane of discontinuity at or near 90°, a maximum reflected signal will return to the transducer. In scanning welds, this is achieved by beams angled into the work, through a lucite wedge and into the work through water, oil or similar couplant material.

When examining groove welds which require ultrasonic testing under the AWS Structural Welding Code, the technician will find the procedures tightly specified in AWS D1.1. The amount of testing to be done and the equipment to be used are specified. The equipment itself must be calibrated using some type of standard reference block, such as the IIW (International Institute of Welding) calibration block. The setup used by technicians (magnitude and location of indications on horizontal and vertical scales of the CRT) are prescribed for both straight beam and angle beam testing.

Advantages *The Best. Deep Penetration 200"*
Equip. Portable

- UT can be used to detect both surface and subsurface discontinuities.
- For pulse-echo testing, access is necessary to only one side of the work.
- The size of flaws and their interface location may be determined quantitatively.
- The method is generally more sensitive for the discovery of planar type discontinuities (which are generally considered more critical) than is radiography.
- Laminations and lamellar tearing can be readily detected using ultrasonic testing.

Limitations *Operator Highly Skilled*
Smooth surface
groove welds > 1/4" Thick

- Welds in some materials are very difficult to examine ultrasonically. For example, welds involving materials and processes which produce large grain size tend to scatter and disperse the sound beam; penetration of the sound beam into these materials is limited and interpretation of the results can be difficult.

- Personnel must be qualified and, in general, they require more training and experience for ultrasonic testing than for most of the other more common NDE methods.
- The scan pattern must be sufficient to pass the projected sound beam through the entire volume of the weld and heat affected zone to permit detection of possible discontinuities.
- For contact testing, the test surface used for scanning with the transducer must be smooth enough so liquid coupling may be obtained. Otherwise, the part may be placed in water and the sound wave transmitted through some length of water path. This is called immersion testing.

Eddy Current (Electromagnetic) Testing (ET)

Eddy current testing requires the part under test to be subjected to the influence of an alternating electromagnetic field. The tests detect surface or subsurface discontinuities in any material that is an electrical conductor. The electromagnetic field induces eddy currents in the part and also establishes magnetic fields if the part is magnetic.

A complete separation of these two effects in magnetic materials is not readily accomplished. However, a high degree of differentiation can be obtained by specialized techniques. Information gathered by probe coils is transmitted to a test circuit and analyzed electronically. Electromagnetic field frequencies are usually in the range of 500 Hz (hertz, or cycles per second) to 5000 Hz.

Uses

Since eddy currents may be induced in any electrical conductor, eddy current testing is employed on magnetic or nonmagnetic materials.

In eddy current tests, the magnitude and direction of the eddy currents are detected by the

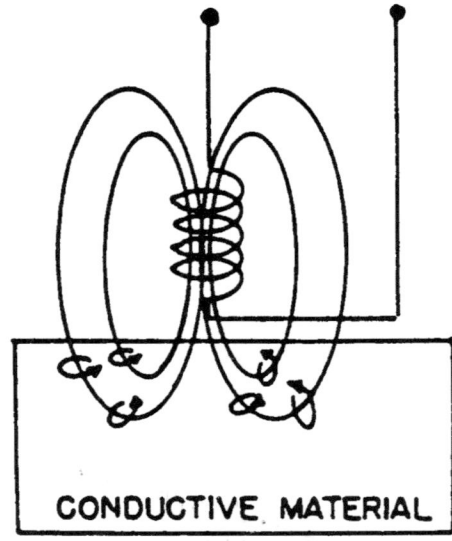

Induced Eddy Currents in Test Objects

coil, or by a separate coil, which acts through electronic circuitry to register the discontinuity.

In testing of magnetic materials, the distribution of magnetic flux is affected by magnetic variables. If the variation in flux is associated with discontinuities in the material, the discontinuities are detected.

Advantages

The four most significant advantages that electromagnetic testing has over other methods are:

- It can, in many cases, be completely automated; thus it provides automatic examination at high speeds and at a relatively low cost.
- Under certain circumstances, the indications produced are proportional to the actual size of the discontinuity. Thus, the tests can be useful for grading and classifying.
- Actual contact of probes with the work is not necessary; close proximity is satisfactory.
- Eddy current testing is capable of detecting many different material characteristics, including: electrical conductivity,

magnetic permeability, thickness, the thickness of a nonconductive coating, alloy content, heat treatment, and the presence of discontinuities at or below the surface.

Limitations *Highly skilled operator*

In preparing the material to be examined, any surface dirt that may be magnetic or electrically conductive must be removed. Before testing is begun, the instrument must be calibrated against a reference standard.

The design of the coil that produces the magnetic field must be appropriate for the shape of the component to be tested and the type of discontinuities that are sought. On welded tubular products, a coil (or coils) is commonly built to surround the material.

The practical limit for penetration of eddy currents in most nonmagnetic metals is approximately 1/4 in. (6 mm) below the surface. The depth of examination beneath the surface of the part depends upon the frequency chosen to excite the electromagnetic field.

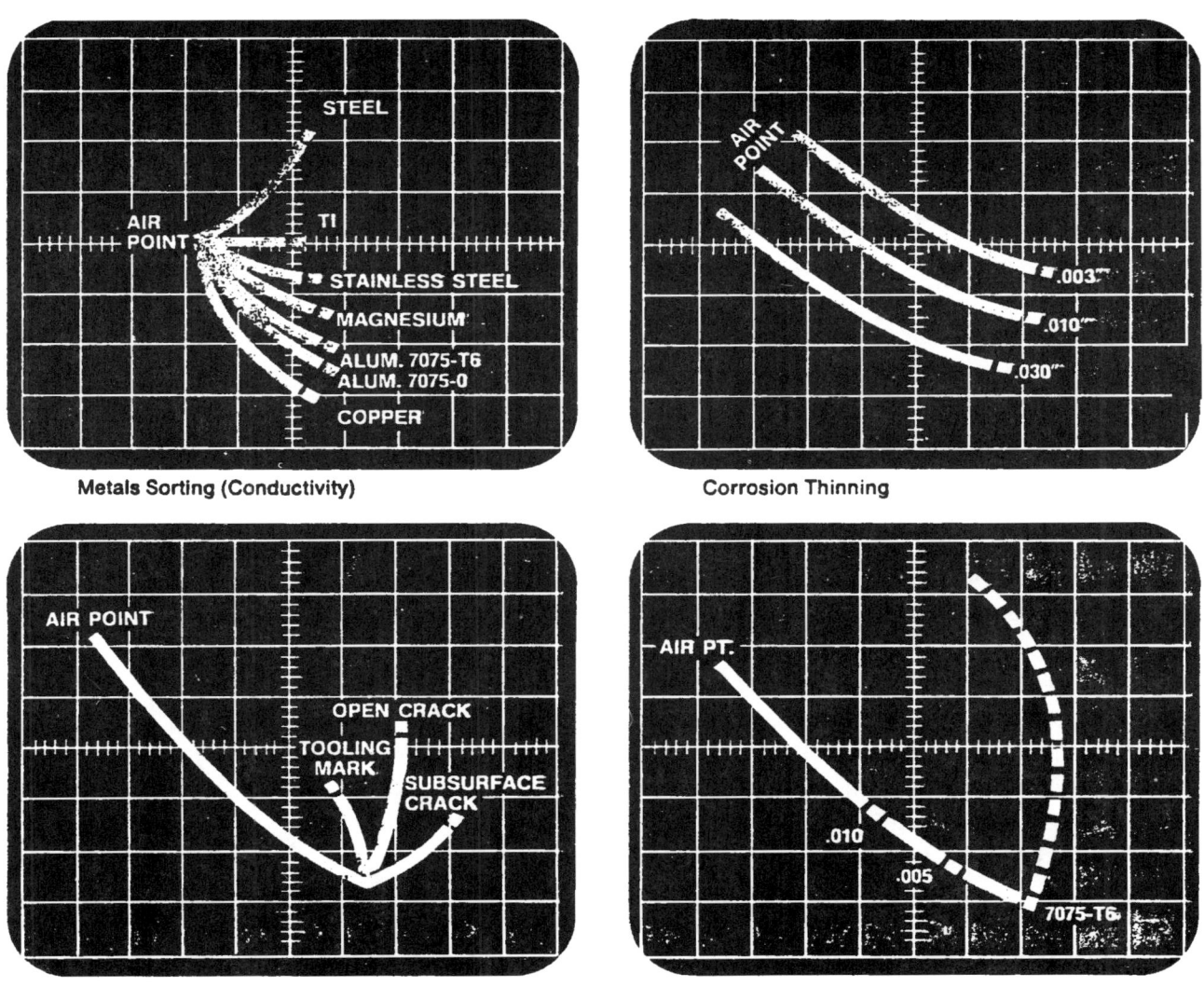

Typical CRT Displays for Eddy Current Testing

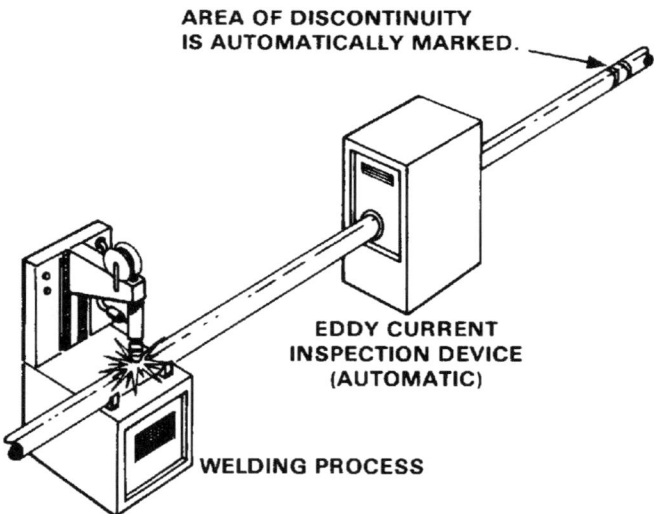

Typical Eddy Current Inspection Arrangement

Acoustic Emission Testing (AET)

Atomic movements leading to cracking are accompanied by sound bursts which may be detected by suitable microphones (piezoelectric ceramic elements). These sounds are emitted in all directions and may be detected from any surface of the vessel. By monitoring the emissions, the weld quality may be assessed during welding and during cooling. Weldments having incomplete joint penetration, incomplete fusion, cracking, porosity, or various imperfections can be detected, and the regions emitting the sound can be located by triangulation from several sensors (using a computer).

After the weldment has cooled, the sounds will cease, and a stimulus (mechanical loading or thermal stress) must be applied to cause further acoustic emission. Stress exceeding the maximum stress previously experienced by the metal causes plastic deformation at the tip of any crack and a tell-tale acoustic burst is emitted. If the discontinuities in the weldment are not affected by loading, they will not be active emitters, and the part will be recognized as structurally "sound" (soundless).

Acoustic emission testing can be a valuable adjunct to hydrostatic pressure testing. Acoustic emissions from a growing crack increase in number and intensity as a function of the applied variable (displacement, load, pressure, or time). Instantaneous interpretation and location of the source can permit a repair to be made before the vessel is harmed by rupturing.

Acoustic emission is also used to monitor in-process welding on production lines and to monitor critical weldments in service.

Leak Tests (LT)

One further indignity may be inflicted upon a well examined weldment: a test for leaks.

Test methods may be as unsophisticated as the pneumatic or gas and soap bubble test. By lightly pressurizing the component and immersing the vessel in water or brushing a soap film over the surface, the formation of bubbles can be observed at any leak.

Open tanks are frequently tested by filling them with water containing fluorescein, which permits ready detection of any seepage under ultraviolet illumination. The hydrostatic (or head) pressure causes the fluorescein to escape through any leaks and be detected.

Another leak test applied to storage tanks (principally floor joints) is referred to as the vacuum box test. Here, the weld surface is

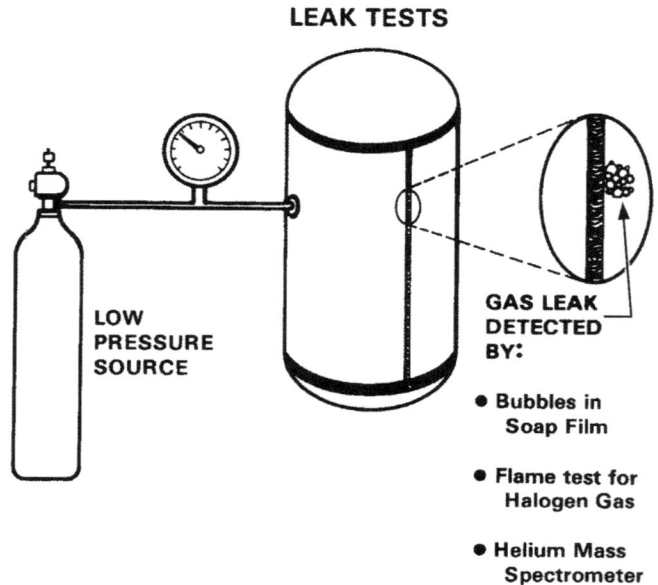

LEAK TESTS

covered with a soap solution and a transparent box with soft rubber seal is placed over a portion of the weld length. Using a vacuum pump or compressed air, a vacuum is created in the box and any leaks will be indicated by the presence of bubbles.

The leak test may use an organic halide gas on one side of the vessel and a halide torch on the other. The torch flame changes color in the presence of halogens.

The ultimate leak test is considered to be the helium leak test which uses helium as a tracer gas. Due to their extremely small size, helium atoms pass through even the smallest leaks. These very small leaks are then detected using special instruments capable of sensing any leakage of the helium gas. (helium spectrometer leak testing)

Details of the more complex tests are available from manufacturers of the equipment used. Once again, the rules for acceptance and rejection must be spelled out in the specifications.

Pressure Tests

A pressure test is one of the family of tests referred to as proof tests (PRT). A pressure test is the "final test" which some of the weldments previously examined must pass to earn a certificate of fitness. Details of how the pressure test is to be conducted should be given in the documents governing each assembly or subassembly.

Here is an important word of caution: In any hydrostatic test of a closed vessel, make certain that the technicians have vented all air trapped within the container before permitting the pressurization to begin. Liquids are essentially noncompressible. If the vessel is entirely filled with liquid, should rupture begin, the first leakage will reduce the pressure dramatically and the fracture will stop. However, if air has been

Complete Sampling Is Commonly Applied To Proof Tests or Final Inspection

trapped, its energy of compression will continue to extend the fracture explosively, with great damage and danger to observers. It is also very important to remember to vent any vessel after the hydrostatic test is completed. When draining test fluids, a vacuum may be created which may be adequate to damage the component.

Magnetic Test for Delta Ferrite

Delta ferrite is an effective crack deterrent in austenitic stainless steels. This phase is magnetic, whereas austenite is nonmagnetic. That property is now used to measure the amount of ferrite in the weld metal, which marks a major advance for weld examination. The alternate methods of measurement are laborious and inaccurate:

- Microscopic examination checks one tiny spot at a time which may not be typical, since ferrite is not distributed uniformly; etching the phase to identify it may distort its size.
- Computations based on chemical composition ignore the treatment effects and are subject to analytical errors.

The American Welding Society recommends that delta ferrite be reported in Ferrite Numbers, a standard measurement based on comparison with

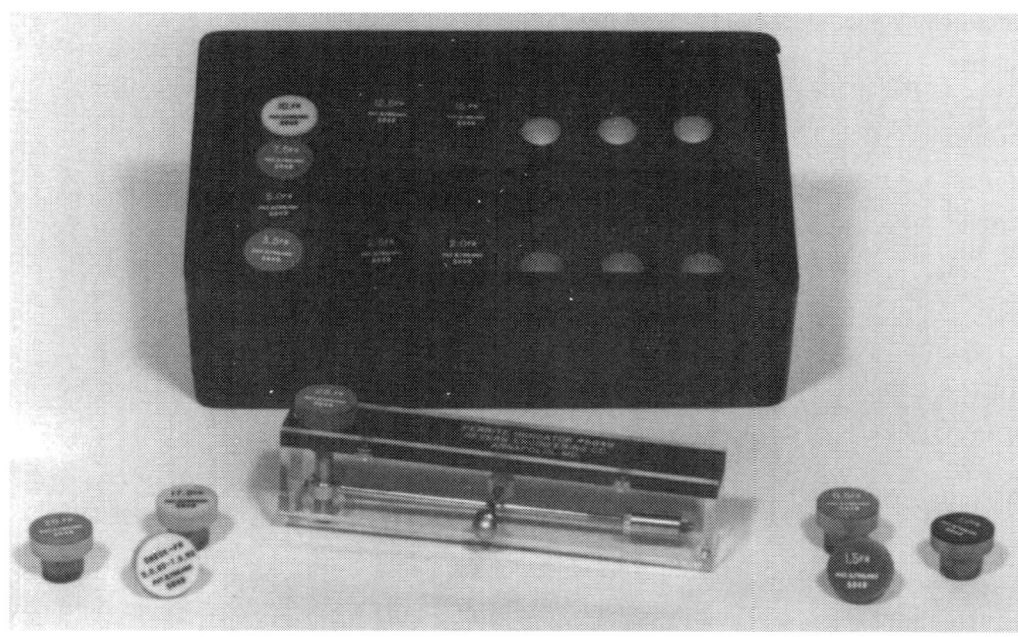
FERRITE INDICATOR (SEVERN GAGE)

magnetic chips obtained from the U. S. National Bureau of Standards (NBS). The testing method is explained in AWS A4.2, *Standard Procedures for Calibrating Magnetic Instruments to Measure the Delta Ferrite Content of Austenitic Stainless Steel Weld Metal.*

Magnetic instruments such as the Magna-Gage and other commercial devices are calibrated by measuring their response to the NBS wafers. The same instrument is then used to measure the desired weld metal.

The observed Ferrite Number (FN) is approximately equal to the percent ferrite that would be found by other techniques, but it has the advantage of being reproducable in any laboratory around the world.

The NBS wafers have been assigned Ferrite Numbers depending on their thickness, which determines their magnetic response. A calibration curve is drawn showing the instrument readings plotted against FN. The curve is then used to interpret readings taken on actual weld specimens.

In the course of inspection work, the welding inspector will no doubt find some contracts specifying delta ferrite in Ferrite Numbers and some specifying percent ferrite. One should endeavor to have all readings measured and reported in Ferrite Numbers, because they can be verified by the NBS standards. However, FN readings could be labeled "percent", with less error than would arise from a micrographic count or from estimating the ferrite using the reported chemical analysis.

Note one difference in the ferrite indications given by the Severn gage. This is a balance beam instrument which presents the opposite ends of a bar magnet simultaneously to a known ferrite standard and to the unknown ferrite content of the weld. The operator notes which of the two metals is magnetically stronger. The test tells only that the weld had more ferrite or less ferrite than the standard. There is no indication of how much the two differed. Other instruments indicate the actual Ferrite Number of the weld, but the Severn gage merely gives an upper or lower limit to the FN.

That will be sufficient for most inspections if the contract calls for, for example, "over 4 FN," or "over 4 percent ferrite." All portions of that weld should pull the magnet away from the 4 FN button. If the contract requires "4 to 10 FN," the welding inspector must also search with a 10 FN button, which should always pull the magnet away from the weld.

If delta ferrite prevents cracks, why is there concern about an upper limit to the Ferrite Number? The answer: High ferrite content increases the probability of forming a brittle sigma phase after heat treatment. Sigma is undesirable in stainless steel welds.

Qualification of Nondestructive Examination Personnel

Nondestructive examination personnel also require qualification and certification. Nondestructive tests are based on indirect indications: typically, the density of a photographic film is interpreted to indicate a discontinuity at the center of the weld, and an adhering line of magnetized iron powder is interpreted to reveal a subsurface crack. These may not be direct revelations of the supposed discontinuity. Some of them may be artifacts extraneous to the subject. The individual interpreting such information must be qualified to accurately interpret test results.

The qualification procedure usually specified is that recommended by the American Society for Nondestructive Testing's (ASNT) Recommended Practice No. SNT-TC-1A. Since this is considered a recommended practice, it only has relevance when a company develops a quality control manual which refers to these recommendations.

This standard provides for three levels of competence and responsibility, as shown in the following table:

NDT Level I permits the individual to operate NDT equipment and interpret results under direct supervision of a person certified for Level II or Level III. NDT Level II clears the individual for all testing and interpretation of indications. NDT Level III qualifies the individual to conduct qualification examinations of persons applying for Levels I and II.

Notice that individuals are not permitted to examine AWS D1.1 Code weldments by NDE methods other than visual unless they hold SNT-TC-1A certification papers. Also, do not mistakenly assume that a Level III individual (able to train and examine personnel at the other levels) would be permitted to make these tests. That person would not be acceptable without Level I or Level II qualification in addition to his Level III certification.

Under the ASME Code rules, the Authorized Inspector verifies the certification of NDE personnel in accordance with the applicable ASME requirements, which most likely reference SNT-TC-1A. When the Authorized Inspector questions the performance of an NDE technician, he has the further authority to audit the program and require requalification.

Summary

Although the CWI will not necessarily be qualified to perform nondestructive examinations other than visual, it is important that that individual at least understands the basic principles of the more common methods. This will assist in developing an understanding of the results reported by those performing the necessary tests.

In addition to review of nondestructive examination results, the CWI may also be responsible for determining if NDE technicians are properly qualified to perform various tests in accordance with contract requirements.

BRIEF ABSTRACT OF SNT-TC-1A LEVELS OF QUALIFICATION

NDT LEVEL I: An individual qualified to perform specific calibrations, specific tests, and specific evaluations according to written instructions and to record the results; the necessary guidance or supervision from a certified NDT Level II or III individual.

NDT LEVEL II: An individual qualified to set up and calibrate equipment and to interpret and evaluate results with respect to applicable codes, standards, and specifications; responsible for on-the-job training and guidance of trainees and NDT Level I personnel; prepares written instructions, and organizes and reports NDT investigations.

NDT LEVEL III: An individual capable of and responsible for establishing techniques; interpreting codes, standards, and specifications; designating the particular test method and technique to be used; responsible for a complete NDT operation; establishes techniques and assists the design engineer in establishing acceptance criteria where none are otherwise available; responsible for the training and examination of NDT Level I and Level II personnel for certification.

REVIEW - CHAPTER 11 NONDESTRUCTIVE EXAMINATION

Q11-1 Which of the following nondestructive examination methods do not usually require electricity?
   a. eddy current
   b. visible dye penetrant
   c. visual
   d. a and b above
   e. b and c above

Q11-2 Which of the following nondestructive examination methods is limited to the detection of surface discontinuities?
   a. visual
   b. penetrant
   c. magnetic particle
   d. all of the above
   e. none of the above

Q11-3 To be most effective, visual inspection should be performed:
   a. before welding.
   b. during welding.
   c. after welding.
   d. all of the above
   e. none of the above

Q11-4 The time during which the penetrant remains on the surface of the part to allow it to be drawn into any discontinuities is called:
   a. waiting time.
   b. penetrating time.
   c. soak time.
   d. dwell time.
   e. none of the above

Q11-5 Which type of magnetizing current provides the best combination of penetrationability and particle mobility?
   a. AC
   b. DC
   c. half-wave rectified DC
   d. b and c above
   e. all of the above

Q11-6 What NDE method will most likely reveal subsurface porosity?
   a. PT
   b. MT
   c. RT
   d. UT
   e. all of the above

Q11-7 Which of the following statements is correct for a radiographic test?
   a. A reduction in thickness will produce a light image on the film.
   b. A low-density discontinuity will produce a light image on the film.
   c. A high-density discontinuity will produce a light image on the film.
   d. a and b above
   e. b and c above

Q11-8 Which of the following discontinuities is almost never detected using RT?
   a. crack
   b. incomplete fusion
   c. undercut
   d. lamination
   e. none of the above

Q11-9 What device is used during radiography to indicate the acquired sensitivity of a radiograph?
   a. rate meter
   b. dosimeter
   c. lead screen
   d. penetrameter
   e. none of the above

Q11-10 Which nondestructive examination method utilizes sound energy as a probing medium?
   a. VT
   b. RT
   c. UT
   d. PT
   e. ET

Q11-11 The process whereby the ultrasonic indications are related to physical distances in a test standard is referred to as:
   a. setup
   b. calibration
   c. standardization
   d. synchronization
   e. none of the above

Q11-12 A test probe containing an alternating current coil is used for which NDE method?
   a. RT
   b. UT
   c. ET
   d. MT
   e. c & d above

Q11-13 Changes in electrical conductivity can be measured using which NDE method?
   a. ET
   b. RT
   c. MT
   d. UT
   e. none of the above

Q11-14 Which of the following NDE methods are suitable for detecting surface cracks?
   a. RT
   b. VT
   c. ET
   d. PT
   e. all of the above

Q11-15 What NDE method is most likely to reveal internal laminations in a rolled plate?
- a. RT
- b. UT
- c. ET
- d. MT
- e. none of the above

Q11-16 PT is limited to the detection of those discontinuities which are?
- a. near the test object surface
- b. open to the test object surface
- c. clean and open to the test object surface
- d. all of the above
- e. none of the above

Q11-17 Visible dye penetrant indications:
- a. must be observed under a black light
- b. don't have to be observed under a black light, but are more sensitive if they are
- c. must be observed under ultraviolet light
- d. must be observed under white light
- e. none of the above

Q11-18 Penetrant can be applied by:
- a. brushing
- b. spraying
- c. dipping
- d. all of the above
- e. none of the above

Q11-19 Fluorescent penetrants are generally more sensitive than visible dye penetrants because:
- a. they can flow into smaller cracks
- b. fluorescent indications are better seen by the human eye
- c. they are subject to greater capillary action
- d. a and c above
- e. b and c above

Q11-20 Which of the following cause decreased sensitivity in PT?
   a. too heavy application of the developer
   b. oily or greasy test object
   c. improper penetrant removal
   d. all of the above
   e. none of the above

Q11-21 PT is limited to test objects which:
   a. are metallic
   b. are porous
   c. are magnetic
   d. are nonporous
   e. have subsurface discontinuities

Q11-22 MT will discover:
   a. surface discontinuities
   b. slightly subsurface discontinuities
   c. underbead cracking
   d. a and b above
   e. all of the above

Q11-23 MT is most sensitive to those discontinuities which are:
   a. within 45° of perpendicular to the lines of flux
   b. within 45° of parallel to the lines of flux
   c. perpendicular to the lines of flux
   d. parallel to the lines of flux
   e. none of the above

Q11-24 MT is limited to test objects which:
   a. are metallic
   b. are porous
   c. are ferromagnetic
   d. are nonporous
   e. have subsurface discontinuities

Q11-25 UT is most sensitive to those discontinuities which are:
   a. within 45° of perpendicular to the sound waves
   b. within 45° of parallel to the sound waves
   c. perpendicular to the sound waves
   d. parallel to the sound waves
   e. none of the above

Q11-26 UT uses frequencies:
   a. below the range of human hearing
   b. within the range of human hearing
   c. above the range of human hearing
   d. beside the range of human hearing
   e. none of the above

Q11-27 In UT the horizontal axis of the CRT screen gives information about:
   a. the distance the sound has traveled in the part
   b. the amount of sound energy reflected
   c. the type of discontinuity
   d. discontinuity orientation
   e. discontinuity cause

Q11-28 RT shows areas of lower density as:
   a. dark regions on the film
   b. light regions on the film
   c. light or dark regions on the film
   d. all of the above
   e. none of the above

Q11-29 RT shows areas of less thickness as:
   a. dark regions on the film
   b. light regions on the film
   c. light or dark regions on the film
   d. all of the above
   e. none of the above

Q11-30 RT shows areas of increased transmission as:
   a. dark regions on the film
   b. light regions on the film
   c. light or dark regions on the film
   d. all of the above
   e. none of the above

Q11-31 Tungsten inclusions generally appear in RT as:
   a. dark regions on the film
   b. light regions on the film
   c. light or dark regions on the film
   d. all of the above
   e. none of the above

Q11-32 Cracks generally appear in RT as:
   a. dark lines on the film
   b. light lines on the film
   c. light or dark lines on the film
   d. all of the above
   e. none of the above

Q11-33 Weld reinforcement generally appears as:
   a. dark regions on the film
   b. light regions on the film
   c. light or dark regions on the film
   d. all of the above
   e. none of the above

Q11-34 Porosity generally appears in RT as:
   a. dark regions on the film
   b. light regions on the film
   c. light or dark regions on the film
   d. all of the above
   e. none of the above

Q11-35 Shallow surface cracks can best be detected in 308 stainless by:
   a. UT
   b. MT
   c. RT
   d. PT
   e. all of the above

Q11-36 Underbead cracks can best be detected by:
   a. ET
   b. MT
   c. UT
   d. PT
   e. all of the above

Q11-37 Porosity in ESW can best be detected by:
   a. UT
   b. MT
   c. RT
   d. PT
   e. all of the above

Q11-38 The vertical axis of the UT CRT screen represents:
   a. distance
   b. time
   c. reflector size
   d. none of the above
   e. all of the above

Q11-39 What NDE method(s) rely on the transmission of sound energy through the test object?
   a. RT
   b. UT
   c. AET
   d. a and b above
   e. b and c above

Q11-40 What NDE method(s) often rely on the application of a hydrostatic pressure to a vessel?
   a. pressure tests
   b. leak tests
   c. proof tests
   d. all of the above
   e. none of the above

Q11-41 What test below is applied to determine the metallurgical effects of welding on austenitic stainless steels?
  a. RT
  b. PRT
  c. ferrite test
  d. UT
  e. PT

Q11-42 Application of a vacuum box to the inside surface of a steel storage tank is one form of which test below?
  a. VT
  b. PRT
  c. LT
  d. all of the above
  e. none of the above

Q11-43 FN is a unit of measurement with which test below?
  a. RT
  b. PRT
  c. ferrite test
  d. UT
  e. PT

Q11-44 NDE personnel are normally qualified in accordance with
  a. ASME Section V
  b. AWS D1.1
  c. ANSI SNT-TC-1A
  d. ASNT SNT-TC-1A
  e. ASME SNT-TC-1A

CHAPTER 12: INSPECTION REPORTS

Introduction

Reporting observations and decisions is the final step for a welding inspector. Many people are waiting for them. The welding inspector may feel that he/she is only responsible to his/her employer, but the employer has responsibilities to many related organizations, be they the customer, fabricator or erector.

Report Contents

A good report begins with good record keeping. Good records not only protect the welding inspector; they also help him/her follow a policy of uniform standards.

The inspection report must be clear and concise so that others will have no difficulty understanding the decisions reached, even months later. Though concise, it must be complete enough to be clear to a reader unfamiliar with the product inspected.

The report must be complete, accurate and have the appropriate signatures. All formal reports and data forms required by the governing standard or code should be completed in ink. Again, it should be noted that erasures are not permitted in these legal documents. Errors are to be crossed through with a single line and not erased. When such corrections are made, they should be accompanied by the inspector's initials and the date of the change. In that way, there will be no question as to who made the change and when it occurred.

Reports must include references to the many other reports that the welding inspector sees and uses as tools in his own inspection process. Among those more commonly used are the following:

- Surveillance Reports
- Certified Material Test Reports
- Quality Reports
- Qualification Records
- Progress Reports
- Production Schedules

The welding inspector of any category should keep a set of records for each contract under surveillance, detailing every inspection made and every joint in the vessel or assembly. For a small job, these records may be contained in a single file, while larger jobs may require that the records be separated into various files. This selection is often a matter of personal preference; however, some codes will stipulate how this information should be organized and maintained. The important feature is to assure that the system used is easily understood by all appropriate individuals, not just the inspector.

Report Format

The inspector, working in accordance with a national code or standard, monitors the quality assurance program of the fabrication through his review of quality reports, qualification records, certificates of compliance, and records of examinations and tests. The inspector for the fabricator is part of the quality control organization and thus assists directly in preparing reports required by the quality assurance program. The governing code or standard will suggest a format for an inspection report.

Structural Welding Code (AWS D1.1)

Suggested forms for an in-house inspector's reports are printed in the Appendix E of AWS D1.1, *Structural Welding Code- Steel*. Copies of these forms are included in the Appendix. They include forms for recording procedure qualifications and welder and welding operator qualifications. Also included are laboratory report forms for three categories of nondestructive examination data.

Variations in the suggested forms are permissible, as long as all of the necessary information is provided. The AWS procedure qualification forms include versions for prequalified welding procedures, welding procedures qualified by actual testing and welding procedures for electroslag and electrogas welding.

ASME Boiler and Pressure Vessel Code

The ASME Boiler and Pressure Vessel Code contains a number of Data Report Forms required of the designer, the owner, the manufacturer and the fabricator of any vessel covered by this code. Under this code, each of these parties must have a quality assurance program (documented in a Quality Assurance Manual) that details manufacturing processes, job titles and responsibilities for inspection. The designated inspector may be expected to complete a Data Report Form for an employer, but will not be able to sign them unless authorized.

The ASME Code places all responsibility on the manufacturer or installer. Only the authorized inspector may sign or initial and date check lists of items personally witnessed on Data Reports. In practice, however, the in-house inspector assembles all of these reports and verifies compliance by the manufacturer.

Form N-1 for nuclear vessels is a typical ASME report. It lists the names of the manufacturer and the purchaser; the type and kind of vessel; dimensions of shell, seams, heads, jacket closure, tube sheets, and tubes; dimensions of any inner chambers (shells, seams, heads); the design pressure of jackets and inner chambers, with test data on impact resistance and pressure tests; data, when applicable, on safety valve outlets, nozzles, inspection manholes, other openings and supports; remarks-, and a certification of design covering the design specifications and stress analysis report (each certified by a professional engineer). The Certificate of Field Inspection (or Field Assembly Inspection) completes this form.

In addition to these data report forms, the welding inspector will also be asked to review and verify information contained on the various forms related to welding procedure and welder qualification. These include: the Welding Procedure Specification (WPS), the Procedure Qualification Record (PQR) and the Performance Qualification Record.

API Report Forms

A typical API form for reporting qualification of procedures or welders for pipeline welding can be found in the Appendix. Although different than those for AWS or ASME, much of the information required is similar, if not identical.

Multiple Inspection: One inspector's reports may well duplicate reports by some of the other inspectors, but that is the nature of the surveillance of major weldments. An illustrative example is the multiple reporting given to the radiograph of a weld.

Many radiographs are evaluated by up to five different people. Each individual records their contribution to the inspection. The report form for radiographic examination might well be used by each of them. Those individuals who might become involved in the radiographic evaluation for a typical scenario are listed below:

1) The *technician* exposes the film and develops it, and records the identification of the weld and the exposure details.

2) The *examiner* or *technician Level I* or *II* in the radiographic department examines the film and reports his interpretation.

3) The *SNT-TC-1A examiner* for the contractor reviews the film for acceptance.

4) The *customer's welding inspector* examines the film and makes his own evaluation.

5) The *ASME Authorized Inspector* (in the employ of one of the insurance companies, or a jurisdiction such as a state) reviews the film and the report for concurrence and acceptance.

Unstructured Reports

Up to now the major emphasis has been on the more formal reports required by the code, standard or specification. However, unstructured reports will be often required of a welding inspector. This is especially true of workmanship opinions. Comments on the visual examination and repair of plate cut edges, on shrinkage and distortion, on dimensional tolerances such as straightness and flatness, camber, warpage and tilt, on the fit and straightness of intermediate stiffeners and bearing stiffeners, on weld profiles, and on the repair of unacceptable welds and difficulties that occurred - all require individual reporting. Such notes are not readily tabulated unless you keep accurate and complete records.

The formality of an unstructured report will vary with the welding inspector's responsibility. An inspector in a small company will be more informal in reporting than will a state inspector. Copies of the reports and attached records should go to all who are entitled to receive them. The inspector should keep a copy for his own files. Even while performing inspections on non-code work, an inspector should keep a proper record of his work. At a minimum, this could be in the form of a complete set of notes.

Report Check List

- After writing the report, the following questions should be asked as a check on the quality of the report:

- Are all the report data forms required by the governing code, standard or specification complete? Accurate? Signed?

- Are all supporting forms, reports and data included or properly referenced?

- Are the facts stated clearly and concisely?

- Can the reader reach the same logical conclusions or make sound decisions from the facts and data in the report today. In two months? In six months?

- Does the overall organization of the report present a total picture to the reader?

- Does the report maintain a logical sequence? For example, does it follow the fabrication process? Procedure inspection? Acceptance process?

- Have the purpose and objectives of the reports been attained?

Conclusion

One should be commended for undertaking a career in the field of inspection. Welding inspection is an important field. It is exciting and demanding. It takes a special kind of person to select this kind of work. Good luck in these future activities.

REVIEW - CHAPTER 12 INSPECTION REPORTS

Q12-1 Once inspections are completed, what important aspect of the inspector's job must be accomplished?
   a. tell the foreman that the weld is acceptable
   b. tell the supervisor that the inspection is complete
   c. fill out an inspection report detailing his findings
   d. all of the above
   e. none of the above

Q12-2 Which of the following is not normally required of inspection reports?
   a. inspector's signature
   b. an indication of only those parts which were acceptable
   c. they should be clear and concise
   d. they should be filled out in ink
   e. none of the above

Q12-3 What authorship is attached to inspection report forms?
   a. the inspector's signature
   b. the welder's signature
   c. the welding supervisor's signature
   d. Forms are anonymously presented.
   e. Forms are not signed.

Q12-4 What handy report forms are available from AWS?
   a. Appendix E of AWS D1.1
   b. Appendix I of AWS D1.1
   c. Appendix O of AWS D1.1
   d. AWS QC 1
   e. ASME N-1 Form

Q12-5 You have made a numerical mistake on a report form. How should it be corrected?
   a. An experienced inspector will use a pencil so such errors can be erased and corrected.
   b. As an inspector-in-training, such errors need not be corrected.
   c. To keep the report legal and credible, the error should be crossed out and the correction added adjacent to the error and noted complete with initials and date of correction.
   d. The report must be completely rewritten.
   e. none of the above

Q12-6 How are errors in writing corrected in written reports?
   a. crossed out
   b. erased with an ink eraser
   c. entire page must be rewritten
   d. crossed out, corrected, initialed and dated
   e. none of the above

Q12-7 Who is authorized to sign off ASME data report forms?
   a. the authorized inspector who performed inspection
   b. an authorized keeper of the code stamp
   c. an officer or manager of the company
   d. any of the above
   e. both a and b above

Q12-8 How are opinions on workmanship or suggestions for repair usually reported?
   a. Comments are written in chalk on the work.
   b. by unstructured reports
   c. Provisions for such comments are contained in structured reports.
   d. Inspectors are forbidden to offer such comments.
   e. none of the above

APPENDIX A - ADDITIONAL RESOURCES & REFERENCES

Addresses of Major Associations of the Welding Fabricating Industry

American Welding Society (AWS)[1]
550 N.W. LeJeune Rd.
P.O. Box 351040
Miami, FL 33135
(305)-443-9353

Aluminum Association (AA)[1]
900 19th St., N.W.
Suite 300
Washington, DC 20006
(202) 862-5100

American Petroleum Institute (API)
1220 L Street, N.W.
Washington, D.C. 20005
(202) 682-8000

American Association of State Highway Officials (AASHTO)
444 N. Capital St., N.W.
Washington, D.C. 20001
(202) 624-5800

Association of American Railroads (AAR)
50 F Street, N.W.
Washington, D.C. 20001
(202) 635-2100

Abrasive Engineering Society (AES)
1700 Painters Run Road
Pittsburgh, PA 15243
(412) 221-0909

American Gas Association (AGA)
1515 Wilson Blvd.
Arlington, VA 22209
(703) 841-8400

American Institute of Mining, Metallurgical and Petroleum Engineering (AIME)
345 East 47th St.
New York, NY 10017
(212) 705-7695

American Institute of Plant Engineers (AIPE)
3975 Erie Ave.
Cincinnati, OH 45208
(513)561-6000

American Institute of Steel Construction (AISC)
400 N. Michigan Ave.
Chicago, Illinois 60611
(312) 670-2400

Association of Iron and Steel Engineers (AISE)
Three Gateway Center
Suite 2350
Pittsburg, PA 15222
(412) 281-6323

American Iron and Steel Institute (AISI)[1]
1000 16th St., N.W.
Washington, D. C. 20036
(202) 452-7236

American Nuclear Society (ANS)
555 North Kensington Ave.
La Grange Park, IL 60525
(312) 352-6611

American National Standards Institute (ANSI)
1430 Broadway
New York, New York 10018
(212) 642-4900

American Railway Engineering Association (AREA)
50 F Street, N.W.
Washington, D.C. 20001
(202) 639-2190

American Society of Civil Engineers (ASCE)
345 East 47th Street
New York, NY 10017
(212) 705-7496

Major Association Addresses continued:

ASM International (ASM)
Route 87
Metals Park, OH 44073
(216) 338-5151

American Society of Mechanical Engineers (ASME)
345 East 47th Street
New York, New York 10017
(212) 705-7722

American Society for Nondestructive Testing (ASNT)
4153 Arlingate Plaza
Columbus, Ohio 43228
(614) 274-6003

American Society for Quality Control (ASQC)
310 W. Wisconsin Ave.
Milwaukee, WI 53203
(414) 272-8575

American Society of Safety Engineers (ASSE)
1800 East Oakton
Des Plaines, IL 60018-2187
(312) 692-4121

American Society for Testing Materials (ASTM)[1]
1916 Race Street
Philadelphia, Pennsylvania 19103
(215) 299-5400

American Welding Institute (AWI)
Knoxville Lab
Route 4, Box 90
Louisville, TN 37777
(615) 970-2150

American Water Works Association (AWWA)
6666 W. Quincy Avenue
Denver, Colorado 80235
(303) 794-7711

American Bureau of Shipping (ABS)
45 Eisenhower Drive
Paramus, New Jersey 07652
(201) 368-9100

Canadian Standards Association (CSA)
178 Rexdale Boulevard
Rexdale, Ontario
Canada M9W 1R3
(416) 744-4000

Canadian Welding Bureau (CWB)
254 Merton Street
Toronto, Ontario M4S 1A9
Canada
(416) 487-5415

Canadian Welding Development Institute (CWDI)
391 Burnhamthorpe Rd.
East Oakville, Ontario L6J 4Z2
Canada
(416) 845-9881

Copper Development Association[1]
2 Greenwich Office Park
Box 1840
Greenwich, CT 06836-1840
(203) 625-8210

Compressed Gas Association (CGA)
1235 Jeff Davis Highway
Arlington, Virginia 22202
(703) 979-0900

Cryogenic Society of America CSA)
1033 South Blvd.
Oak Park, IL 60302
(312) 383-7053

Edison Welding Institute (EWI)
1100 Kinnear Rd.
Columbus, OH 43212
(614) 486-9400

Fabricators & Manufacturers' Association International (FMA)
7811 N. Alpine Rd.
Rockford, IL 61111
(815) 654-1902

Grinding Wheel Institute (GWI)
712 Lakewood Center North
14600 Detroit Ave
Cleveland, OH 44107
216-226-7700

Industrial Accident Prevention Association (IAPA)
100 Front Street West
Royal York Hotel Arcade
Toronto, Ontario M5J1R3
Canada
(416) 366-3711

Major Association Addresses continued:

Industrial Safety Equipment Association (ISE)
1901 N. Moore St.
Arlington, VA 22209
(703) 525-1695

International Institute Of Welding (IIW)
550 N. W. LeJeune Road
P.O. Box 351040
Miami, FL 33135
(305) 443-9353

International Organization for Standardization (ISO)
(See American National Standards Institute)

International Oxygen Manufacturers' Association (IOMA)
P.O. Box 16248
Cleveland, OH 44116
(216) 228-2166

Material Handling Institute (MHI)
8720 Red Oak Blvd.
Suite 201
Charlotte, NC 28210
(704) 522-8644

National Association of Corrosion Engineers (NACE)
Box 218340
Houston, TX 77218
(713) 492-0535

National Board of Boiler and Pressure Vessel Inspectors (NBBPVI)
1055 Crupper Avenue
Columbus, Ohio 43229
(614) 888-8320

National Electrical Manufacturers' Association (NEMA)
2101 L. St., N.W.
Washington, DC 20037
(202) 457-8400

National Fire Protection Association (NFPA)
Batterymarch Park
Quincy, Massachusetts 02269
(617) 770-3000

Naval Publication and Forms Center[2]
5801 Taber Avenue
Philadelphia, Pennsylvania 19120
(215) 697-2000

Steel Tank Institute (STI)
P.O. Box 4020
North Brook, IL 60065
(312) 498-1980

Superintendent of Documents [3]
U. S. Government Printing Office
Washington, DC 20402
(202) 783-3238

Titanium Development Association
P.O. Box 2307
Dayton, OH 45401
(513) 223-8432

Ultrasonic Industry Association (UIA)
P.O. Box 5126
Old Bridge, NJ 08857
(201) 521-4441

Underwriters Laboratories, Inc. (UL)
333 Pfingsten Road
Northbrook, Illinois 60062
(312) 272-8800

Uniform Boiler and Pressure Vessel Laws Society (UBPVLS)
2838 Long Beach Road
Oceanside, New York 11572
(516) 536-5485

Welding Research Council (WRC)
345 East 47th St.
Suite 1301
New York, NY 10017
(212) 705-7956

Welded Steel Tube Institute (WSTI)
522 Westgate Tower
Cleveland, OH 44116
(216) 333-4550

Note: For a detailed explanation of the agencies listed herein refer to AWS Welding Handbook, Vol 1, Chapter 13, "Codes and Standards"

[1] Number Assigner for "UNS"
[2] Military Specifications
[3] Federal Specifications

NDI Selection Guide

Equipment Needs	Applications	Advantages	Limitations
Visual			
Magnifier, color enhancement, projectors, other measurement equipment, i.e., rulers, micrometers, optical comparators, light source.	Welds which have discontinuities on the surface.	Economical, expedient, requires relatively little training and relatively little equipment for many applications.	Limited to external or surface conditions only. Limited to the visual acuity of the observer or inspector.
Liquid Penetrant			
Fluorescent or dye penetrant, developers, cleaners (solvents, emulsifiers, etc.). Suitable cleaning gear. Ultraviolet light source if fluorescent dye is used.	Weld discontinuities open to surface, i. e., cracks, porosity.	Portable, relatively inexpensive equipment. Expedient inspection results. Results are easily interpreted. Requires no electrical energy except for light sources.	Surface films such as coatings, scale, smeared metal mask or hide rejectable defects. Seepage from weld porosity at the surface can also mask indications. Parts must be cleaned before and after inspection.
Magnetic Particle			
Prods, yokes, coils suitable for inducing magnetism into the weld. Power source (electrical). Magnetic powders. - some applications require special facilities and ultraviolet lights.	Most weld discontinuities open to the surface - some large voids slightly subsurface. Most suitable for cracks.	Relatively economical and expedient. Inspection equipment is considered portable. Unlike dye penetrants, magnetic particle can detect some discontinuities slightly below the surface.	Applicable only to ferromagnetic materials. Parts must be cleaned before and after inspection. Thick coatings may mask rejectable discontinuities. Some applications require parts to be demagnetized after inspection. Magnetic particle inspection requires use of electrical energy for most applications.
Radiography (Gamma)			
Gamma ray sources, gamma ray camera projectors, film holders, film, lead screens, film processing equipment, film viewers, exposure facilities, radiation monitoring equipment	Welds which have voluminous discontinuities such as porosity, incomplete joint penetration, corrosion, etc. Lamellar type discontinuities such as cracks and incomplete fusion can be detected with a lesser degree of reliability. May also be used in certain applications to evaluate dimensional requirements such as fit-up, root conditions, and wall thickness.	Generally not restricted by type of material or grain structure. Surface and subsurface inspection capability. Radiographic images aid in characterizing discontinuities. Provides a permanent record for future review.	Planar discontinuities must be favorably aligned with radiation beam to be reliably detected. Radiation poses a potential hazard to personnel. Cost of radiographic equipment, facilities, safety programs, and related licensing is relatively high. A relatively long time between exposure process and availability of results. Accessibility to both sides of the weld required.
Radiography (X-Rays)			
X-ray sources (machines, electrical power source, same general equipment as used with gamma sources (above)	Same application as above.	Adjustable energy levels. Generally produces higher quality radiographs than gamma sources. Also, same advantages as above.	High initial cost of X-ray equipment. Not generally considered portable. Also, same limitations as above.
Ultrasonic			
Pulse-echo instrument capable of exciting a piezoelectric material and generating ultrasonic energy within a weld, and a suitable cathode ray tube scope capable of displaying the magnitudes of received sound energy. Calibration standards, liquid couplant.	Most weld discontinuities including cracks, slag, and incomplete fusion Can also be used to verify weld thickness.	Most sensitive to planar type discontinuities. Test results known immediately. Portable. Most ultrasonic flaw detectors do not require an electrical power outlet. High penetration capability.	Surface condition must be suitable for coupling of transducer. Couplant (liquid) required. Small, thin welds may be difficult to inspect. Reference standards are required. Requires a relatively skilled operator or inspector.
Eddy Current			
An instrument capable of inducing electromagnetic fields within a weld and sensing the resulting electrical currents (eddy) so induced with a suitable probe or detector. Calibration standards.	Weld discontinuities open to the surface, (i.e., cracks, porosity, incomplete fusion) as well as some subsurface discontinuities. Alloy content, heat treatment variations.	Equipment used with surface probes is generally lightweight and portable. Painted or coated welds can be inspected. Can be partially or completely automated for a high speed, relatively inexpensive test.	Relatively shallow depth of inspection. Many material and test variables can affect the test signal.

APPENDIX B - SAMPLE FORMS

CONTENTS	PAGE

AWS D1.1
Form E-1 (Front) Welding Procedure Specification (WPS) B-2
Form E-1 (Back) Procedure Qualification Record (PQR) B-3
Form E-3 Welding Procedure Qualification Test Record For Electroslag and Electrogas B-4
Form E-4 Welder, Welding Operator or Tack Welder Qualification Test Record................................. B-5
Form E-7 Report of Radiographic Examination of Welds.. B-6
Form E-8 Report of Magnetic Particle Examination of Welds B-7

ASME B31.1 SECTION IX
Form E00006 (Front) QW-482 Suggested Format For Welding Procedure Specifications (WPS)......... B-8
Form E00006 (Back) QW-482 .... B-9
Form E00007 (Front) QW-483 Suggested Format For Procedure Qualification Record (PQR)........... B-10
Form E00007 (Back) QW-483 .... B-11
Form E00008 QW-484 Suggested Format For Manufacturer's Record Of Welder Or Welding
Operator Qualification Tests (WPQ) B-12

API 1104
Procedure Specification Form.. B-13

AWS QC-1
Certified Welding Inspector Certificate B-14

B-2

WELDING PROCEDURE SPECIFICATION (WPS) YES ☐
PREQUALIFIED_____ **QUALIFIED BY TESTING**_____
or PROCEDURE QUALIFICATION RECORDS (PQR) YES ☐

Company Name _____
Welding Process(es) _____
Supporting PQR No.(s) _____

Identification # _____
Revision _____ Date _____ By _____
Authorized by _____ Date _____
Type - Manual ☐ Semi-Automatic ☐
       Machine ☐ Automatic ☐

JOINT DESIGN USED
Type: Single ☐ Double Weld ☐
Backing Yes ☐ No ☐
Backing Material _____
Root Opening _____ Root Face Dimension _____
Groove Angle _____ Radius (J-U) _____
Backgouging Yes ☐ No ☐ Method _____

POSITION
Position of Groove _____ Fillet _____
Vertical Progression Up ☐ Down ☐

BASE METALS
Material Spec. _____
Type or Grade _____
Thickness: Groove _____ Fillet _____
Diameter (Pipe) _____

ELECTRICAL CHARACTERISTICS
Transfer Mode (GMAW)
Short-Circuiting ☐ Globular ☐ Spray ☐
Current: AC ☐ DCEP ☐ DCEN ☐ Pulsed ☐
Other _____
Tungsten Electrode (GTAW):
Size _____
Type _____

FILLER METALS
AWS Specification _____
AWS Classification _____

TECHNIQUE
Stringer or Weave Bead _____
Multi-pass or Single Pass (per side) _____
Number of electrodes _____
Electrode Spacing: Longitudinal _____
                   Lateral _____
                   Angle _____
Contact Tube to Work Distance _____
Peening _____
Interpass Cleaning _____

SHIELDING
Flux _____ Gas _____
              Composition _____
Electrode-Flux (Class) _____ Flow Rate _____
_____ Gas Cup Size _____

PREHEAT
Preheat Temp. Min. _____
Interpass Temp. Min. _____ Max. _____

POSTWELD HEAT TREATMENT
Temp. _____
Time _____

WELDING PROCEDURE

| Pass or Weld Layer(s) | Process | Filler Metals | | Current | | Volts | Travel Speed | Joint Details |
		Class	Diam.	Type & Polarity	Amps or Wire Feed Speed			

Form E-1 (Front)

(Source AWS D1.1-94)

B-3

PROCEDURE QUALIFICATION RECORD (PQR)# _____
TEST RESULTS

TENSILE TEST

Specimen no.	Width	Thickness	Area	Ultimate tensile load, lb.	Ultimate unit stress, psi	Character of failure and location

GUIDED BEND TEST

Specimen no.	Type of bend	Result	Remarks

VISUAL INSPECTION
Appearance _____
Undercut _____
Piping porosity _____
Convexity _____
Test date _____
Witnessed by _____

Other Tests

Radiographic-ultrasonic examination
RT report no: _____ Result _____
UT report no: _____ Result _____

FILLET WELD TEST RESULTS
Minimum size multiple pass   Maximum size single pass
Macroetch   Macroetch
1. _____ 3. _____   1. _____ 3. _____
2. _____   2. _____

All-weld-metal tension test

Tensile strength, psi _____
Yield point/strength, psi _____
Elongation in 2 in., % _____
Laboratory test no. _____

Welder's name _____ Clock no. _____ Stamp no. _____
Tests conducted _____ Laboratory
                         Test number _____
                         Per _____

We, the undersigned, certify that the statements in this record are correct and that the test welds were prepared, welded, and tested in accordance with the requirements of sections 5, Part B of ANSI/AWS D1.1, (_____) Structural Welding Code- Steel.
                                                                            year

Signed _____
               Manufacturer or Contractor
By _____
Title _____
Date _____

Form E-1 (Back)   (Source: AWS D1.1-94)

WELDING PROCEDURE QUALIFICATION TEST RECORD
FOR ELECTROSLAG AND ELECTROGAS WELDING

PROCEDURE SPECIFICATION

Material specification _____
Welding process _____
Position of welding _____
Filler metal specification _____
Fillet metal classification _____
Filler metal _____
Flux _____
Shielding gas _____ Flow rate _____
Gas dew point _____
Thickness range this test qualifies _____
Single or multiple pass _____
Single or multiple arc _____
Welding current _____
Preheat temperature _____
Postheat temperature _____
Welder's name _____

VISUAL INSPECTION (9.25.1)
Appearance _____
Undercut _____
Piping porosity _____
_____
Test date _____
Witnessed by _____

TEST RESULTS

Reduced-section tensile test

Tensile strength, psi
1. _____
2. _____

All-weld-metal tension test
Tensile strength, psi _____
Yield point/strength, psi _____

Side-bend tests
1. _____ 3. _____
2. _____ 4. _____

Radiographic-ultrasonic examination
RT report no. _____
UT report no. _____

Impact tests
Size of specimen _____ Test temp ____
Ft·lb: 1. _____ 2. _____ 3. _____ 4. ____
5. _____ 6. _____ Avg. _____
High _____ Low _____
Laboratory test no. _____

WELDING PROCEDURE

Pass no.	Electrode size	Welding current		Joint detail
		Amperes	Volts	

Guide tube flux _____
Guide tube composition _____
Guide tube diameter _____
Vertical rise speed _____
Traverse length _____
Traverse speed _____
Dwell _____
Type of molding shoe _____

We, the undersigned, certify that the statements in this record are correct and that the test welds were prepared, welded, and tested in accordance with the requirements of section 4, Part E, and section 5, Part B of ANSI/AWS D1.1 (_____) Structural Welding Code- Steel.
    year

Procedure no. _____ Manufacturer or contractor _____
Revision no. _____ Authorized by _____
                                                                      Date _____

Form E-3

(Source: AWS D1.1-94)

B-5

WELDER, WELDING OPERATOR OR TACK WELDER QUALIFICATION TEST RECORD

Type of Welder _____
Name _____ Identification No. _____
Welding Procedure Specification No. _____ Rev _____ Date _____

Variables	Record Actual Values Used in Qualification	Qualification Range
Process/Type (5.16.2)		
Electrode (single or multiple)		
Current/Polarity		
Position (5.16.5)		
Weld Progression (5.16.7)		
Backing (YES or NO) (5.16.18)		
Material/Spec. (5.16.1)	_____ to _____	
Base Metal		
Thickness: (Plate)		
Groove		
Fillet		
Thickness: (Pipe/tube)		
Groove		
Fillet		
Diameter: (Pipe)		
Groove		
Fillet		
Filler Metal (5.16.3)		
Spec. No.		
Class		
F-No.		
Gas/Flux Type (5.16.4)		
Other		

VISUAL INSPECTION (5.12.6 or 5.12.7)
Acceptable YES or NO _____

Guided Bend Test Results (5.28.1/5.29.1)

Type	Result	Type	Result

Fillet Test Results (5.28.2/5.28.3; 5.39.3/5.29.4)
Appearance _____ Fillet Size _____
Fracture Test Root Penetration _____ Macroetch _____
(Describe the location, nature, and size of any crack or tearing of the specimen.)

Inspected by _____ Test Number _____
Organization _____ Date _____

RADIOGRAPHIC TEST RESULTS (5.28.4/5.39.2)

Film Identification Number	Result	Remarks	Film Identification Number	Result	Remarks

Interpreted by _____ Test Number _____
Organization _____ Date _____

We, the undersigned, certify that the statements in this record are correct and that the test welds were prepared, welded, and tested in accordance with the requirements of Section 5, Part C or D of ANSI/AWS D1.1, (_____) Structural Welding Code- Steel.   year

Manufacturer or Contractor _____
Autorized by _____
Date _____

Form E-4   (Source: AWS D1.1-94)

B-6

REPORT OF RADIOGRAPHIC EXAMINATION OF WELDS

Project _____
Quality requirements - section no. _____
Reported to _____

WELD LOCATION AND IDENTIFICATION SKETCH

Technique _____
Source _____
Film to Source _____
Exposure time _____
Screens _____
Film type _____

(Describe length, width, and thickness of all joints radiographed)

Date	Weld identification	Area	Interpretation		Repairs		Remarks
			Accept.	Reject	Accept.	Reject	

We, the undersigned, certify that the statements in this record are correct and that the welds were prepared and tested in accordance with the requirements of the American Welding Society ANSI/AWS D1.1 (_____) Structural Welding Code- Steel.
                                                                                                    year

Radiographer(s) _____   Manufacturer or contractor _____
Interpreter _____   Authorized by _____
Test date _____   Date _____
Form E-7

(Source AWS D1.1-94)

B-7

REPORT OF MAGNETIC PARTICLE EXAMINATION OF WELDS

Project _____
Quality requirements - section no. _____
Reported to _____

WELD LOCATION AND IDENTIFICATION SKETCH

Quantity: _____ Total Accepted: _____ Total Rejected: _____

Date	Weld identification	Area Examined		Interpretation		Repairs		Remarks
		Entire	Specific	Accept.	Reject	Accept.	Reject	

PRE-EXAMINATION

Surface Preparation: _____

EQUIPMENT

Instrument Make: _____ Model: _____ S. No.: _____

METHOD OF INSPECTION

☐ Dry ☐ Wet ☐ Visible ☐ Fluorescent
How Media Applied: _____
☐ Residual ☐ Continuous ☐ True Continuous
☐ AC ☐ DC ☐ Half-Wave
☐ Prods ☐ Yoke ☐ Cable Wrap ☐ Other _____
Direction for Field: ☐ Circular ☐ Longitudinal
Strength of Field: _____
(Amper turns, field density, magnetizing force, number, and duration of force application.)

POST EXAMINATION

Demagnetizing Technique (if required): _____
Cleaning (if required): _____ Marking Method: _____

We, the undersigned, certify that the statements in this record are correct and that the welds were prepared and tested in accordance with the requirements of the American Welding Society ANSI/AWS D1.1 (_____) Structural Welding Code- Steel.
    year

Inspector _____ Manufacturer or contractor _____
Level _____ Authorized By _____
Test date _____ Date _____

Form E-8
© 1991 by American Welding Society. All rights reserved.
(Source AWS D1.1-94)

QW-483 (Back)

Tensile Test (QW-150)   PQR No. _____

Specification No.	Width	Thickness	Area	Ultimate Total Load lb	Ultimate Unit Stress psi	Type of Failure & Location

Guided-Bend Tests (QW-160)

Type and Figure No.	

Toughness Tests (QW-170)

Specimen No.	Notch Location	Notch Type	Test Temp.	Impact Values	Lateral Exp. % Shear	Lateral Exp. Mils	Drop Weight Break	Drop Weight No Break

Fillet-Weld Test (QW-180

Result - Satisfactory: Yes _____ No _____   Penetration into Parent Metal: Yes _____ No _____
Macro - Results _____

Other Tests

Type of Test _____
Deposit Analysis _____
Other _____

- -

Welder's Name _____ Clock No. _____ Stamp No. _____
Tests Conducted by: _____ Laboratory Test No. _____
We certify that the statements in this record are correct and that the test welds were prepared, welded, and tested in accordance with the requirements of Section IX of the ASME Code.

Manufacturer _____

Date _____ By _____
(Detail of record of tests are illustrative only and may be modified to conform to the type and number of tests required by the Code.)

Source ASME B31.1, SEC IX

B-9

QW-482 (Back)

WPS No. _____ Rev. _____

POSITIONS (QW-405)
Position(s) of Groove _____
Welding Progression: Up _____ Down _____
Position(s) of Fillet _____

PREHEAT (QW-406)
Preheat Temp. Min. _____
Interpass Temp. Max. _____
Preheat Maintenance _____

(Continuous or special heating where applicable should be recorded.)

POSTWELD HEAT TREATMENT (QW-407)
Temperature Range _____
Time Range _____

GAS (QW-408)

Percent Composition

| | Gas(es) | (Mixture) | Flow Rate |

Shielding _____ _____ _____
Trailing _____ _____ _____
Backing _____ _____ _____

ELECTRICAL CHARACTERISTICS 9QW-409)
Current AC or DC _____ Polarity _____
Amps (Range) _____ Volts (Range) _____

(Amps and volts range should be recorded for each electrode size, position, and thickness, etc. This information may be listed in a tabular form similiar to that shown below.)

Tunsten Electrode Size and Type _____
(Pure Tungsten, 2% Thoriated, etc.)
Mode of Metal Transfer for GMAW _____
(Spray arc, short circuiting arc, etc.)
Electrode Wire feed speed range _____

TECHNIQUE (QW-410)
String or Weave Bead _____
Orfice or Gas Cup Size _____
Initial and Interpass Cleaning (Brushing, Grinding, etc.) _____

Method of Back Gouging _____
Oscillation _____
Contact Tube to Work Distance _____
Multiple or Single Pass (per side) _____
Multiple of Single Electrodes _____
Travel Speed (Range) _____
Peening _____
Other _____

Weld Layer(s)	Process	Filler Metal		Current		Volt Range	Travel Speed Range	Other (e.g., Remarks, Comments, Hot Wire Addition, Technique, Torch Angle, Etc.)
		Class	Dia.	Type Polar.	Amp. Range			

(Source ASME B31.1, SEC IX)

QW-483 SUGGESTED FORMAT FOR PROCEDURE QUALIFICATION RECORD (pqr)
(See QW-200.2, Section IX, ASME Boiler and Pressure Vessel Code)
Record Actual Conditions Used to Weld Test Coupon.

Company Name _____
Procedure Qualification Record No. _____ Date _____
WPS No. _____
Welding Process(es) _____
Types (Manual, Akutomatic, Semi-Auto.) _____

JOINTS (QW-402)

Groove Design of Test Coupon
(For combination qualifications, the deposited weld metal thickness shall be recorded for each filler metal or process used.)

BASE METALS (QW-403)	POSTWELD HEAT TREATMENT (QW-407)
Material Spec. _____	Temperature _____
Type or Grade _____	Time _____
P-No. _____ to P-No. _____	Other _____
Thickness of Test Coupon _____	
Diameter of Test Coupon _____	
Other _____	

GAS (QW-408)

                    Percent Composition
          Gas(es)   (Mixture)   Flow Rate
Shielding _____ _____ _____
Trailing _____ _____ _____
Backing _____ _____ _____

FILLER METALS (QW-404)
SFA Specification _____
AWS Classification _____
Filler Metal F-No. _____
Weld Metal Analysis A-No. _____
Size of Filler Metal _____
Other _____

Weld Metal Thickness

ELECTRICAL CHARACTERISTICS (QW-409)
Current _____
Polarity _____
Amps _____ Volts _____
Tungsten Electrode Size _____
Other _____

POSITION (QW-405)
Position of Groove _____
Weld Progression (Uphill, Downhill) _____
Other _____

TECHNIQUE (QW-410)
Travel speed _____
String or Weave Bead _____
Oscillation _____
Multipass or Single Pass (per side) _____
Single or Multiple Electrodes _____
Other _____

PREHEAT (QW-406)
Preheat Temp. _____
Interpass Temp. _____
Other _____

This form (E00007) may be obtained from the Order Dept., ASME, 22 Law Drive, Box 2300, Fairfield, NJ 07007-2300

(Source: ASME B31.1, SEC IX)

B-11

QW-483 (Back)

Tensile Test (QW-150)   PQR No. _____

Specification No.	Width	Thickness	Area	Ultimate Total Load lb	Ultimate Unit Stress psi	Type of Failure & Location

Guided-Bend Tests (QW-160)

Type and Figure No.	

Toughness Tests (QW-170)

Specimen No.	Notch Location	Notch Type	Test Temp.	Impact Values	Lateral Exp. % Shear	Lateral Exp. Mils	Drop Weight Break	Drop Weight No Break

Fillet-Weld Test (QW-180

Result - Satisfactory: Yes _____ No _____ Penetration into Parent Metal: Yes _____ No _____
Macro - Results _____

Other Tests

Type of Test _____
Deposit Analysis _____
Other _____

- -

Welder's Name _____ Clock No. _____ Stamp No. _____
Tests Conducted by: _____ Laboratory Test No. _____
We certify that the statements in this record are correct and that the test welds were prepared, welded, and tested in accordance with the requirements of Section IX of the ASME Code.

Manufacturer _____

Date _____ By _____
(Detail of record of tests are illustrative only and may be modified to conform to the type and number of tests required by the Code.)

Source ASME B31.1, SEC IX)

QW-484 SUGGESTED FORMAT FOR MANUFACTURER'S RECORD OF WELDER OR WELDING OPERATOR QUALIFICATION TESTS (WPQ)
See QW-301, Section IX, ASME Boiler and Pressure Vessel Code

Welder's name _____ Clock number _____ Stamp no. _____
Welding process(es) used _____ Type _____
Identification of WPS followed by welder during welding of test coupon _____
Base material(s) welded _____ Thickness _____

Manual or Semiautomatic Variables for Each Process (QW-350)	**Actual Values**	**Range Qualified**
Backing (metal, weld metal, welded from both sides, flux etc.) (QW-402)		
ASME P-No. _____ to ASME p-No. (QW-404)		
( ) Plate ( ) Pipe (enter diameter, if pipe)		
Filler metal specification (SFA): _____ Classification (QW-404)		
Filler Metal F-No.		
Consumable insert for GTAW or PAW		
Weld deposit thickness for each welding process		
Welding position (1G, 5G, etc.) (QW-405)		
Progression (uphill/downhill)		
Backing gas for GTAW, PAW, GMAW; fuel gas for OFW (QW-408)		
GMAW transfer mode (QW-409)		
GTAW welding current type/polarity		

Machine Welding Variables for the Process Used (QW-360)	**Actual Values**	**Range Qualified**
Direct/remote visual control		
Automatic voltage control (GTAW)		
Automatic joint tracking		
Welding position (1G, 5G, etc.)		
Consumable insert		
Backing (metal, weld metal, welded from both sides, flux, etc.)		

Guided Bend Test Results

Guided Bend Test Type   ( ) QW-462.2 (Side) Results   ( ) QW-462.3(a)(Trans. R & F) Type   ( ) QW-462.3(b) (Long, R & F) Results

Visual examination results (QW-302.4) _____
Radiographic test results (QW-304 and Qw-305) _____
(For alternative qualification of groove welds by radiography)
Fillet Weld - Fracture test _____ Length and percent of defects _____ in.
Macro test fusion _____ Fillet leg size _____ in. x _____ in. Concavity/covexity _____ in.
Welding test conducted by _____
Mechanical tests conducted by _____

We certify that the statements in this record are correct and that the test coupons were prepared, welded, and tested in accordance with the requirements of Section IX of the ASME Code.

Organization _____

Date _____   By _____

This form (E00008) may be obtained from the Order Dept., ASME, 22 Law Drive, Box 2300, Fairfield NJ 07007-2300

(Source ASME B31.1, SEC IX)

B-13

Reference: API Standard 1104, 2.2

PROCEDURE SPECIFICATION NO. _____

For _____ Welding of _____ Pipe and Fittings
Process _____
Material _____
Diameter and wall thickness _____
Joint design _____
Filler metal and no. of beads _____
Electrical or flame characteristics _____
Position _____
Direction of welding _____
No. of welders _____
Time lapse between passes _____
Type and removal of lineup clamp _____
Cleaning and/or grinding _____
Preheat stress relief _____
Shielding gas and flow rate _____
Shielding flux _____
Speed of travel _____
Sketches and tabulations attaches _____

Tested _____ Welder _____
Approved _____ Welding supervisor _____
Adopted _____ Chief engineer _____

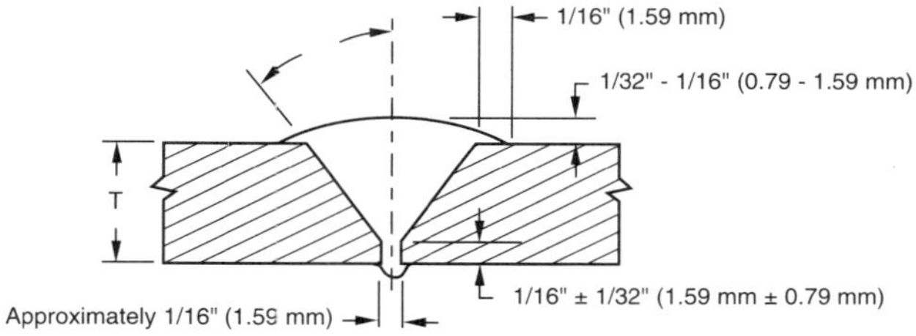

Standard V-Bevel Butt Joint

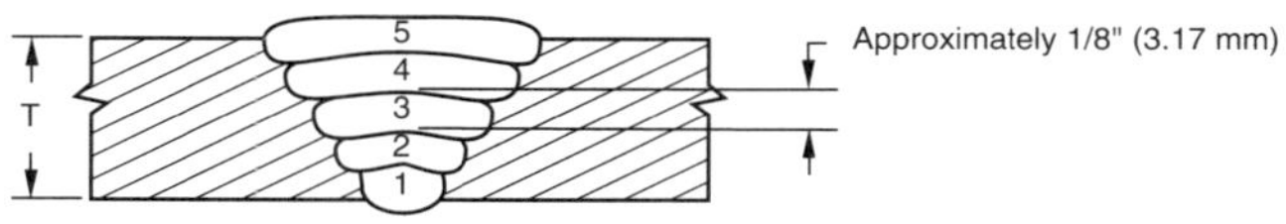

Sequence of Beads

Note: Dimensions are for reference only.

ELECTRODE SIZE AND NUMBER OF BEADS

Bead Number	Electrode Size and Type	Voltage	Amperage and Polarity	Speed

(Source API Standard 1104)

American Welding Society

Certifies that Welding Inspector

C. W. Eye

has complied with the requirements of Section 6.1 of the AWS Standard for Qualification and Certification of Welding Inspectors QCI-88

00000001
CERTIFICATE NUMBER

March 1994
VALID DATE

March 1997
EXPIRATION DATE

PRESIDENT AWS

CHAIRMAN Q & C COMMITTEE

© 1984 American Welding Society

APPENDIX C - ANSWER KEY REVIEW QUESTIONS CHAPTERS 1 - 12

Chapter 1
Q1-1   d
Q1-2   d
Q1-3   c
Q1-4   e
Q1-5   d
Q1-6   a
Q1-7   b
Q1-8   e
Q1-9   e
Q1-10 e
Q1-11 e
Q1-12 a
Q1-13 e
Q1-14 b
Q1-15 a

Chapter 2
Q2-1   d
Q2-2   d
Q2-3   b
Q2-4   a
Q2-5   a
Q2-6   d
Q2-7   a
Q2-8   c
Q2-9   b
Q2-10 a
Q2-11 b
Q2-12 d
Q2-13 d
Q2-14 b
Q2-15 d

Chapter 3
Q3-1   d
Q3-2   e
Q3-3   a
Q3-4   c
Q3-5   b
Q3-6   d
Q3-7   b
Q3-8   c
Q3-9   d

Chapter 4
Q4-1   c
Q4-2   d
Q4-3   e
Q4-4   c
Q4-5   d
Q4-6   c
Q4-7   e
Q4-8   b
Q4-9   c
Q4-10 e
Q4-11 d
Q4-12 c
Q4-13 a
Q4-14 b
Q4-15 c
Q4-16 b
Q4-17 c
Q4-18 d
Q4-19 a
Q4-20 c
Q4-21 b
Q4-22 e
Q4-23 d
Q4-24 c
Q4-25 a
Q4-26 d
Q4-27 e
Q4-28 e
Q4-29 b
Q4-30 b
Q4-31 c
Q4-32 a
Q4-33 e
Q4-34 b
Q4-35 e
Q4-36 d
Q4-37 a
Q4-38 b
Q4-39 d
Q4-40 a
Q4-41 a
Q4-42 b
Q4-43 c
Q4-44 d
Q4-45 c
Q4-46 a
Q4-47 a
Q4-48 b
Q4-49 e
Q4-50 e
Q4-51 d
Q4-52 b
Q4-53 e
Q4-54 c
Q4-55 b
Q4-56 d
Q4-57 e
Q4-58 d
Q4-59 b
Q4-60 a
Q4-61 a
Q4-62 b
Q4-63 e
Q4-64 a
Q4-65 d
Q4-66 b
Q4-67 b
Q4-68 a
Q4-69 d

Chapter 5
Q5-1   c
Q5-2   d
Q5-3   c
Q5-4   a
Q5-5   b
Q5-6   a
Q5-7   e
Q5-8   c
Q5-9   a
Q5-10 b
Q5-11 e
Q5-12 b
Q5-13 d
Q5-14 d
Q5-15 e
Q5-16 d
Q5-17 a
Q5-18 c
Q5-19 e
Q5-20 d
Q5-21 d
Q5-22 c

Chapter 5 (cont.)
Q5-23 a
Q5-24 c
Q5-25 b
Q5-26 d
Q5-27 a
Q5-28 a
Q5-29 d
Q5-30 b
Q5-31 c
Q5-32 a
Q5-33 b
Q5-34 a
Q5-35 b
Q5-36 c
Q5-37 b
Q5-38 e
Q5-39 c
Q5-40 b
Q5-41 c
Q5-42 c
Q5-43 d
Q5-44 a
Q5-45 b
Q5-46 d
Q5-47 d
Q5-48 c
Q5-49 b
Q5-50 a
Q5-51 c
Q5-52 c
Q5-53 a
Q5-54 b
Q5-55 d
Q5-56 e

Chapter 6
Q6-1 e
Q6-2 a
Q6-3 c
Q6-4 b
Q6-5 e
Q6-6 e
Q6-7 b
Q6-8 b
Q6-9 e
Q6-10 e
Q6-11 c
Q6-12 a
Q6-13 b
Q6-14 b
Q6-15 b
Q6-16 d
Q6-17 e
Q6-18 a
Q6-19 c
Q6-20 e
Q6-21 b
Q6-22 c
Q6-23 e
Q6-24 c

Chapter 7
Q7-1 d
Q7-2 a
Q7-3 a
Q7-4 d
Q7-5 b
Q7-6 c
Q7-7 c
Q7-8 e
Q7-9 c
Q7-10 e
Q7-11 e
Q7-12 a
Q7-13 a
Q7-14 d
Q7-15 c
Q7-16 d
Q7-17 b
Q7-18 a
Q7-19 b
Q7-20 d
Q7-21 b
Q7-22 e
Q7-23 b
Q7-24 a
Q7-25 e
Q7-26 c
Q7-27 c
Q7-28 a
Q7-29 b
Q7-30 e
Q7-31 c
Q7-32 c
Q7-33 a
Q7-34 e
Q7-35 a

Chapter 8
Q8-1 c
Q8-2 e
Q8-3 c
Q8-4 d
Q8-5 e
Q8-6 e
Q8-7 a
Q8-8 c
Q8-9 d

Chapter 9
Q9-1 b
Q9-2 e
Q9-3 c
Q9-4 a
Q9-5 b
Q9-6 e
Q9-7 d
Q9-8 a
Q9-9 d
Q9-10 d
Q9-11 b
Q9-12 d
Q9-13 a
Q9-14 c
Q9-15 a
Q9-16 c
Q9-17 c
Q9-18 c
Q9-19 c
Q9-20 c
Q9-21 b
Q9-22 c
Q9-23 b
Q9-24 b
Q9-25 d
Q9-26 b
Q9-27 c
Q9-28 d
Q9-29 a
Q9-30 d
Q9-31 e
Q9-32 e
Q9-33 c
Q9-34 c
Q9-35 d
Q9-36 b
Q9-37 a
Q9-38 e
Q9-39 b
Q9-40 a
Q9-41 e
Q9-42 a
Q9-43 d
Q9-44 a
Q9-45 a
Q9-46 e
Q9-47 d
Q9-48 d
Q9-49 e
Q9-50 e
Q9-51 d
Q9-52 c
Q9-53 d
Q9-54 c
Q9-55 e
Q9-56 e
Q9-57 d
Q9-58 d
Q9-59 c
Q9-60 d
Q9-61 b
Q9-62 b
Q9-63 e
Q9-64 d

Chapter 9 continued
Q9-65 a
Q9-66 a
Q9-67 e
Q9-68 c
Q9-69 b
Q9-70 c
Q9-71 a
Q9-72 e
Q9-73 d

Chapter 10
Q10-1   d
Q10-2   d
Q10-3   d
Q10-4   c
Q10-5   d
Q10-6   c
Q10-7   b
Q10-8   d
Q10-9   b
Q10-10 c
Q10-11 a
Q10-12 b
Q10-13 e
Q10-14 b
Q10-15 c
Q10-16 b
Q10-17 c
Q10-18 e
Q10-19 a
Q10-20 b

Chapter 11
Q11-1   e
Q11-2   d
Q11-3   d
Q11-4   d
Q11-5   c
Q11-6   c
Q11-7   c
Q11-8   d
Q11-9   d
Q11-10 c
Q11-11 b
Q11-12 e
Q11-13 a
Q11-14 e
Q11-15 b
Q11-16 c
Q11-17 d
Q11-18 d
Q11-19 b
Q11-20 d
Q11-21 d
Q11-22 d
Q11-23 c
Q11-24 c
Q11-25 c
Q11-26 c
Q11-27 a
Q11-28 a
Q11-29 a
Q11-30 a
Q11-31 b
Q11-32 a
Q11-33 b
Q11-34 a
Q11-35 d
Q11-36 c
Q11-37 c
Q11-38 c
Q11-39 e
Q11-40 d
Q11-41 c
Q11-42 c
Q11-43 c
Q11-44 d

Chapter 12
Q12-1   d
Q12-2   b
Q12-3   a
Q12-4   a
Q12-5   c
Q12-6   d
Q12-7   a
Q12-8   b

INDEX

A
abbreviations used in references, A-1-3
acceptability, 11-1
acoustic emission testing, 11-14
air carbon arc cutting, 9-24
aluminum alloys, 6-17
annealing, 6-10
ANSI standards, 3-5
answers to review questions, (see Appendix C)
API Standard, 3-5
    report forms, see Appendix B
approval of tests, 8-8
arc strikes, 10-24
argon shielding, 9-7, 9-10
ASME Boiler & Pressure Vessel Code, 3-2
    requirements, 3-3, 8-8
    reports, 12-1
ASNT SNT-TC-IA, 11-17
attitude, 1-2
austenite, 6-8, 6-15
available welding talent, 9-3
AWS filler metals, 3-6
AWS Structural Welding Code, 2-7, 3-2, 8-8
    reports, 12-2
    welding talent

B
back side accessibility, 9-2
bainite, 6-8
base metal defects, 2-5
base metals, 8-2
body centered cubic (BCC), 6-8 - 6-9
body centered tetragonal (BCT), 6-9
Boiler Code—(see ASME)
brazing, 9-21
bridges, 6-12

C
carbide precipitation, 6-16
carbon equivalent C.E., 6-15
carbon steel, 6-7
    soaking temperature, 6-5
cathode ray tube, 11-19
cementite, 6-8
certificate, B-14
certification, 1-5
changes in procedure, 8-9

chemical analysis, 2-4
chemistry (welding) 6-12, 6-15, 9-1, 9-5, 9-8 - 9-10, 9-13, 9-14, 9-17, 9-19, 9-20, 9-22
chipping, 5-18, 9-25
cleaning, 2-7, 2-10, 8-3
Code of Ethics, 1-5
code inspector, 3-3, 12-2
Codes, Standards, & Specifications, 3-1
communication, 1-7
conformity to drawings, 2-15
conflict of interest, 1-6
consumables, 9-3
contamination, 6-14
contents, welding procedure, 8-1
conversions, 7-27
cooling rates, 6-10
corner joint example, 4-2
corrosion attack, 6-16
cost nondestructive testing, 11-2
    destructive testing, 11-2
cracking, 6-3, 6-14, 6-16, 10-2, 10-19-10-24, 10-25
crystalline structures, 6-8 - 6-9
    BCC, 6-8 - 6-9
    BCT, 6-9
    FCC, 6-9
    HCP, 6-9
current, type and range, 8-3
cutting, 9-23

D
defective material, 2-4
delamination 10-18
delayed cracking, 10-19
delta ferrite, 6-16, 11-17
designating joint design of welds, 2-1
destructive testing, 7-1
diffusion, 6-3
diffusion brazing, 9-22
dimensional accuracy, 2-12
dimensions of material, 9-2
dip brazing, 9-22
discontinuities, 10-1 -10-3
ductility, 7-8 - 7-9
duties of inspector, 2-1
dye penetrant inspection, 11-4
dynamic loading, 6-12

E

economics, 9-4
edge preparation, 4-1
eddy current testing, 11-12
education and training, 1-4
electroslag welding, 9-16
engineer, welding, 1-9
    design/project, 1-9
equipment check, 2-6
equipment availability, 9-3
erosion, 9-23
essential variables, 8-9
eutectoid composition, 6-10
evaluation of test results, 8-8
expansion (thermal), 6-5
eye protection, 9-4

F

face centered cubic, (FCC), 6-9
fatigue, dynamic loading, 6-12
ferrite, 6-8, 11-16
    delta, 6-16
    number, FN, 6-15
ferrule, 9-20
filler material storage, 2-5
filler metal, 2-1, 2-5, 3-6
fillet weld examples, 4-4, 4-10, 4-12, 5-7, 5-8
fillet weld gage (photo), 2-14
finished product inspection, 2-11
fissures, 10-24
flux, 9-5, 9-9, 9-14
flux cored arc welding, 9-9
fluorescent penetrant inspection, 11-4
furnace brazing, 9-21

G

gamma radiation, 11-8
gas metal arc welding, 9-7
gas tungsten arc welding, 9-10
gouging, 9-24
grinding, gouging, 2-11, 9-25
groove welds, 4-2, 4-4, 4-8, 5-3, 5-5

H

halide torch leak test, 11-15
heat-affected zone, 4-9
heat input, 6-11, 8-6
helium spectrometer leak test, 11-15
hexagonal close packed (HCP), 6-9
holding oven, 2-6
hydrogen contamination, 6-14
hydrostatic testing, 11-14

I

inclusions, 10-8
    nonmetallic, 10-8
    tungsten, 10-11
    slag, 10-9, 10-10
incomplete fusion, 10-11, 10-12, 10-14 (see discontinuities chapter 9)
incomplete joint penetration, 10-2, 10-11
induction brazing, 9-22
infrared brazing, 9-22
insufficient throat, (see underfill)
integrity, 1-5
interpass temperature, 2-11
interpretation, penetrant indications, 11-5
    test data, 2-15 (see also chapters 3 & 8)

J - K

L

lamellar tearing, 10-2, 10-18, 10-20
lamination, 10-18
leak tests, 11-14
lens shades, 9-4
limitations of inspection, 11-1
liquid penetrant inspection, 11-4, A-4
location of discontinuity, 9-6, 9-8, 9-10, 9-12, 9-14, 9-18, 9-21
longitudinal cracks, 10-19

M

macrofissure, 10-24
Magna-Gage, 11-17
magnetic particle inspection, 11-5, A-4
magnetic test for ferrite, 11-16
marking, 8-6
martensite, 6-8, 6-15
Master Chart of Welding Processes, 9-1
materials, 2-1
mechanical cutting, 9-25
mechanical properties, 2-4
melting and freezing, 6-4
metallurgy, 6-1
metric system, 7-25
microfissure, 10-24
microstructure, 6-11, 10-17
multiple inspection, 12-2

N

NBS ferrite standards, 11-16
NDT level I, II, III, 11-17
nuclear components, report form N-1, 12-2

O

operator qualification test, B-12
overlap, 10-16
oxyacetylene welding, 9-18
oxygen cutting, 9-23

P

pascal, 7-25
pearlite, 6-8, 6-10
peening, 6-7, 8-6
penetrameter, 11-8
penetrant inspection, 11-4
personnel to be qualified, 3-4
phase transformations, 6-9
physical condition, 1-2
pipe, 7-7, 8-16, 8-17
plant manager, 1-8
plasma arc cutting, 9-24
plate welding, 8-11
porosity, 10-2 - 10-8
position of welding, 8-3, 8-5, 8-15 - 8-19
postheating, 8-6
power boilers, 3-4, 8-9
preparation, edge, 4-1
preheating, 2-8
preplaced brazing filler metal, 9-21
prequalified process limitation, 9-1
prequalified procedure specification, 8-7, Appendix B
pressure tests, 11-15
pressure vessels (see also ASME), 3-2
procedures, 3-4
procedure specifications (WPS), Appendix B
processes, inspection, Appendix A
  welding, chapter 9
progress of inspection, 2-1
project engineer, 1-9
proof tests, (see pressure tests)
public statements, 1-6
pulsed arc, 9-8
purchase specification check, 2-2

Q

qualifications to become inspector, 1-2
qualifications of inspector, 1-2, 11-18
quality level, 9-4
quenching, 6-8

R

radiographic inspection, 11-7
radiographs, (illustrations, chapter 10)
records, 1-4, (see chapter 12)
requalification, 8-10

resistance brazing, 9-22
rolling direction, 8-10 - 8-12, 8-14
root cracks, 10-20, 10-24
root opening, 4-3, 5-4
root requirements, 9-2

S

safety, 9-4
  radiation hazard, 11-9
  toxic fumes, 9-22
sample forms, Appendix B
sampling, 7-24
scope, welding procedure, 8-2
seams and laps, 10-18
selecting inspection method, 11-1
selecting test samples, 7-24
semiautomatic welding, 9-7
sequence of welding, 4-11, 5-20
Severn gage, 11-16
shielded metal arc welding, 9-5
shielding, 6-13, 9-7, 9-9, 9-10, 9-13, 9-14, 9-17
short-circuiting transfer, 9-7
shrinkage stress, 6-6
S.I., International System of Units, 7-25
sigma phase, 11-16
signoff, 12-1, Appendix B
size of discontinuity, 10-25
    weld, chapter 5
skill level, 9-3, 9-4
slag inclusions, 2-10
solicitation of employment, 1-6
solids vs liquids, 6-2
specifications, 3-1
spray transfer, 9-7
stabilization, 6-16
stainless steel, 6-15
standardization of tests, 8-20
standards, 3-1, 3-4, 3-5
steels, 6-4, 6-15
strength, 7-10
stress relief, 6-6
stud arc welding, 9-20
submerged arc welding, 9-14
subsurface cracks, 11-6
supervisor, 1-7
superintendent (field/shop), 1-8
surface cracks, 11-6
symbols, chapter 5

T

tack welding, 8-3, B-5
tempering, 6-10
terms and definitions, welding, 1-2, chapter 4

tests, mechanical properties, 2-4
    qualification welds, chapter 8
thermal cutting, 9-23
throat, 10-19, 10-20, 10-23
titanium alloys, 6-18
toe cracks, 10-19, 10-20, 10-23, 10-24
tools, visual inspection, 11-4
    UT 11-9
torch brazing, 9-21
toughness, 6-9
transverse cracks, 10-20, 10-24
tungsten inclusions, 10-8, 10-11

U

ultrasonic inspection, 11-9
unauthorized practice, 1-7
underbead cracks, 10-20, 10-24
undercut, 10-2, 10-17
unstructured reports, 12-3
underfill, 10-2, 10-11, 10-12

V

variables, 8-9
verification, inspection, 2-12
    job material, 2-2
vision, 1-2
visual inspection, 11-2
voids, 9-22

W

warpage, 6-5
weldability, 6-1
"Weldability of Steels," 6-2
welders, 1-8, 8-1

X

x-ray, 11-8, - 11-10

Z

zirconium alloys, 6-18